ENVIRONMENTAL SCIENCE, ENGINEERING AND TECHNOLOGY

RADIOACTIVE WASTE: SOURCES, TYPES AND MANAGEMENT

ENVIRONMENTAL SCIENCE, ENGINEERING AND TECHNOLOGY

Additional books in this series can be found on Nova's website under the Series tab.

Additional E-books in this series can be found on Nova's website under the E-books tab.

ENVIRONMENTAL SCIENCE, ENGINEERING AND TECHNOLOGY

RADIOACTIVE WASTE: SOURCES, TYPES AND MANAGEMENT

SATOSHI YUAN
AND
WENXU HIDAKA
EDITORS

Nova Science Publishers, Inc.
New York

For permission to use material from this book please contact us:
Telephone 631-231-7269; Fax 631-231-8175
Web Site: http://www.novapublishers.com

Additional color graphics may be available in the e-book version of this book.

Library of Congress Cataloging-in-Publication Data

Radioactive waste : sources, types and management / Satoshi Hidaka and Wenxu Yuan.
p. cm.
Includes index.
ISBN 978-1-62100-188-1 (hbk.)
1. Radioactive wastes--Management. 2. Radioactive waste disposal. I. Hidaka, Satoshi, 1976- II. Yuan, Wenxu, 1973-
TD897.85.R338 2011
621.48'38--dc23

2011032411

Published by Nova Science Publishers, Inc. † New York

CONTENTS

Preface		**vii**
Chapter 1	DMNR: A New Concept for Real-Time Online Monitoring of Short and Medium Term Radioactive Waste *Paolo Finocchiaro*	**1**
Chapter 2	Radioactive Waste from Research Reactor Operation and Decommissioning *J. G. Marques and A. Kling*	**41**
Chapter 3	Promising Analytical Techniques for HLW Analysis *A. V. Legin, D. O. Kirsanov, V. A. Babain, L. N. Gall and N. R. Gall*	**77**
Chapter 4	Treatment of High-Level Radioactive Waste Arising from Pyroprocessing and Its Waste Form *Ippei Amamoto*	**97**
Chapter 5	Sources of Radioactive Contamination of Frozen Taiga Landscapes in Southern Yakutia *A. P. Chevychelov and P. I. Sobakin*	**127**
Chapter 6	Cementitious Wasteforms for Immobilization of Low-Activity Radioactive Wastes *Dawn M. Wellman, Chase C. Bovaird, Shas V. Mattigod, Kent E. Parker, Libby N. Clayton, Laura Powers and Marcus I. Wood*	**145**
Chapter 7	Highly Radioactive Waste Disposal: A Global Concern *Kenneth A. Rogers*	**185**
Index		**213**

PREFACE

Radioactive wastes are wastes that contain radioactive material and are usually by-products of nuclear power generation and other applications of nuclear fission or nuclear technology, such as research and medicine. Radioactive waste is hazardous to human health and the environment and is regulated by government agencies. This book presents topical research in the study of the sources, types and management of radioactive waste, including DMNR real-time online monitoring of short and medium term radioactive waste; radioactive waste from research reactor operation and decommissioning; treatment of high-level radioactive waste arising from pyroprocessing and its waste form and highly radioactive waste disposal.

Chapter 1 - One of the main topics in decommissioning and storage of radioactive waste is its monitoring in the short, medium and long term. Whereas for the long term waste the geological disposal is foreseen, or better the use of new "cleaner" reactors, the short and medium term waste is currently packed into drums and either kept inside the production site or transported to dedicated storage sites. In any case special care must be devoted to the prompt detection of possible leaks due to structural weakness or accident. Whichever the topological arrangement or the physical structure of the drums, the goal would be measuring the radioactivity around each single drum. This could be done by means of ordinary geiger-Muller counters, but the cost of such a solution would be prohibitive. Therefore, what is usually done is a periodic check by operators (or robots) plus some overall monitoring in fixed locations inside the repository itself. Detector Mesh for Nuclear Repositories is a project for a prototype demonstrator of the online monitoring of short-medium term radioactive waste, currently under development at INFN-LNS Catania. Such a system is planned to be distributed, fine-grained, robust, reliable, and based on low-cost components. With the main purpose of counting gamma radiation, a new kind of mini-detector was implemented that behaves like a scintillating geiger-Muller counter. It has to be replicated in shape of a grid around each single waste drum. This system could also open new perspectives on the modality of waste packaging and storage. What was proposed is cheap scintillating detectors, readout by means of SiPM photosensors capable of detecting the tiny light signals of few photons. This way one can easily place grids of 5-10 radiation sensors around each waste drum, and record continuously the measured activity in order to know its instant rate and also its history. In order to build such a system one needs to perform a characterization of the photosensors to be employed, and this was done for several families of SiPM in terms of gain, photon resolution, overall efficiency, noise, cross-talk. The cross talk evaluation, in particular, is crucial in order to determine the lowest threshold attainable. This is why a simple yet powerful analytical method was developed to measure it. Front-end electronics and an FPGA-based counting system were developed, in order to handle the data flow coming from the field sensors. Such a system also deals with the redundant data transmission toward a console with a graphical user interface and a data storage system. The redundant transmission is to be done simultaneously on differential cables and wireless, using powerful low-cost units. A robotic arm is being developed for remote ad-hoc inspection, both visual and by using high resolution radiation detectors; other operations around the waste drums should also be possible using the same robotic arm. Test results with radioactive sources showed very encouraging performance in terms of sensitivity, therefore a small demonstrator system will be placed around real radioactive waste drums quite soon.

Chapter 2 - Research reactors are important tools for radioisotope production, education and training, material structure studies, neutron radiography, silicon doping, neutron activation analysis, and neutron capture therapy. More than 600 research reactors were built worldwide since Fermi's graphite pile in 1942, of which about 200 are currently in operation. Research reactors present a large variety of designs, with different arrangements of moderator, reflector and coolant materials, fuel enrichment up to 93% and a wide power range, from sub-critical assemblies up to 250 MW. In this chapterwe will review the information available on the production of radioactive waste during the operation of research reactors as well as during decommissioning and dismantling of the installations.

Chapter 3 - Reprocessing of spent nuclear fuel requires thorough analytical control at every stage of the process to avoid alarm situations, environmental failures and economical losses. There are only limited possibilities for this monitoring nowadays – the standard in the area are de facto expensive and time-consuming techniques like ICP-AAS and ICP-MS. There is a growing interest to the developing of alternative approaches for HLW analysis in the last several years. One of the very attractive fields is the use of MS ERIAD – a version of the well-known electrospray-mass-spectroscopy. This method seems being a very promising one not only for the analysis of complexes in solution, but for elemental analysis at the same time. The advantages of the method are extremely small sample volumes (several microliters), high sensitivity and very simple sample preparation. Furthermore, it is possible to design MS ERIAD interface for standard mass- spectrometers. An extensive growth of the use of chemometric techniques in analytical chemistry through the last decade has lead to the appearance of several other promising analytical alternatives. One of them is the employment of arrays of non-specific, cross-sensitive chemical sensors in combination with chemometric tools. Although not competing in preciseness with ICP-based methods this approach allows for fast and simple quantification of rare earth elements in complex mixtures in the presence of major components like uranium and thorium. Another approach brought into life by chemometrics is the use of classical optical spectroscopy methods in UV, Vis and IR ranges in combination with modern multivariate regression algorithms. This chapter is devoted to the detailed description of these promising analytical techniques for HLW analysis.

Chapter 4 - The molten salt, 3LiCl(59.5mol%)-2KCl(40.5mol%), as the medium used in the pyroreprocessing by the electrorefining method will be contaminated by the accumulation of various actinoid elements (An) and fission products (FP) due to prolonged electrolytic operation. Though the An can beremoved at the pyrocontactor step by extraction using cadmium melt, the FP remain in the eutectic medium, which is regarded as a spent electrolyte at certain stage, can cause the rising of the melting point of the electrolytic bath and /or lowering the current efficiency. Some measures e.g., its replacement by a virgin medium, etc.

should be taken to maintain its stable condition. The constant replacement of the electrolyte, however, could lead to the generation of enormous volume of high-level radioactive waste (HLW). In terms of environmental load reduction and economical improvement, it is desirable to have the spent electrolyte purified for recycling by removing its FP. Some technical developments on spent electrolyte treatments have been carried out in several countries. One of them is the zeolite sorption of FP following An removal which is being developed in the United States, Japan, etc. Meanwhile, the contaminants precipitation methods by converting contaminants to insoluble compounds are undertaken in Russia and a few other countries and are anticipated to reduce the total volume of HLW. In the case of the Russian process, its purpose is to remove contaminants such as the minor actinoids (MA) and rare earth elements (RE) from the medium before its disposal. In the authors' case, the main objective is to recycle the purified medium, delaying its disposal for as long as possible. It is with this objective in mind that this study was undertaken. Author et al. have introduced a simple filtration method to remove RE particles which were formed due to the conversion of RE chlorides to phosphates. Here, the iron phosphate glass system is used as a filtration medium for the removal of FP particles. However, some soluble FP such as compounds of alkali metals, alkaline earth metals, etc. still remain in the medium. Next, authors attempted to apply the iron phosphate glass system as a sorbent instead, to remove the soluble FP. Authors have obtained some positive results and have intention to incorporate it into the spent electrolyte recycle process as a part of the FP separation and immobilization system. Although this method is still on a laboratory scale level, it is a potential rational method to use the iron phosphate glass form which possesses a larger loading capacity for FP than the conventional glass form.

In this manuscript, the author intends to describe the progress hitherto of the iron phosphate glass system functioning as a FP filter medium and sorbent as well as being the waste form itself.

Chapter 5 - According to the data of Far Eastern division of Federal Service for Nuclear Supervision of Russia (FSNS) after 25 years of mining in this area over 1 million tons of Uranium ore have been mined and stored on the surface as mine dumps. The total content of uranium in this amount of ore is approximately 2000 tons. This suggests high levels of radioactive contamination in this territory. Moreover, because these radioactive mine dumps have been around for nearly 40 years, one can suppose that anthropogenically contaminated territories, characterized by high levels of natural radioactive nuclides in soil, vegetation cover and surface waters, occupy large areas. It was revealed that the territory of Yuzhnoye deposit was the most contaminated one.

Chapter 6 - Solidification of low-activity wastes with cementitious materials is a widely accepted technique that contains and isolates waste from the hydrologic environment. The radionuclides ^{129}I, ^{75}Se, ^{99}Tc, and ^{238}U are identified as long-term dose contributors. The anionic nature of these radionuclides in aqueous solutions allows them to readily leach into the subsurface environment. Any failure of concrete encasement may result in water intrusion and consequent mobilization of radionuclides from the waste packages via mass flow and/or diffusion into the surrounding subsurface environment. In order to assess the long-term performance of waste grouts for encasement of radionuclides, the diffusion of radionuclide species must be measured when contacted with vadose zone porewater or groundwater. A series of sediment-concrete half-cell diffusion experiments were conducted under unsaturated conditions (4%, 7%, and 15% by weight moisture content) to quantify the diffusion of radionuclides from surrounding subsurface sediment into concrete wasteforms. The data from this investigation provides valuable information for understanding the speciation, behavior, and fate of radionuclides immobilized within concrete wasteforms under vadose zone conditions. Compilation of this data with previous data quantifying the diffusion of radionuclides from concrete wasteforms into the surrounding subsurface environment will allow quantification of the complete diffusion profile and predictive simulation of the long-term performance of concrete wasteforms and the fate of contaminants immobilized in cementitious waste forms.

Chapter 7 - Global inventories of spent nuclear fuel are expected to increase substantially in future years. Because of the widely accepted international principle of whoever generates the waste also is responsible for its safe and secure disposal, the focus for the long-term management of highly radioactive waste has been on national disposal efforts. However, it is unlikely that any national repository for highly radioactive waste will be ready to accept nuclear waste before 2020. As a result, there has been an increased realization that radioactive waste is a global concern that requires more international cooperation. There are a variety of international actors involved in the management of highly radioactive waste. Clearly, the countries that generate the nuclear waste play a critical role in the management of the waste since they ultimately are responsible for constructing and operating disposal facilities. International conventions help to monitor and regulate national radioactive waste management efforts and establish international norms. International governmental organizations and non-governmental organizations are playing increasingly important roles since they promote research and development, share information, and lobby for various policies affecting nuclear power and highly radioactive waste disposal.

While the international community has made some progress in addressing the issue of safely managing the increasing inventories of highly radioactive waste, much still needs to be done. Ultimately, the safe and effective long-term management of highly radioactive waste must be handled on a global scale in order to reduce safety, environmental and security concerns.

In: Radioactive Waste: Sources, Types and Management ISBN: 978-1-62100-188-1
Editors: Satoshi Yuan and Wenxu Hidaka

Chapter 1

DMNR: A New Concept for Real-Time Online Monitoring of Short and Medium Term Radioactive Waste

Paolo Finocchiaro
Istituto Nazionale di Fisica Nucleare, Laboratori Nazionali del Sud, Catania, ITALY

Abstract

One of the main topics in decommissioning and storage of radioactive waste is its monitoring in the short, medium and long term. Whereas for the long term waste the geological disposal is foreseen, or better the use of new "cleaner" reactors, the short and medium term waste is currently packed into drums and either kept inside the production site or transported to dedicated storage sites. In any case special care must be devoted to the prompt detection of possible leaks due to structural weakness or accident.

Whichever the topological arrangement or the physical structure of the drums, the goal would be measuring the radioactivity around each single drum. This could be done by means of ordinary geiger-Muller counters, but the cost of such a solution would be prohibitive. Therefore, what is usually done is a periodic check by operators (or robots) plus some overall monitoring in fixed locations inside the repository itself.

Detector Mesh for Nuclear Repositories is a project for a prototype demonstrator of the online monitoring of short-medium term radioactive waste, currently under development at INFN-LNS Catania. Such a system is planned to be distributed, fine-grained, robust, reliable, and based on low-cost components.

With the main purpose of counting gamma radiation, a new kind of mini-detector was implemented that behaves like a scintillating geiger-Muller counter. It has to be replicated in shape of a grid around each single waste drum. This system could also open new perspectives on the modality of waste packaging and storage. What was proposed is cheap scintillating detectors, readout by means of SiPM photosensors capable of detecting the tiny light signals of few photons. This way one can easily place grids of 5-10 radiation sensors around each waste drum, and record continuously the measured activity in order to know its instant rate and also its history.

In order to build such a system one needs to perform a characterization of the photosensors to be employed, and this was done for several families of SiPM in terms of gain, photon resolution, overall efficiency, noise, cross-talk. The cross talk evaluation, in particular, is crucial in order to determine the lowest threshold attainable. This is why a simple yet powerful analytical method was developed to measure it.

Front-end electronics and an FPGA-based counting system were developed, in order to handle the data flow coming from the field sensors. Such a system also deals with the redundant data transmission toward a console with a graphical user interface and a data storage system. The redundant transmission is to be done simultaneously on differential cables and wireless, using powerful low-cost units.

A robotic arm is being developed for remote ad-hoc inspection, both visual and by using high resolution radiation detectors; other operations around the waste drums should also be possible using the same robotic arm.

Test results with radioactive sources showed very encouraging performance in terms of sensitivity, therefore a small demonstrator system will be placed around real radioactive waste drums quite soon.

1. Introduction

The term radioactive waste (or shortly radwaste) indicates waste containing radioactive material. It is usually produced as a residue of nuclear processes, such as nuclear fission, even though industrial processes and medical applications may also produce radioactive waste. Contrary to chemical waste dangerousness, radioactivity decreases with time, therefore one could in principle seal and isolate the waste for some time until its radioactivity poses no relevant risk. Unfortunately "some time" can range from a few hours to years for medical and industrial radwaste, and up to tens of thousands of years for high-level waste from nuclear fission reactors and nuclear weapons decommissioning [1].

As to radwaste disposal, the IAEA classification of Radioactive Waste [2] defines:

- high level waste (HLW);
- low and intermediate level waste (LILW);
- waste that can be cleared from nuclear regulatory control.

LILW is further separated into short lived and long lived waste according to the typical decay time of its radioactivity. Operational waste from nuclear power plants typically falls into the category of short lived LILW if it cannot be cleared from nuclear regulatory control. Spent fuel would be HLW if declared as waste, whereas most of the currently produced radwaste belongs is LILW.

The main approaches pursued so far to manage radwaste are segregation and storage for short-lived wastes, near-surface disposal for low and some intermediate level wastes, deep "geological" burial or transmutation for the long-lived high-level wastes.

A summary of the amounts of radioactive wastes and management approaches for most developed countries is presented and reviewed periodically as part of the IAEA Joint Convention on Safety of Spent Fuel Management and on the Safety of Radioactive Waste Management. In particular, during the meeting held in Vienna on May 2009, all of the representatives "*recognized that safety of spent fuel and radioactive waste management is a crucial and difficult topic, and that there are considerable areas for improvement*", and "*recommended that the safety of spent fuel and radioactive waste management be taken into account from the very beginning"*. This means that so far there is no final decision worldwide concerning the disposal of radwaste, and that there is room for R&D, improvement and innovation.

Decommissioning radioactive facilities produces different categories of radioactive waste. How much waste is generated in each category depends on the timing of dismantling operations. Postponing it may reduce the amounts of intermediate level waste (ILW) and increase amounts of low level waste (LLW) and other waste that can be cleared from regulatory control. This will influence disposal arrangements and costs. Recently several countries have introduced the category of very low level waste (VLLW), that allows to strongly reduce the cost of LLW disposal. However, disposal facilities do not exist everywhere and therefore, if no appropriate disposal facilities for the amounts and categories of waste are available, there are two possibilities:

- maintain the installation in safe enclosure mode or
- dismantle the installation, condition the waste and place it into suitable temporary storage sites [3].

Countries without waste disposal sites apply long term safe enclosure policies for the shutdown of their nuclear facilities. Even countries with some waste disposal options may not have disposal options for all kinds of decommissioning waste. Materials (or equipment) removed from a decommissioned nuclear installation typically fall into one of the following four classes, even though with differences from country to country:

- materials that are to be stored (or disposed of) under controlled and monitored conditions;
- materials that can be released for a specific restricted use outside the nuclear industry (e.g. as the foundation for an airport runway);
- materials authorized for reuse within the nuclear industry;
- materials that can be cleared for unrestricted reuse or disposal [3].

In the following it will be briefly described the typical kinds of radioactive waste, their radioactivity and the rough effect of shielding, the advantages of a continuous monitoring; then the proposed solution will be described in detail.

2. The Radioactive Waste

The radioactivity of nuclear waste decreases with time, as every radioactive species decays following the well known exponential law of Eq.1

$$N(t) = N_0 e^{-\frac{t}{\tau}} \qquad (1)$$

where N(t) is the population at the time t, N_0 is the radioisotope population at a chosen initial time, τ is the "decay constant" characteristic of each nuclear species. If we imagine to count the number of nuclei of a given radioisotope at the time t=0, and then to count it again after a time τ, we will find that its population is reduced by a factor

$$\frac{N(\tau)}{N_0} = \frac{1}{e} \approx 0.37 \qquad (2)$$

Therefore the decay constant τ is the time needed to reduce the population of the given radioisotope to *1/e* (37%) of the initial value. Quite frequently people make use of the "half-life" $T_{1/2}$ that is the time needed to reduce the initial population to 50%. From Eq.1 it follows that the two constants are bound by the relation

$$T_{1/2} = \tau \cdot \ln 2 \qquad (3)$$

The new nuclear species produced by a radioactive decay can be, in its turn, radioactive as well and undergo further decay governed by a different decay constant. This gives rise to the so called "decay chains". I do not go into the mathematical details of the combined decay processes, however the overall conclusions are that:

- The longer the half-life the smaller the radioactivity will be, and vice versa.
- The observed decay rate behavior in radioactive materials is dominated by the longest half-life of its radioactive components.
- Eventually all radioactive waste decays into non-radioactive elements, but this can take seconds for some species and up to million years for others.

For instance, some radioactive elements such as ^{239}Pu in "spent" fuel will remain hazardous to living creatures for hundreds of thousands of years. Conversely other short-lived radioisotopes, such as ^{131}I which has an 8-day half-life, will be very active and dangerous initially, but with a quickly decreasing hazardousness in matter of weeks or months. Type and energy of the ionizing radiation emitted by the decay of each radioisotope represent important factors to determine how dangerous it is. On top of this, the chemical properties of the radioisotope, and of its possible "parents" and/or "children" along the decay chain, are additional elements playing a relevant role in the effective hazardousness.

The main sources of radwaste are typically by-products of:

- Nuclear fuel cycle, including mining, enrichment, reprocessing, decommissioning.
- Nuclear weapons decommissioning.
- Legacy radwaste, that is radioactive waste produced decades ago and stored or disposed of in several locations.
- Medical diagnostics and treatment.
- Industrial applications, like high-energy gamma radiography and oil well logging by means of neutron emitters.
- Processing naturally radioactive material, like ^{40}K that is contained in water and all the minerals containing natural Potassium; remarkably, even oil, gas and coal contain radioactive elements.

The main radioactivity types are alpha, beta and gamma, also called "ionising radiation". In addition we can list nuclear fission and neutron emission.

- Alpha decay, typically occurring in heavy elements, consists in the emission of an ^{4}He nucleus, also called "alpha particle", with a given kinetic energy. Due to its mass and electric charge it is not penetrating and is typically stopped within few micrometers of matter.
- The beta decay can be positive or negative, depending on whether a positron (e+, positive electron) or an electron (e-) is emitted. In both cases an additional neutral particle is emitted, neutrino or antineutrino, which shares the available kinetic energy with the beta particle. For this reason the beta energy of a given decaying species is variable from 0 to a maximum value. The beta radiation is typically stopped within few millimeters in matter or few centimeters in air, whereas the neutrinos basically do not interact at all.
- The gamma decay consists in the emission of a high-energy quantum of electromagnetic radiation (also called "gamma photon") by a nucleus in an excited state. The well defined excitation energy value is imparted to this gamma photon, that is very penetrating. Depending on the energy, stopping gamma rays can require from centimeters to meters of high-density materials (e.g. lead). X rays, also produced in radioactive decays, are simply lower-energy gamma rays.
- Nuclear fission is the process where a heavy nucleus splits into two fragments, generally releasing at the same time a few neutrons. The two fragments are typically radioactive according to one or more of the previous decay modes and, being quite heavy, are stopped immediately in matter. Conversely, the neutrons carry no electric charge and as a consequence they are highly penetrating and do not produce direct ionization. However, in order to lose kinetic energy thay have to collide directly with nuclei, hence they will produce indirect ionization even at relevant distance from their production point. In order to stop neutrons one needs special materials and geometries.

All the listed types of radiation are emitted by the radwaste, in absolute and relative amounts that depend on the particular composition. This is why each kind of waste might need to be embedded in a suitable matrix, then surrounded by some shielding envelope and finally inserted into some sort of mechanical assembly (typically a drum). Figure 1 shows a simple sketch of the radiation types and their penetration level in ordinary radwaste.

According to the recommendations of IAEA, storage is an option that has to be considered in any waste management strategy [4]. Proper storage should be available at all stages in waste processing for isolation and environmental protection, and to facilitate retrieval for subsequent steps. This is particularly important for LILW, which may be stored to allow its radioactivity to decrease to levels that permit authorized discharge or reuse. This is the case, for instance, with waste abundant in radionuclides of very short half-life (few hundred days). Thus, when designing storage sites, sufficient additional capacity should be foreseen as a reserve capacity for waste generated in any incidents or in the event that the waste cannot be transferred off the site because, for example, a disposal facility is not available.

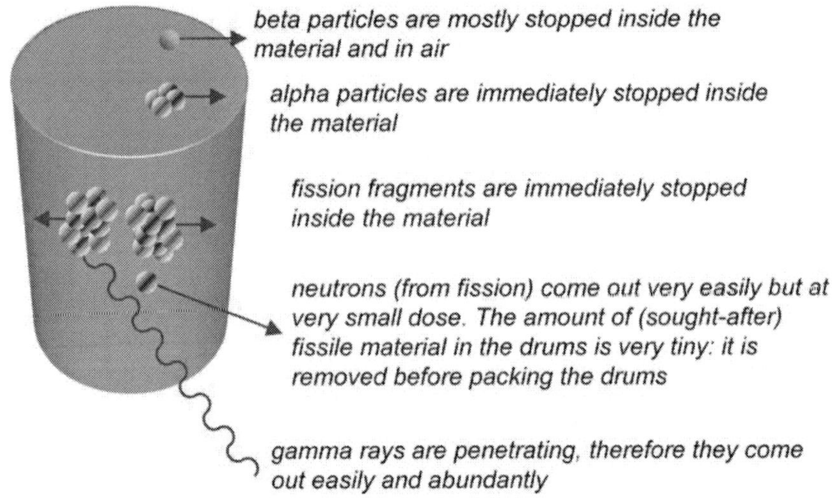

Figure 1. Sketch of the radiation types, and their typical penetration level, in ordinary radwaste.

An important point to be considered is the possibility that significant volumes of liquid waste be generated, in which case one should foresee special collecting tanks instead of drums. Such tanks should be made of chemically resistant material and surrounded by secondary containment vessels, to prevent the spread of contamination in the event of any leakage, and by adequate shielding layers.

3. Current Monitoring Methods

The stored waste packages should be kept at the same integrity level as when received, until retrieved for further treatment, conditioning or disposal. IAEA again recommends that the design of the facility should permit radiation monitoring and inspection, including visual or other examination of the waste packages, to obtain early indications of any physical deterioration or signs of leakage [4]. Radiation monitoring and visual inspection is expected to be performed whenever the waste is handled or moved (placed into storage, retrieved or transported off the site), in order to protect workers handling the waste, to help preventing the accidental spread of contamination and to provide an additional check of the record keeping system. Therefore the basic operations to be routinely performed are [5]:

- Monitoring at source and in the environment.
- Monitoring the status of the containment for the radioactive waste in the storage location.
- Monitoring changes in the characteristics of the radioactive waste, in particular if storage is continued for extended periods, by means of inspection and regular analysis.

The monitoring procedures officially employed at the moment mostly everywhere are based on some preliminary characterization of the waste during the packaging phase, and then on visual inspection and dose counting during its transport to the storage place. Periodical checks by means of handheld radiation detectors are scheduled, ambient radiation monitors are installed in predefined locations inside the storage site, and environmental monitors are installed outside. Ambient monitors usually feature several alarms, set at appropriate levels, to ensure the detection, location and assessment of any leakage from the containment. These monitors, however, only provide a rough overall indication, whereas localized leaks could go undetected still being extremely dangerous for an operator who moves near the leaking package for routine operations.

The characterization process, hopefully to be performed for all steps in radioactive waste management, should include the measurement of physical and chemical parameters, the identification of radionuclides and the measurement of activity content. This would allow to keep track of the history of the radioactive waste packages through the stages of conditioning, storage and disposal, and to maintain logbooks for the future. Not all of these measurements are suitable to be realistically performed at all the stages, however. In several cases the packages can be periodically checked by moving them to a characterization station, still inside the storage site, even though this displacement procedure can in itself produce additional risks of accidents.

The results of all these measurements must be recorded and retained and retrievable for an appropriate period of time, along with all the relevant details on radioactive waste, on waste packages, and on the contents of waste stores. At any time during storage it should be possible to determine from the records the type, activity and characteristics of the waste stored in each specified location [5].

Even though there exist rules, recommendations and laws regulating the management of radwaste, there is a worldwide diffuse distrust toward it. From a scientific standpoint the situation, at least for the short and medium term waste, is clear because the most relevant part of the problem is the radioactivity that is well known and handled. However, the technological implementation of the solution generates distrust, and quite often fears, that can be sometimes justified and sometimes not. By surfing the web with the help of a search engine one can immediately realize that the problem is not settled at all.

In the general conception of the population the words "nuclear" or "radioactivity" immediately engender fears somehow connected with terrible weapons and the end of world war two. Unfortunately the spread of psychological (or real) fears is much faster than the slow persuasion by means of the development of knowledge and awareness.

Figure 2. A cartoon illustrating the question posed by radwaste monitoring information: to know or not to know?

A possible way out, in my opinion, could be represented by the information: in other words, if the data concerning the radwaste monitoring could be measured in real time and somehow made available online, this could represent a double achievement. On the one hand the diffusion of information could provide a contribute to the overall knowledge of the general public thus attenuating the fears; on the other hand it would prevent possible accidents to go undetected (or worse, possible data manipulation aimed at hiding the real danger connected with radwaste). The back face of this medal is that the data could also be misused to spread false alarms and fears.

The question, jokingly posed in Figure 2, is therefore: to know or not to know?

4. Real-Time Monitoring and Data Availability

From a technical point of view a real-time monitoring of radwaste packages was already possible years ago, even though with some limitations. As it was said in the previous section, ambient radiation monitors are typically installed in several locations inside the storage sites. What one should do is replicating these systems in form of a mesh of detectors (i.e. geiger counters) to be deployed all over the storage site around the packages and interconnected with an overall transmission and control system. This could work in principle but posing a set of operational issues, first of all the cost. The main features one would require to the radiation sensors of such a system are:

- A mesh of detectors to be installed onto standard platforms hosting radwaste drums.
- Radiation hardness in order to withstand the expected radiation doses for nominally 10-100 years.
- Mechanical robustness and flexibility, so that one can easily modify their shape without the need of fabricating different models for each different arrangement.
- Low intrinsic efficiency, in order to cope with the high and very high activity reasonably expected near a leaking package, without being saturated.
- Linear sensor shape and size of the order of 1-2 meters, in order to "watch" at least one full dimension of a drum (height or circumference).
- Possibility to freely choose at design stage the sensor geometrical efficiency, by simply choosing a different length or width of the standard sensor.
- High sensitivity, required to detect radiation depositing only a small amount of energy.
- Reliability.
- Possible position sensitivity, which could improve the leak detection features.
- Ease of handling and parts replacement.
- Low cost.

Moreover, the system should feature a robot, or robotic arm, providing remote visual inspection of the drums and capable of high resolution radioactivity measurements by means of a more sophisticated and specialized radiation detector. The robot should also be able to replace failing sensors and/or electronic modules and batteries.

As for the acquired data, the system should exploit some redundancy on the transmission from the sensors to the control station, possibly making use of batteries to make up for power failures and to keep the monitoring in operation also during possible displacements of the platforms.

Under these conditions it could be possible to achieve a real-time three-dimensional radioactivity map, capable of monitoring the expected decrease of radioactivity over time on a drum-by-drum basis. Alarm levels and leak detection would be an additional benefit of such a configuration.

Downstream from the hardware and the data acquisition software, the system should feature a high-level software for graphical user interface (GUI) and a powerful database engine for archiving/retrieval of all the information related to the waste packages. For instance the history of monitored counting rates could be integrated in the database with the data needed for waste characterization and for a national inventory of waste; moreover, the records from the control processes for treatment, packaging and conditioning could also be included, as well as the audit records for individual containers and packages. Accidents or non-compliances with the specifications for waste packages, and the respective consequent actions, should also be stored along with the monitoring records [4].

The database, which should be safely redundant (i.e. duplicated or triplicated in different physical locations and protected against power failures), could then be interrogated at any moment in order to extract and analyze data under many different correlation conditions. This would constitute a powerful analysis tool for the detection of anomalous behavior, leaks, damage, wrong operations, procedural defects or human mistakes. Finally, it could to a large extent automate the production of the periodical reports to be submitted to the regulatory bodies in compliance with the conditions of authorization.

5. The Proposed System: DMNR

A step further in this waste monitoring concept, I believe, comes with the Detector Mesh for Nuclear Repositories (DMNR) project, currently under development at INFN-LNS Catania with the goal of constructing a proof-of-principle prototype demonstrator of the online monitoring of short-medium term radioactive waste [6],[7].

As shown in Figure 1 the main radioactivity coming out of a waste package is typically gamma, even though in some cases one could have to deal with neutrons from fission. This is why the focus was placed on the development of a new kind of mini-detector that behaves like a cheap and simple scintillating geiger counter. The system is based on arrays of scintillating optical fibers readout by means of SiPM photosensors capable of detecting the tiny light signals of few photons. Simulations showed that even employing a thin fiber, this sensor can detect gamma rays. Moreover, it was also proved that the same detection principle, coupled to a neutron converter like ^{6}LiF, could be usefully exploited to detect thermal neutrons.

For the first prototype round fibers (1mm diameter) and square sensitive area SiPM (1mm x 1mm, produced by SensL) were chosen, because the round fiber does not have any preferential orientation with respect to the incident radiation, whereas a square SiPM is better tolerant to slight misalignments with the fiber.

Some front-end electronics and an FPGA-based counter set were also developed, in order to handle the counting rate data flow coming from the fiber sensors. Such a system also deals with the redundant data transmission toward a console with a graphical user interface and a data storage system. The redundant transmission is done simultaneously on differential cables and wireless, using powerful low-cost and low-consumption units also powered by rechargeable batteries.

A robotic arm is currently being finalized for remote ad-hoc inspection, both visual and by using high resolution radiation detectors; additional operations around the waste drums will likely be possible using the same robotic arm. An overall sketch of the DMNR system is shown in Figure 3.

It is worth to be noted that, due to the system modularity, the possible configurations range from a complete monitoring system to a sort of small "black box" to be used as a mobile monitoring system during displacements or transportation of radwaste.

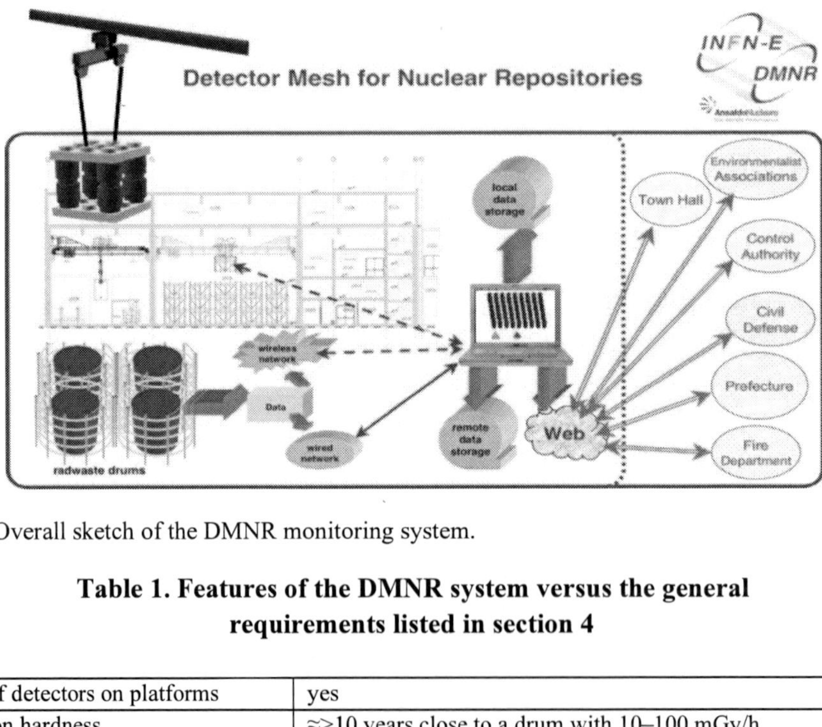

Figure 3. Overall sketch of the DMNR monitoring system.

Table 1. Features of the DMNR system versus the general requirements listed in section 4

mesh of detectors on platforms	yes
radiation hardness	≳>10 years close to a drum with 10–100 mGy/h
robustness and flexibility	yes, plastic scintillators; SiPM not damaged by ambient light exposure
low intrinsic efficiency	yes, 1mm thick round plastic fibers
linear sensor shape, 1-2m long	yes
geometrical efficiency can be chosen by length or thickness	yes, fibers can be freely cut from spool
high sensitivity	yes, SiPM can detect down to single photons
reliability	yes, simple system
possible position sensitivity	yes, by crossing fibers
ease of handling	yes, only plastic cases
low cost	yes (for mass production)

The global features of the DMNR system as compared to the general requirements listed in the previous section, are reported in Table 1.

6. THE BASIC SENSOR UNIT

6.1. The Scintillating Fiber

Based on the previously listed general constraints, and considering other overall safety and redundancy requirements and concepts, a possible detector unit principle and an ensuing overall setup of the system were sketched.

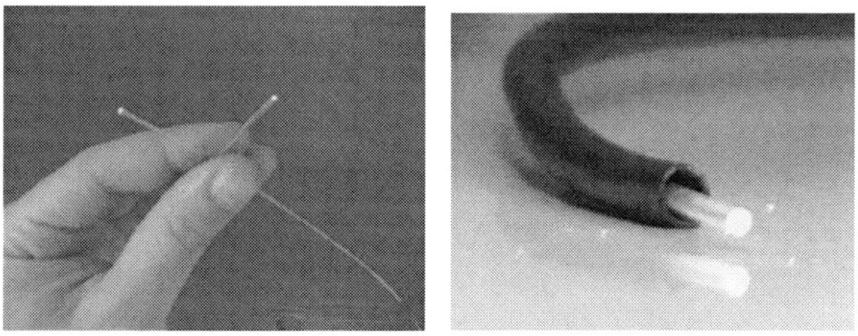

Figure 4. A scintillating fiber. On the right it is shown inside a black sheath acting as first mechanical protection and ambient light shield.

Table 2. Properties of the scintillating fibers currently employed in the DMNR project

Fiber type	BCF-20
Core material	Polystyrene
Cladding material	Acrylic
No. of H atoms per cc (core)	4.82×10^{22}
No. of C atoms per cc (core)	4.85×10^{22}
No. of electrons per cc (core)	3.4×10^{23}
Core refractive index	1.60
Cladding refractive index	1.49
Density	1.05
Numerical aperture	0.58
Cladding thickness, round fibers	3% of fiber diameter
Cladding thickness, square fibers	4% of fiber size
Trapping efficiency, round fibers	3.44% minimum, at each end
Trapping efficiency, square fibers	4.4% minimum, at each end
Emission color	Green
Wavelength of emission peak (nm)	492
Decay time (ns)	2.7
1/e attenuation length (m)	>3.5
Light yield (photons/MeV)	~8000
Operating temperature	-20°C to +50°C
Vacuum compatible	Yes

The idea is to employ tiny plastic scintillators that are poorly sensitive to gamma radiation and, if used in reduced size, would be yet less efficient. For the prototype system it was decided to use plastic scintillating fibers (Figure 4), because they have the further advantage of coming in a predefined shape that is also suitable for immediate geometric modifications: one can cut them to a desired length and arrange them around the source at will, even flexing or looping them. The main properties of the fibers currently being employed are listed in Table 2.

Each detector consists of a scintillating optical fiber, 1-2 m long and 1 mm diameter, which can be arranged around each drum in longitudinal and/or ring geometry (Figure 5). If both geometries are adopted, the most likely leak position can be determined by combining the increased activity measured by a longitudinal and a ring fiber.

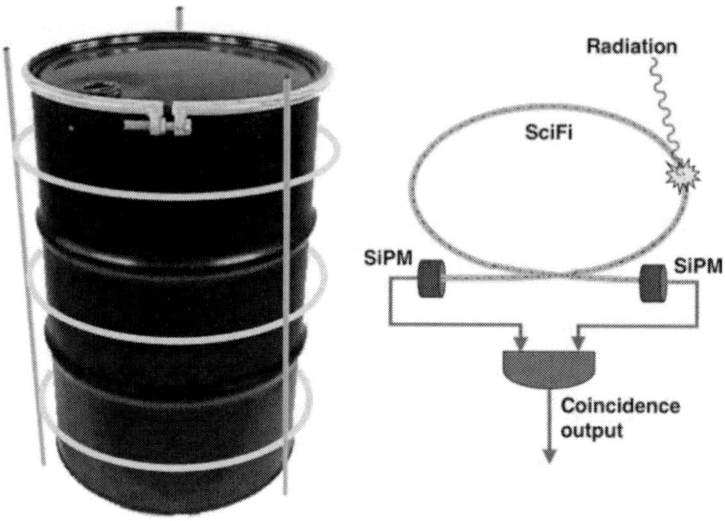

Figure 5. A possible arrangement of the fibers around a radwaste drum (left), and a sketch of the sensor working principle (right).

Each fiber can intercept radiation coming off the drum wall, mostly gamma rays, so that the energy released inside its active volume is converted into scintillation light, that propagates to both ends of the fiber itself. The scintillation efficiency of the fibers is about 8000 photons/MeV. Considering that (i) the expected gamma rays from nuclear waste have energy below 2 MeV; (ii) the average energy (excluding zeroes) released in a 1 mm thick fiber is roughly around 100 keV or higher, as resulting from GEANT numerical simulations and shown in Figure 6; (iii) at least 6% of the produced scintillation light is conveyed to each fiber end [8]; (iv) the photon detection efficiency, PDE, of a photodetector can be assumed of the order of ≈20% in the emission wavelength range of the fiber (as shown below); as a consequence the number of photons to be detected can be as low as few units, too low to be managed by ordinary photomultiplier tubes. Moreover, photomultipliers would also pose additional issues like cost, non-ruggedness, fragility, high-voltage power supply, damage from accidental exposure to ambient light. Therefore one needs a rugged photodetector to be coupled to each fiber end, with a sensitivity down to the single photon and capable of fast timing, in order to select events in strict coincidence between the two fiber sides. When a pulse is detected on both sensors within few nanoseconds, it is recognized as an ionising event and can then be counted and recorded to compute the overall activity. The coincidence between the two fiber ends is required in order to reduce the spurious background counts due to dark noise of the photosensors.

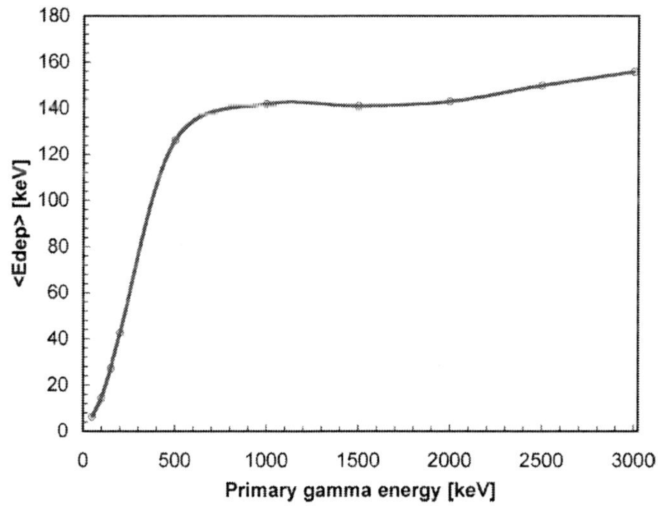

Figure 6. Average energy deposited in a 1mm diameter plastic fiber as a function of the incident gamma ray energy, as resulting from numerical simulations performed with GEANT Monte Carlo code. The line joining the simulated points was drawn as a guide for the eye.

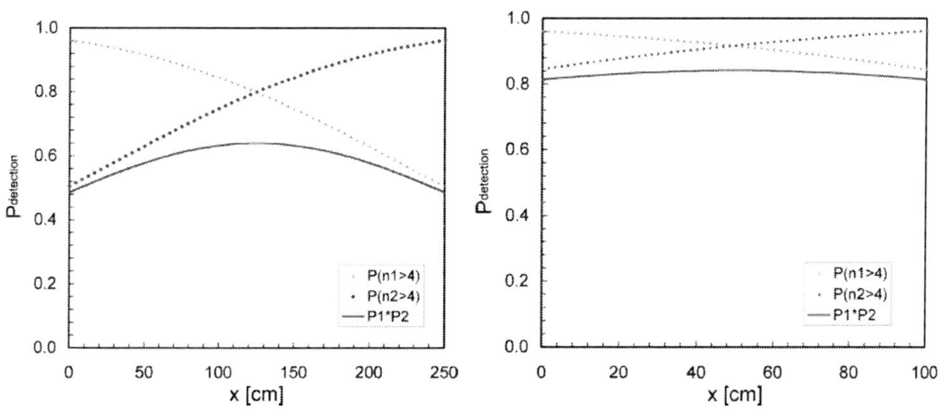

Figure 7. Detection probability as a function of the impact position, for the individual photosensors and combined, respectively for a 2.5m long fiber (lefthand plot) and for a 1m long fiber (righthand plot). The average deposited energy was assumed 100keV (Figure 6), the detection threshold on each photosensor was assumed 4 photons. Under these conditions one can see that the expected sensor response is flat for the 1m fiber, and its non-uniformity for the 2.5m fiber is not dramatic (20%).

Figure 7 shows a rough calculation of the detection probability as a function of the impact position for a gamma ray depositing 100 keV onto a fiber, using the parameters indicated above in (i-iv). Assuming a Poisson distribution for the scintillation photons, and a detection threshold of 4 photons on each fiber end, the plots report the individual detection probabilities at each end (P1 and P2, dotted lines) and the coincidence probability for two fibers 2.5m and 1m long.

As to the radiation hardness, Acosta et al. [9] measured a 0.75 reduction factor of the light transmission efficiency after irradiation with a 95kGy dose of gamma rays from a ^{60}Co source. In the proximity of a radwaste drum one can reach a dose rate of the order of 10-100mGy/h [10], that corresponds to a time span around 100-1000 years. From this point of view fibers made from glass scintillator, rather than plastic, would be still harder. However, the long decay time of their scintillation light (≈3ms) represents a limitation for pulse counting activity measurement.

Using plastic scintillating fibers fast neutrons, beta and gamma rays can be detected, whereas other high-energy charged particles, even though detectable with high efficiency, never reach the fibers in a radiation waste environment. In order to make them sensitive also to thermal neutrons, such fibers could be suitably doped or covered with some material that is able to convert low energy neutrons ($E < 1eV$) into higher energy radiation, such as ^{6}LiF. Further details on this technique, that go beyond the scope of the present subject, can be found in ref.[11].

6.2. The Silicon Photomultiplier

In order to detect the scintillation photons produced by the fibers, the choice fell on the recently born silicon photomultiplier, somewhere named SiPM, GAPD, SPM or MPPC, depending on the manufacturer. In the following it will be denoted as SiPM. This device consists of an array of identical photodiode cells, operated slightly above their breakdown voltage in the geiger avalanche regime, with output terminals connected together. This way the common output signal is the analog sum of those from many independent identical cells, and the SiPM operates as a quasi-digital photon counter (Figure 8). Its peculiarity is the sensitivity to single photons (see Refs. [12] and [13] and the references therein).

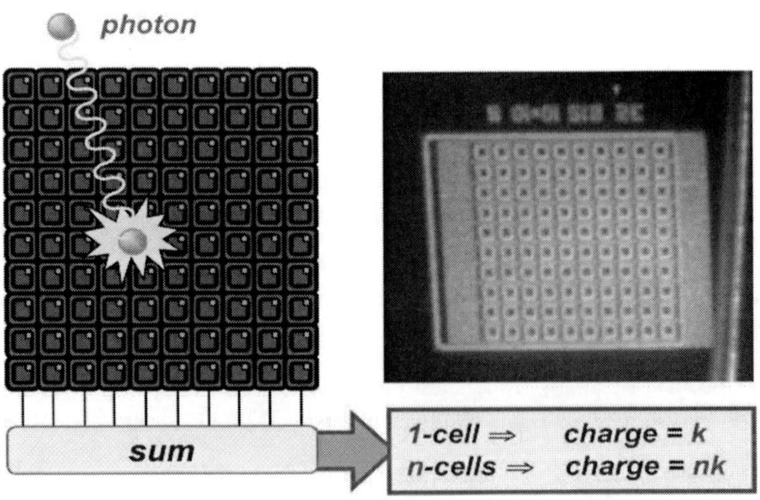

Figure 8. Sketch of the SiPM operating principle: the output is a charge signal directly proportional to the number of elementary cells fired. The picture on the righthand side represents a close-up view of a prototype SiPM, produced by ST Microelectronics, featuring 10x10 cells on a 0.5mm x 0.5mm area.

With these two choices, scintillating fiber and SiPM, one should be able to meet the efficiency and sensitivity issues. Unfortunately the SiPM is affected by intrinsic dark noise, in the form of counts due to a single cell randomly triggered for thermal reasons and not distinguishable from true photons. Moreover, it is also affected by cross-talk: whenever a cell is triggered, the electrons in the avalanche can themselves generate a few visible photons [12], and these photons have a finite probability to reach adjacent cells and trigger them. Therefore, in order to reduce (or suppress) the spurious counts, it was decided: (i) to employ two SiPMs, one at each end of the fiber, operated in coincidence; (ii) to set a suitable detection threshold on each SiPM equivalent at least to 4 triggered cells.

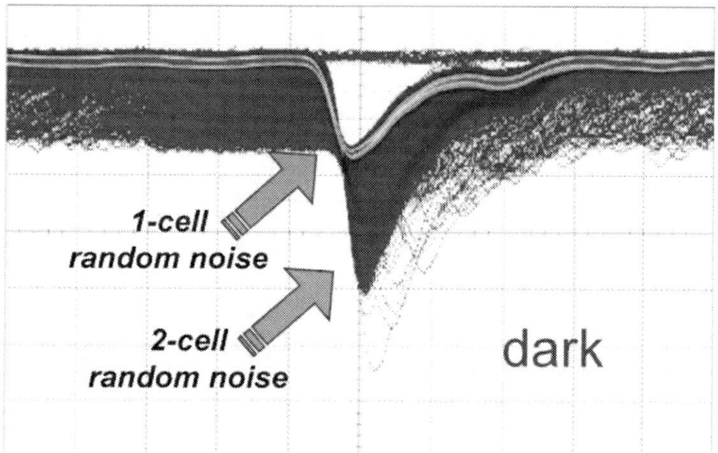

Figure 9. Snapshot of a persistence plot on a digital scope, showing a collection of dark count signals from a prototype SiPM not affected by cross-talk (the same non-commercial device shown in Figure 8, produced by ST Microelectronics).

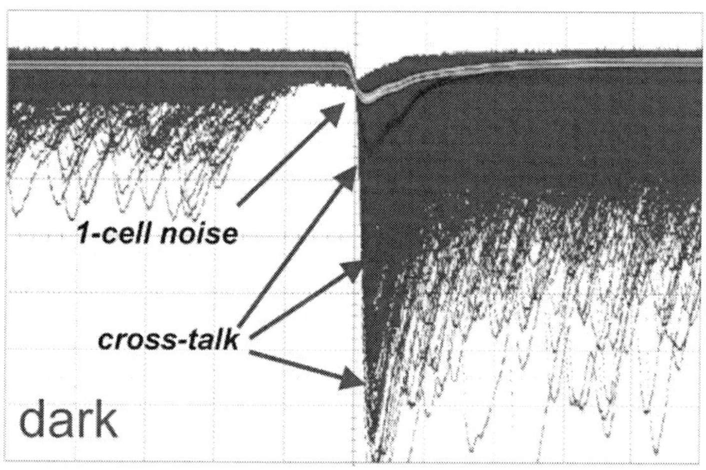

Figure 10. Snapshot of a persistence plot for another SiPM clearly affected by cross-talk. It is immediately evident the need of two sensors, one at each fiber end, with suitable thresholds in order to reduce the spurious counts.

Figure 9 shows a persistence plot, made using a digital scope, with a collection of dark count signals in a non-commercial prototype SiPM not affected by cross-talk (the same device shown in Figure 8, produced by ST Microelectronics). Clearly visible are the random noise signals involving two cells. In Figure 10 the same plot is shown for another SiPM clearly affected by cross-talk. From the figure it is immediately evident the need of two sensors, one at each fiber end, with suitable thresholds in order to reduce the spurious counts.

Figure 11 shows again the same kind of plot, but this time with the SiPM illuminated by light pulses produced by a fast pulsed laser. Several bands are visible, each one corresponding to a discrete number of photons detected.

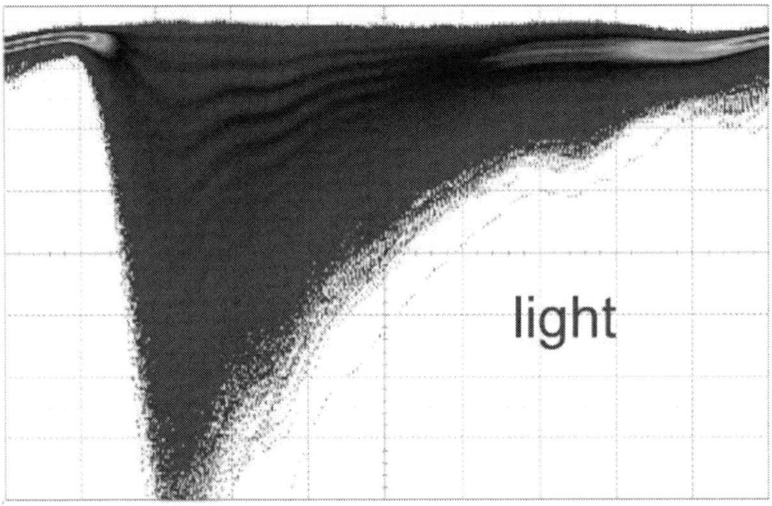

Figure 11. Snapshot of a persistence plot with the SiPM illuminated by light pulses produced by a fast pulsed laser. Several bands are visible, each one corresponding to a discrete number of photons detected.

The ideal charge output of a SiPM should approximate digital information, i.e., a signal which is an integer multiple of the elementary cell output, depending on the number of detected photons. This, of course, only holds in the ideal case because of several reasons:

- Small differences between individual cells may result in fluctuations of the output signal.
- Electrical noise, uncorrelated dark counts, afterpulses, and crosstalk can perturb the output signal shape.
- A photon interacting with a cell while it is recharging gives rise to a smaller signal.
- A photon interacting in a boundary region between cells or at the wrong depth may give rise to a smaller signal.

For all of the aforementioned reasons, what one expects as a SiPM charge spectrum when it is illuminated with a low-intensity time-coherent light pulse is an overall Poisson distribution modulated with gaussian peaks related to integer numbers of detected photons. And this is actually what one finds, as shown in Figure 12 and Figure 13 where each peak corresponds to an integer number of cells fired (i.e. photons detected).

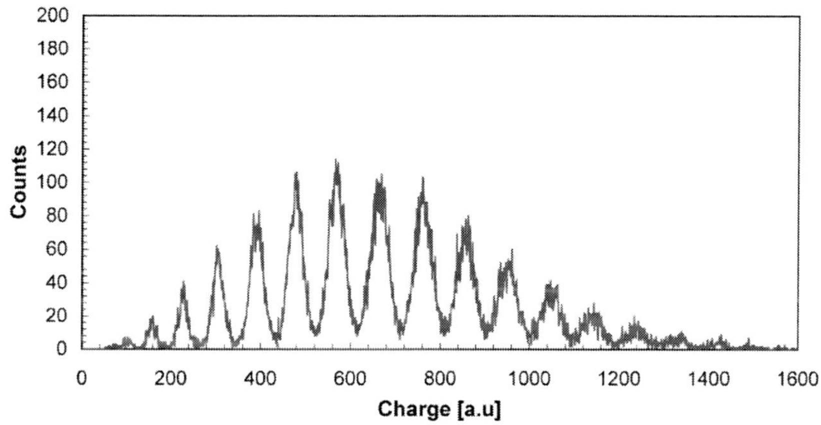

Figure 12. Charge spectrum of a SiPM (ST Microelectronics, 100 cells) illuminated with low-intensity light pulses produced by a fast pulsed laser. An ORTEC FTA810B linear amplifier was employed.

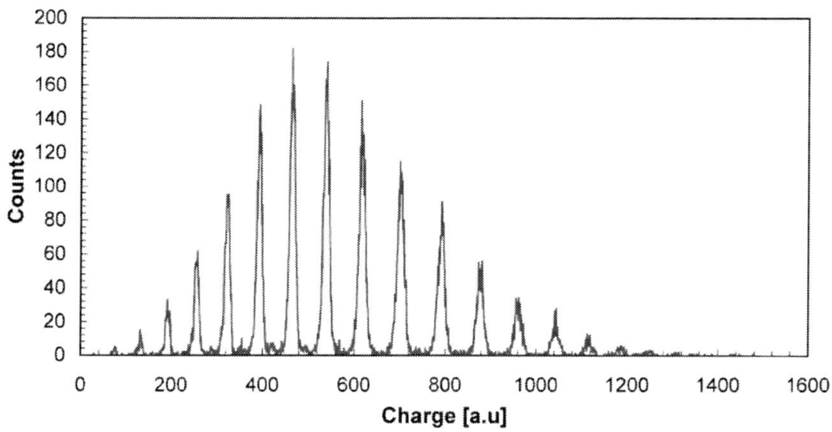

Figure 13. The charge spectrum of Figure 12, under the same conditions, but using a linear amplifier developed at LNS. The already excellent photon resolving power visible in Figure 12 is further improved.

As mentioned earlier, the distribution of the charge signals from the SiPM, when illuminated with low-intensity laser pulses, is a Poisson one that in principle should be discrete. However, due to the mentioned effects, each single value is smeared in a gaussian fashion and its width increases with the order *n* of the peak, following Eq.4, where σ_1 is the width of the first gaussian ($n=1$).

$$\sigma^2(n) = n\sigma_1^2 \tag{4}$$

The spectrum shape is then expected to be the convolution of a Poisson distribution with several equally spaced gaussian peaks, as in Eq.5.

$$F(x) = A\frac{dP}{dx} = A \cdot \sum_{n=1}^{\infty} \frac{\mu^n}{n!} e^{-\mu} \cdot \frac{1}{\sigma(n)\sqrt{2\pi}} \cdot e^{-\frac{[x-c(n)]^2}{2\sigma^2(n)}} \tag{5}$$

A is a normalization constant, equal to the integral of the measured spectrum.

μ is the average value of the Poisson distribution.

$c(n)$ represents the x-coordinate of the centroid of the nth peak.

Additional details about the fit procedure and slight deviations from this behavior can be found in ref.[12].

The SiPM gain is defined as the number of elementary electrons created in response to the interaction of one photon. Since each cell of the device operates in geiger mode, the interaction of one photon produces an electron-hole pair, followed by an avalanche multiplication. The avalanche multiplication factor is the gain, and it obviously depends on the bias voltage. Within the operating range the gain is expected to grow linearly with the bias voltage ([14],[15]). In order to evaluate the gain for each bias value, one has to measure the spacing between peaks in terms of QDC (charge-to-digital converter) channels, and then to scale it by the known QDC yield in terms of charge/channel and by the known gain of the employed amplifier.

The dark noise figure is characteristic of each SiPM model, as it depends on the manufacturer (different technologies), on the geometry of the elementary cell, on the geometrical arrangement of the cells, on their number and overall size. In order to make a reasonable comparison between the noise figure of several different SiPM models, the dark count rate as a function of the counter threshold level was explored: a given value of the threshold will prevent all pulses with lower amplitude from being counted. What one expects, when reporting the data onto a plot, is a stepwise descending line that corresponds to progressively removing from the counts the lower order peaks of the spectra of Figure 12 and Figure 13. For each device tested (and for each bias value) the x-axis of the plot was normalized to the position of the 1–photon peak, and the y-axis to the maximum count rate of the device itself. This way all the data could be plotted on the same graph, with the number of fired cells on the x-axis and relative counting rate on the y-axis. On Figure 14 this graph is shown for the SiPM models/bias values listed in Table 3. The horizontal segments shown in the lower left part of the plot indicate the statistical dark counting rate expected at 2–photon level in case of no cross talk; the arrows are a guide to the eye for comparing these values to the corresponding observed ones.

Table 3. Main features of the SiPM photosensors tested with respect to their dark noise

SiPM producer	size (mm^2)	n. of cells	bias voltage (V)
SensL	1x1	400	31
SensL	1x1	400	32
Hamamatsu	1x1	100	69.4
Hamamatsu	3x3	14400	72.5
ST Microelectronics	0.5x0.5	100	32.5
FBK IRST	1x1	100	31.5

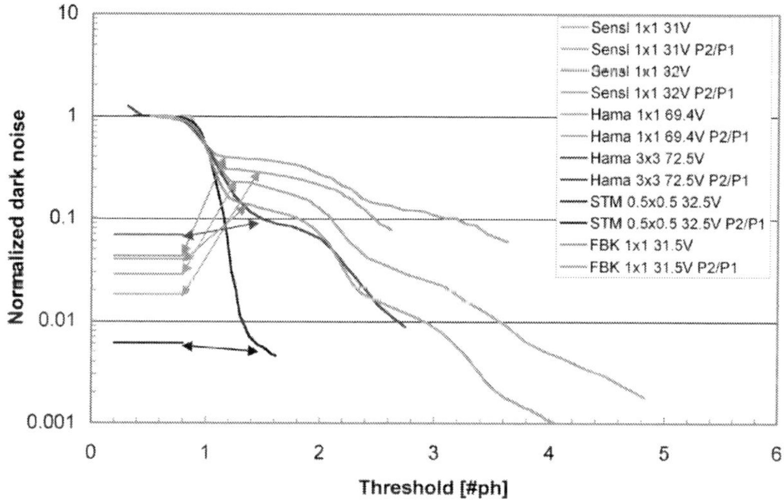

Figure 14. Dark counting rate as a function of the threshold, expressed in number of equivalent photons or fired cells, for the SiPM sensors of Table 3. Each data series was normalized to its own maximum value. The horizontal segments on the left indicate the counting rate expected at 2–photon level in case of no cross talk; the arrows are a guide to the eye for comparing these values to the corresponding observed ones.

By making use of the properties of the Poisson distribution one can easily derive the cross-talk features of the tested devices. Indeed one can observe the overall counting rate, assume it comes from purely statistical behavior and then call μ_{micro} the probability of occurrence of a dark count per unit time (1ns, for instance). This holds in the hypothesis of micro uniformity over time: there is no correlation between dark counts, and therefore one can easily calculate the expected poissonian probability of observing more than one count in a given time interval. Conversely, by analyzing the spectrum as in Figure 13 one can directly measure the probability of having more than one count in the same time interval (the time interval is the duration of the integration signal on the data acquisition electronics). In this second case one is observing a macroscopic Poisson distribution: the pulses come uniformly distributed over time, but every now and then there might be some burst of two or more correlated pulses. μ_{macro} is the average probability of a dark count per unit time in this second case. Once known the μ_{micro} and μ_{macro} values, it is straightforward to obtain the probability of cross-talk by computing the difference between the observed probability of multiple counts and the expected one (Eq.6).

$$P(\mu_{macro};k>1)-P(\mu_{micro};k>1)=P_{crosstalk} \quad (6)$$

Indeed what we here called cross-talk should be more correctly called “correlated noise”, therefore including also afterpulses; however, in this framework afterpulses play a minor role than cross-talk. Figure 15 shows the three terms of Eq.6 for the devices/conditions listed in Table 3. Notice that the non-commercial SiPM prototype produced by ST Microelectronics shows no cross-talk, as expected, due to a technological process of opto-insulation around each elementary cell which prevents secondary photons from reaching neighboring cells.

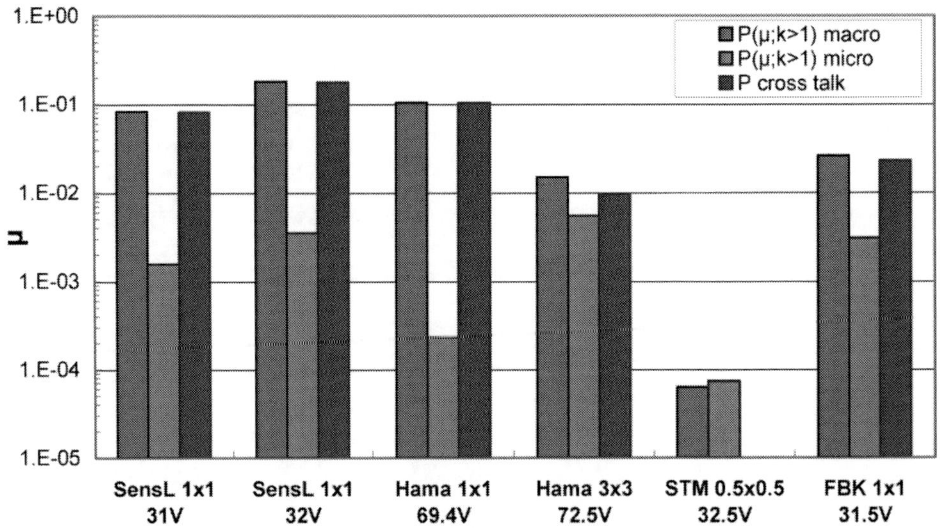

Figure 15. The three terms of Eq.6 for the devices/conditions listed in Table 3: observed probability of multiple dark counting (blue), expected probability of random multiple dark counting (green), cross-talk probability (red).

Another issue which has to be addressed is photon detection efficiency (PDE) of the photosensors. The typical method employed makes use of a Xenon lamp and a set of diffraction gratings and wavelength selection grids, in order to produce monochromatic light. This light is then sent into an integrating sphere, that is a hollow sphere coated with high-reflectivity white material that guarantees a perfectly randomized distribution of the photons. On the surface of the sphere one places a reference photosensor (namely a calibrated photodiode) and the sensor under measurement (Figure 16). Knowing the geometrical characteristics of the two sensors, one can determine the photon flux impinging on the reference sensor and rescale it to the other one. Therefore people used to measure the photocurrent on the SiPM, scale it down by the known gain and obtain the PDE. Unfortunately this method is not suitable for silicon photomultipliers: the avalanche multiplication also multiplies the correlated noise in an unpredictable mode. This is why the manufacturers report on their datasheets PDE values which explicitly include the cross-talk and afterpulses. How to measure the photocurrent getting rid of the correlated noise? Taking advantage of the single photon sensitivity of the SiPM we devised a single photon counting method. Instead of measuring a current we count pulses, making sure to keep the photon flux so low that the probability of having more than one photon impinging on the sensor at the same time is negligible [16], [17]. The method can be summarized into the following steps:

- The counter threshold is set on the 1-photon plateau (0.5-0.75 ph on Figure 14).
- The SiPM and the reference photodiode watch into an integrating sphere (Figure 16).
- The monochromatic light intensity is kept low (no 2-photon events on the SiPM).
- The digital pulse to the counter is long enough to miss afterpulses.
- Pulses are counted with and without light, signal counts are obtained by subtraction.
- The deadtime correction is applied.

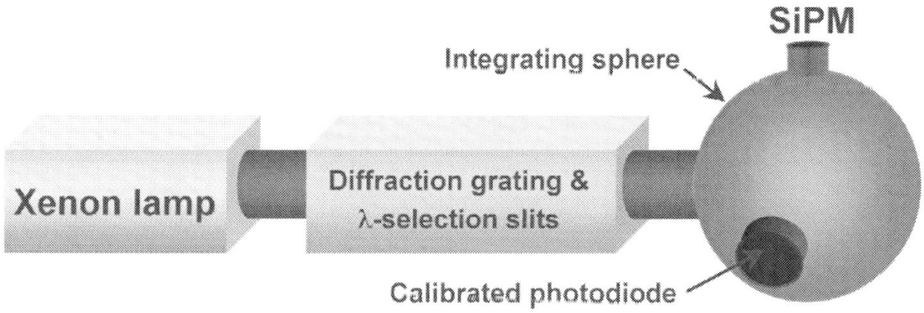

Figure 16. Sketch of the equipment used for the measurement of the Photon Detection Efficiency of the SiPM.

In Figure 17 we show a sample plot with the result of PDE measurements done with this method on two of the SiPM sensors listed in Table 3. The difference in the measured behavior can be accounted for by the better fill-factor, i.e. geometrical efficiency, of the Hamamatsu sensor (78% versus 36%) and by its known better efficiency below 450nm.

At the moment at LNS several families of SiPM are being characterized with respect to gain, photon resolution, PDE, noise, cross-talk.

The mechanical arrangement of the SiPM with respect to the fiber in the aforementioned radwaste monitoring environment is crucial: one needs to devise a solution that is at the same time simple, robust and cheap (Table 1). With the aim of proving that this could be done, not claiming it as a final solution, the opto-mechanical holder shown in Figure 18 was designed. The basic sensor unit is then assembled by installing two SiPM sensors at the ends of a fiber using the coupler and the mechanical support structure of Figure 19. The plastic tube containing fiber and shield, shown on the left in Figure 19, can be chosen either rigid or flexible according to the longitudinal or circular arrangement near a waste drum. Remarkably, most of the mechanical parts of the sensor assembly (Figure 19), including the hooks to push-clip them onto the platform, are off-the-shelf cheap materials easily available at any hardware store.

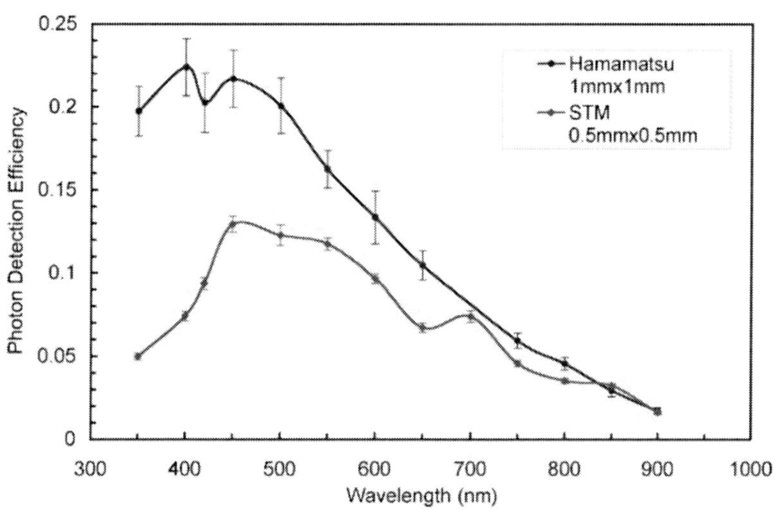

Figure 17. Sample plot of Photon Detection Efficiency measured at several wavelengths for two of the sensors of Table 3. The lines are only meant to guide the eye.

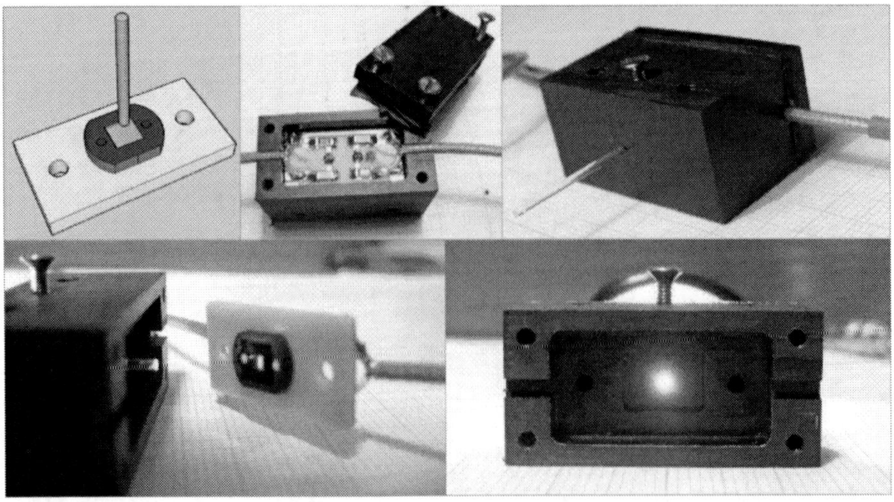

Figure 18. Mechanical coupler between fiber and SiPM. From left to right, top to bottom: coupling scheme; overall setup from the backside; the closed coupler box; the fiber inserted into the coupler and the SiPM soldered onto its electronic mini-board; front view of the light coming from the fiber, as seen by the SiPM. The black tubular sheath of Figure 4 was not yet installed here.

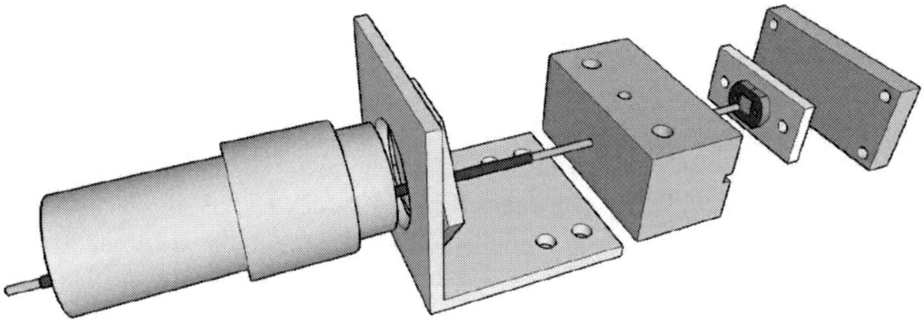

Figure 19. Exploded view of the prototype opto-mechanical coupler between fiber and SiPM, and of the overall mechanical support structure for the sensor unit.

6.3. Front-End Electronics

Despite the physical amplification due to the avalanche multiplication, the output signal of a SiPM is generally small. If we consider that the typical gain is around 10^5–10^6, the 1-photon signal on an oscilloscope with 50Ω termination would result in an amplitude around 0.1mV. The first tests were done using a commercial high-level linear amplifier, namely ORTEC FTA810B, featuring a linear gain around 200 with a bandwidth of 350MHz. After such an amplification the same signal would be seen on the oscilloscope with an amplitude of a few 10mV. Extensive measurements were done using this amplifier, and a sample high-quality charge spectrum is shown in Figure 12. It was then decided to design a dedicated amplifier, that was developed at LNS in discrete components SMD technology, featuring a linear gain around 200 and a bandwith of 2GHz. A sample charge spectrum, taken under the same conditions as those of Figure 12, is shown in Figure 13. The 3σ photon resolving power in this configuration is actually astounding, about 116 versus 18 for the previous amplifier.

Employing this amplifier allowed a much more comfortable handling of the analog to digital electronics, without recourse to a dedicated ASIC (Application Specific Integrated Circuit). In particular a discrimination circuit was developed, in order to generate a digital output pulse whenever an input pulse overcomes a predefined threshold value, along with a fast digital coincidence circuit that only generates a digital output pulse when its two digital inputs are true.

Each of the two SiPM outputs from a fiber is fed to an amplifier, whose output goes into a discriminator. The outputs of the two discriminators are combined by the coincidence unit, that will thus produce an output pulse only when a predefined amount of light is detected at both fiber ends.

Figure 20 shows a sketch of the front-end electronics architecture for one fiber. The output signal shape and type (analog or digital and polarity) are represented near each component. It is worth to remark that a single FPGA can handle several fibers at once.

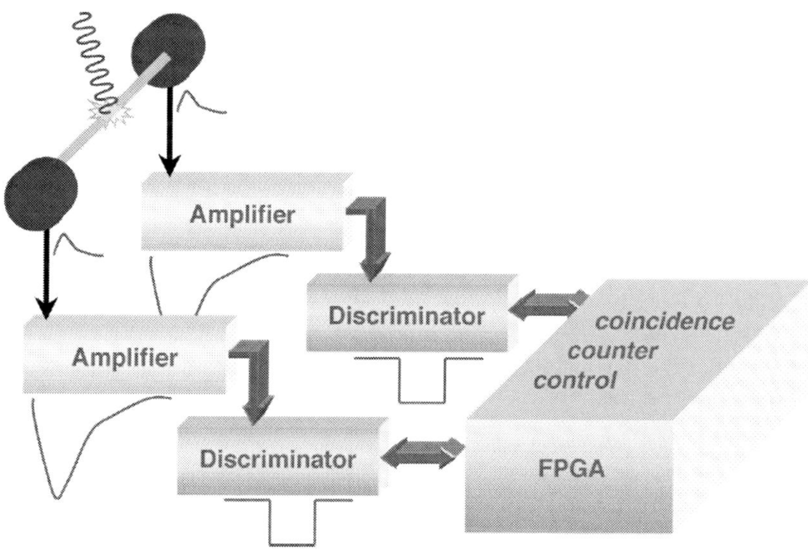

Figure 20. Scheme of the front-end electronics for one fiber. The output signal shape and type (analog/digital and polarity) are represented near each component. It is worth to remark that a single FPGA will handle several fibers at once.

6.4. Response to Gamma Radiation

In order to confirm the expectation about the fiber sensitivity to gamma rays, some tests, simulations and precision measurements were perfomed. The first test concerned the rough sensitivity, and it was done by simply placing a low-activity ^{60}Co laboratory source in contact with a sensor unit at three different positions, respectively denoted left, middle, right (^{60}Co emits two gamma rays with energy of 1.333 and 1.173 MeV). By recording on an event-by-event basis the time difference between the arrival of the two light pulses at the two SiPM sensors, a histogram was built whose x–coordinate is proportional to the coordinate along the fiber of the radiation interaction point [18]. Figure 21 shows the plots obtained in the three cases, which prove that indeed gamma rays from the source were being detected.

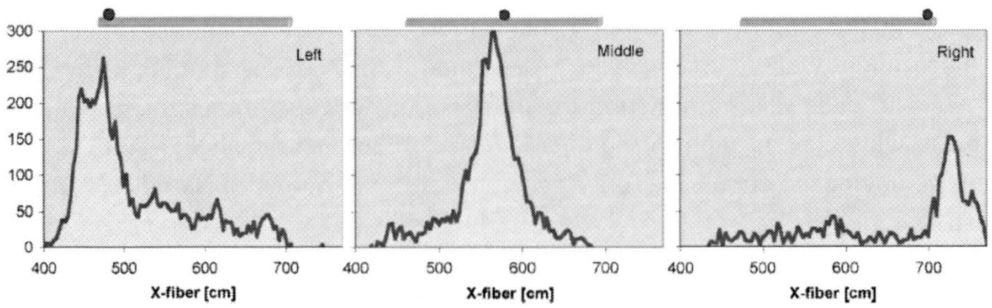

Figure 21. Proof of gamma sensitivity of the fiber+SiPM sensor unit. A ^{60}Co source was placed in three positions along the fiber, and the corresponding position histograms were built by recording the time difference between the arrival of the two light pulses at the two SiPM sensors.

Afterwards simulations were performed using the GEANT code, which included the fiber material and geometry, the poissonian production of the scintillation photons, the properties of the SiPM. Not being able to estimate the real properties of the optical contact between the fiber and the SiPM face, it was assumed it contributed in a systematical way. In other words, one expected to find a proportionality between the simulated and the measured data, with a scaling coefficient independent of the gamma source type.

Due to the weakness of the available gamma sources, it was required to perform precision measurements with the source placed at small distances from the fiber itself. This was done by means of a micrometric translation stage holding the source. The measurements were done using ^{60}Co (1.333 and 1.173 MeV) and ^{137}Cs (0.662 MeV) sources. Figure 22 reports the results of simulations and measurements; the simulated data points were rescaled by a unique systematical constant, to account for the unknown effect of the optical contact, of the effective light yield of the particular fiber (in the simulation the nominal value was used), and of the SiPM detection efficiency [19].

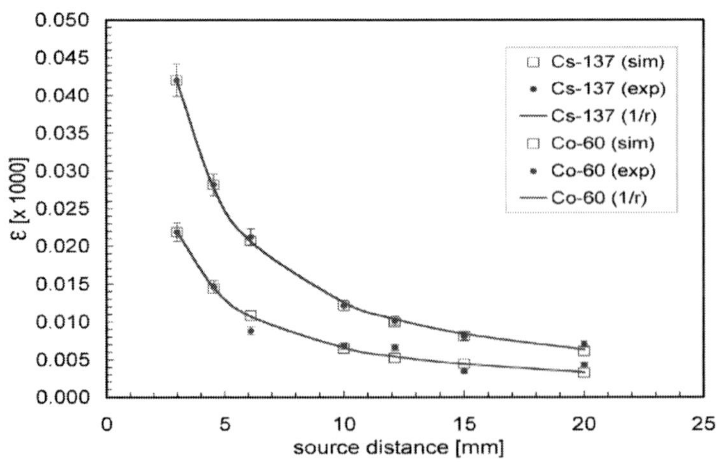

Figure 22. Efficiency of the fiber sensor to gamma radiation, as a function of the source distance, for two different sources: ^{60}Co and ^{137}Cs. Open squares: simulation; full circles: measured data; line: calculation assuming a *1/r* dependence.

Moreover, by assuming a point–like source and an infinite fiber length, one can easily calculate the fiber solid angle as seen from the source, either by solving a simple integral or even by elementary geometrical considerations. The geometrical efficiency thus resulting, i.e.

the ratio between the fiber solid angle and the 4π total solid angle, is $d/2\pi r$ where d is the fiber diameter and r is the distance from the source. Two curves proportional to $1/r$ were reported on the same plot (Figure 22) after normalization to the first data point of each series. The conclusion, as can be easily verified on the plot, is that both the simulation and the $1/r$ behavior reproduce pretty well the measured data.

7. Counting, Readout and User Interface

An FPGA-based counting unit was developed which was programmed to implement a set of logic coincidence circuits, with a corresponding set of counters, and to handle the data and control communications with the main console. FPGA (Field Programmable Gate Array, Figure 23) is a programmable component that allows to build complex custom logic circuits by means of software programming [20]. Its main advantages with respect to an ASIC (Application Specific Integrated Circuit) are:

- Speed of prototyping, as one does not need to physically build any electronics, and every elementary component is programmed.
- Protection of the investment on software against the obsolescence of the hardware.
- Low cost.
- Use of a standard hardware description language (VHDL), which can be migrated across different FPGA models and manufacturers.

Figure 23. The prototype data acquisition and control unit based on FPGA, implemented on an Altera Cyclone II FPGA development board.

The system chosen for the test implementation was an Altera Cyclone II FPGA development board [21]. Each FPGA unit in the DMNR scheme (Figure 24), is meant to handle one platform with all of its fibers, even though this is not a limitation due to the hardware but rather a choice to keep a better modularity. Indeed each platform is planned to be self-consistent, e.g. hosting on-board all of its sensors and related electronics.

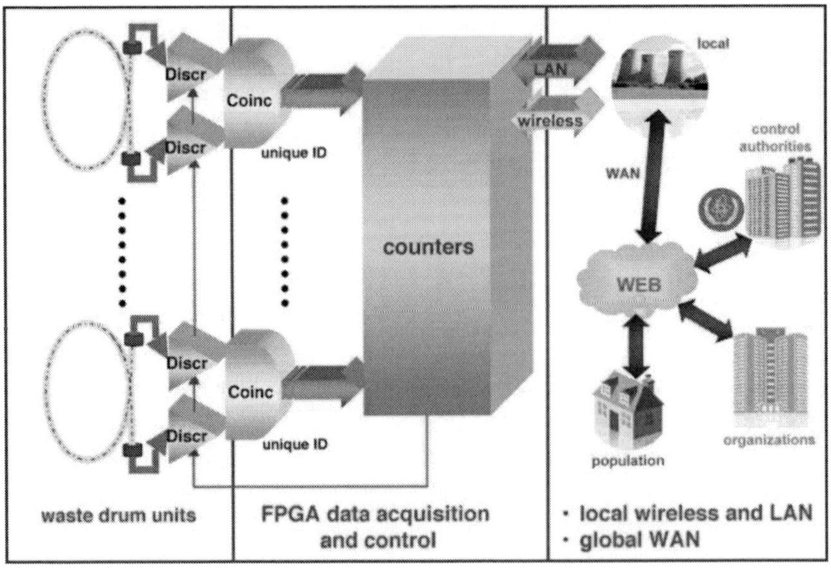

Figure 24. General scheme of the monitoring data acquision and transmission. Inside the storage site there is a redundant transmission path over a Local Area Network and wireless (ZigBee). From there, a connection to the Wide Area Network provides access to/from the web control authorities, other organizations, private citizens.

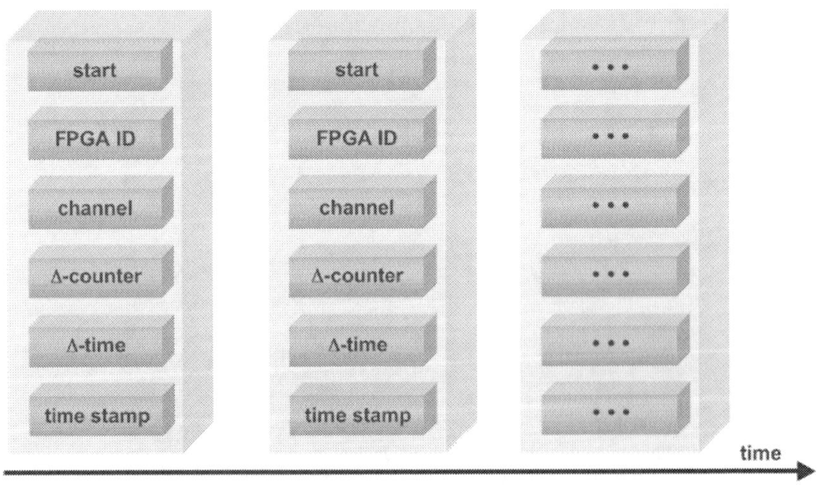

Figure 25. The output data stream from each FPGA is made of frames. Each frame, related to a fiber, has a start word, followed by the FPGA identifier and channel pinning down the fiber; then follow the number of registered counts and the counting time interval. The last word is a global time stamp.

Each coincidence circuit looks after one fiber: whenever a digital signal is produced by one of the two discriminators, the coincidence circuit waits a predefined time window (typically 50ns) for the possible companion signal coming from the other side of the fiber. If this is the case, a digital counter dedicated to that fiber is incremented by one unit. After a predefined counting time interval, the content of all the counters is sent to the console (Figure 25) along two redundand physical paths: a wired connection and a wireless one. Actually an alternative solution was also implemented where the coincidence units are realized in hardware, with the FPGA only taking care of the counters and controls. The choice between the two solutions will be a tradeoff between cost and performance: software coincidence units will require faster and more expensive FPGA's, whereas employing cheaper FPGA's will require additional external coincidence hardware.

The basic features of the developed counting and readout prototype are listed below:

- The counting rate is measured on a fiber by fiber basis.
- Two readout policies are exploited at the same time: periodic and on-request.
- The counters are 32-bit deep, therefore in the hypothesis of a 1MHz counting rate a counter could be operating up to 1.2 hours before each readout. However, periodic readout intervals of seconds or minutes are automatically foreseen.
- The maximum input count rate is about 50MHz (in the current version).
- Scalability and modularity.

As already said, the coincidence between the signals from the two ends of a fiber is needed to reject spurious (and noise) counts. Moreover, as the fiber length is of the order of 1-2m and the speed of light inside the fiber is 20cm/ns, the two signals must be overlapping within ≈10ns. A 20ns duration for the coincidence window should be enough, but for the moment it was chosen to set 50ns to stay on the safe side. However, this duration can be programmed in the range 20–100ns. Any signal pair falling outside the coincidence window is disregarded as it does not represent a valid count.

The data transmission protocols implemented so far are UART RS232/RS422 (or RS485 multidrop), being the migration from one to another easily achieved via software. A standard industrial bus and protocol will be advisable for a real implementation (ProfiBus, ProfiNet, CanBus, Ethernet, etc.). As for the wireless transmission the activity is currently focused on the implementation on ZigBee modules, which are rather cheap, easily programmed and have a very low power consumption [22].

Once the data reach the console, on the one hand they are passed on to a storage process, on the other hand they undergo an online analysis which allows to set up an interactive synoptic table useful to the operators. The storage process, still under development, will consist in the creation of a double copy logfile, one local and one remote. Moreover, it will feed a suitable database that will be concurrently accessible by many processes at the same time to retrieve and correlate hystorical data. The database, under well defined constraints, will be browsable via web by the properly authorized subjects and entities.

As to the online analysis a test data structure was developed just to show the potentiality of the application. Figure 26 illustrates the hierarchical data structure for a repository in the DMNR architecture. Remarkable is that this structure can be dynamically modified and/or rearranged, e.g. by redefining the number of fibers for certain drums, or the number of drums in a platform. The top level is represented by the plant, which means that within the same architecture one can handle many storage plants. Inside each plant there will be several sectors, each one related for instance to a specific region of the repository, or related to a different class of waste, or to a different storage setup. Inside each sector one can have several groups, again exploiting possible physical or logical classifications. Finally, each group is made of platforms, made of drums made of fibers. The fiber is the lowest logical structure in the hierarchy.

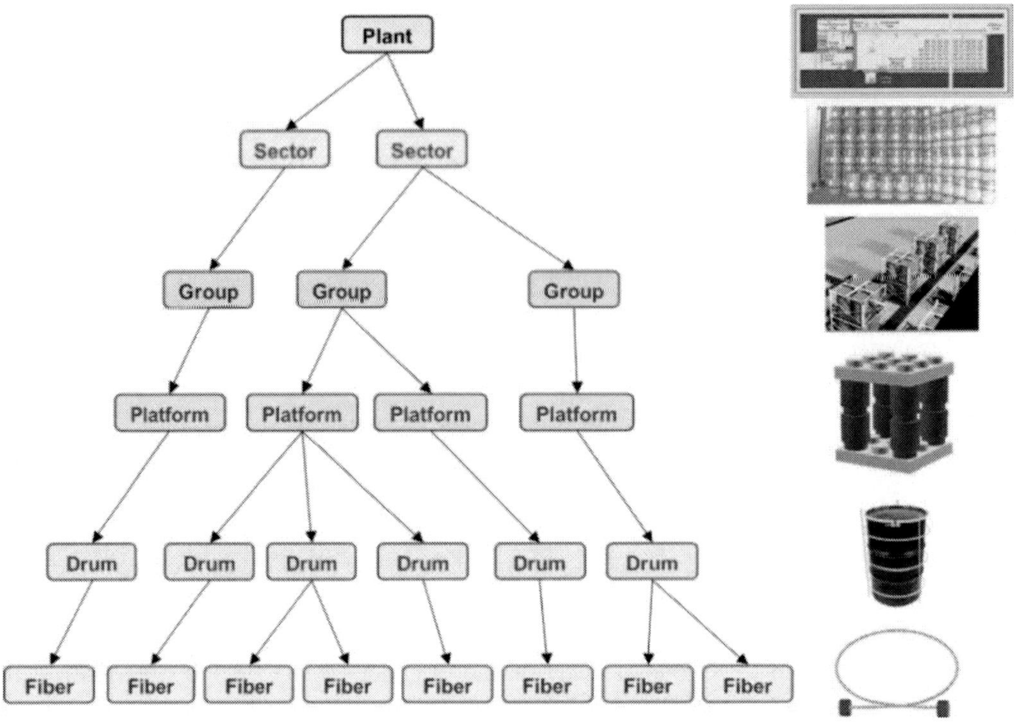

Figure 26. The hierarchy structure for a repository in the DMNR architecture. Remarkable is that this structure can be dynamically modified and/or rearranged, e.g. by redefining the number of fibers for certain drums, or the number of drums in a platform, etc.

The online monitor keeps checking the counting rate fiber by fiber, and on operator request can open a dedicated window to display data at several levels of detail (Figure 27). The instant counting rate of each fiber is displayed in both numerical and graphical fashion, along with additional information concerning the deviation from the previous running average. Moreover, there is a set of alarm levels, with corresponding thresholds, which can be programmed in terms of sigma units. In the tests with sources 3, 4 and 5 sigma units were adopted, respectively corresponding to a random deviation probability of $2.7{\cdot}10^{-3}$, $6{\cdot}10^{-5}$, $6{\cdot}10^{-7}$. Every now and then there can be transient deviations, due to statistical fluctuations or operations like displacements of drums, triggering temporary alarms. In case of a persistent alarm the amplitude of the deviation is an indication of the severity, and it means there was a relevant change in some condition like a new geometrical arrangement of some drum or a real accident like a leak. Every alarm is automatically back-propagated through the hierarchy, i.e. each unit in alarm will also set an alarm in its upper level unit (Figure 26).

An additional feature currently being studied is the determination of the absolute position of the anomaly causing the alarm, based on crossed information from several different fibers.

The high level software, running under Microsoft Windows operating system, was developed using the Object Oriented Programming Language C#. The graphical user interface was developed in Windows Presentation Foundation (WPF) and Silverlight, that simplifies the implementation of special graphical functions, animations, 3D drawings.

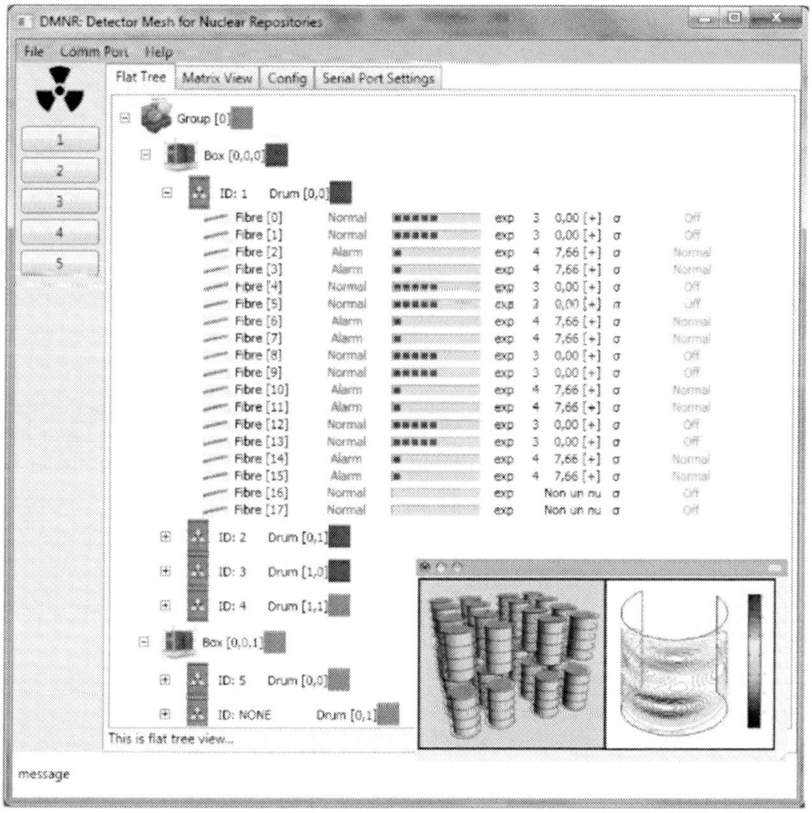

Figure 27. Screen snapshot of the main control panel of the prototype DMNR graphical user interface.

Extending the proposed data architecture with the use of a suitable database will allow to easily meet many of the IAEA specifications concerning quality assurance, as listed in ref.[4]. Quality records could be established and maintained for each waste package, based on information to be automatically retrieved and assembled into periodical reports. The results of such reviews could be archived and available online for a long time, even longer than the formal requirements, also allowing to be easily used for long term statistical analysis. Moreover, the availability of the whole history of a waste package may be important for demonstrating the compliance of the waste package with the specifications, also ensuring the traceability of its characteristics from the collection to the processing and storage.

Examples of the information that can be stored on a drum by drum basis are:

- The initial drum characterization data at production time, along with the key process parameters during its pretreatment, treatment and conditioning.
- The calibration records for equipment and systems for process control.
- The characterization of the waste form and the associated container (e.g. material certificates for the container and its lid and welds or seals, including quality control tests and their records).
- The values of significant monitoring parameters, together with a history of the counting rates, occurred displacements, storage locations.

The data could be easily combined into graphs and histograms, and eventually included into the periodical reports. In addition one could easily aggregate and compare data in order to produce reports for several drums together from a uniform process batch (groups) as well as for parts of a repository (sectors) or for entire repositories (plants).

8. Mechanical Arrangement: Platform and Robotic Arm

8.1. The Platform

In order to test the effective mechanical properties (overall size, load capacity, mechanical stability, etc.) of a platform it was decided to design and construct a prototype holding up to four waste drums (Figure 28). This preliminary version still lacks alignment structures that should guarantee the self-alignment when stacking them, as well as the electrical connection to a ground based power supply socket and data network. The current version hosts eight fibers: four straight ones are located on the inside edge of the four vertical struts (and can be spotted in Figure 28 right), and four circular ones below the phantom drums. The fiber sensors are clamped to the structure by simply pushing them into hook clips preinstalled on the platform. The electronics could be hosted at the center of the base frame or fixed to the side of a strut. Other geometries are currently under study.

The platforms are foreseen to be loaded, unloaded, placed and displaced by hooking and lifting them with an overhead travelling crane, equipped with suitable sensors and a motion control system in order to guarantee the required mechanical precision. We remark that the drums shown in Figure 28 are just hollow cylindrical phantoms, whereas the real drums should be surmounted by some handle to allow hooking them individually with the crane as well.

Figure 28. The prototype platform developed at INFN-LNS to test the effective mechanical properties (overall size, load capacity, mechanical stability, etc.). It hosts eight fibers: four straight ones are located on the inside edge of the four pillars, and four circular ones below the phantom drums. The electronics could be hosted at the center of the base frame or fixed to the side of a strut. Other geometries are currently under study.

The basic features of the final platform should be:

- The fiber-based radiation sensors with the related electronics and battery should be hosted on-board and, if possible, easily accessible and replaceable.
- The platform should have some additional sensors like an inclinometer and a thermometer.
- It should have a modular structure, with possibility of stacking at least three units.
- Robust electro-mechanical spring connectors should be available for self-alignment and power/data connection chaining.
- The data transmission system should be redundant, i.e. connected via cable and low-consumption wireless modules.
- Each single drum must be equipped with tags, barcode and/or RFID, suitably installed in predefined positions on its external surface.
- The platform should provide for robot operations like remote visual inspection and handling, RFID and barcode identification, analysis with specialized radiation detectors.

Figure 29 shows a pictorial view of a possible arrangement of platforms inside a waste repository, while the overhead travelling crane is lowering a robotic arm for drum inspection.

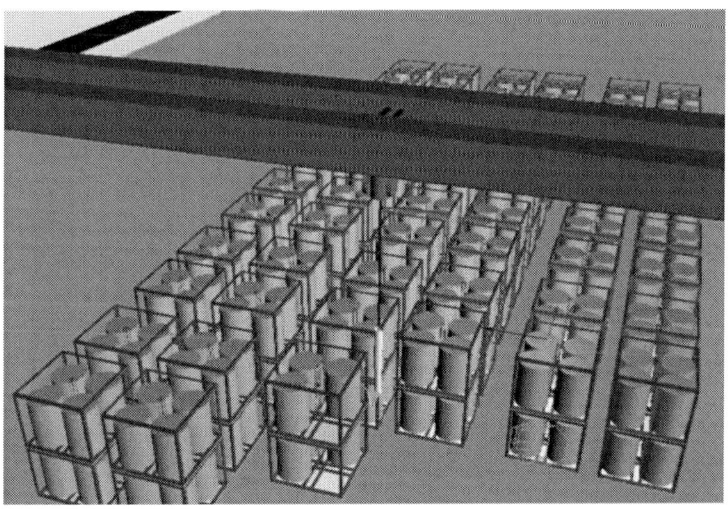

Figure 29. Pictorial view of a possible arrangement of platforms inside a waste repository. Visible is the overhead travelling crane while lowering a robotic arm for drum inspection.

8.2. The Robotic Arm

In such an environment a system for remote inspection becomes a must-have, and then one has to think about a robot system. In most cases a robot will prevent human operators to be unnecessarily exposed to radiation doses that, even though low, would contravene the

ALARA[1] principle. Moreover, an inspection procedure by robot allows a higher degree of automation and repeatability, and facilitates archiving the inspection steps and the results.

A cart or even a walking robot is not advisable, essentially for space constraints due to two main reasons: (i) the free room inside a storage site is limited because of cost-effectiveness; (ii) a drum arrangement as tight as possible acts as a self-shield for radiation, thus screening most of the radiation coming from the inner drums. Therefore a viable solution is a robotic arm carried by the overhead travelling crane, which can be positioned over the destination platform and then lowered down to the required height. The navigation technique will likely be based on 3D site maps generated by means of CAD files, and will make use of active or passive landmarks. Such an arm can be made simple, reasonably cheap, reliable, and can be easily integrated into preexisting structures and equipment. It must be slim but extensible up to about one meter, and should be able to bear a video camera and, when needed, a lightweight radiation detector.

Figure 30 shows a 3D sketch of the robotic arm prototype currently under construction. It features five degrees of freedom: vertical sliding, arm and forearm rotation in the vertical plane, up-down and left-right rotation of the end effector. The end effector will bear a compact color video camera and an additional 1kg payload, which is reasonably destined to a small radiation detector and an RFID reader. The support electronics, along with the arm-control computer, will be hosted on the body of the system where it does not exert any relevant torque. In order to stabilize the whole structure with respect to the platform to be inspected, a small electromagnet is available on the arm front, which clamps to the metal frame of the platform itself, along with a tiny video camera quite helpful during the docking phase (Figure 31).

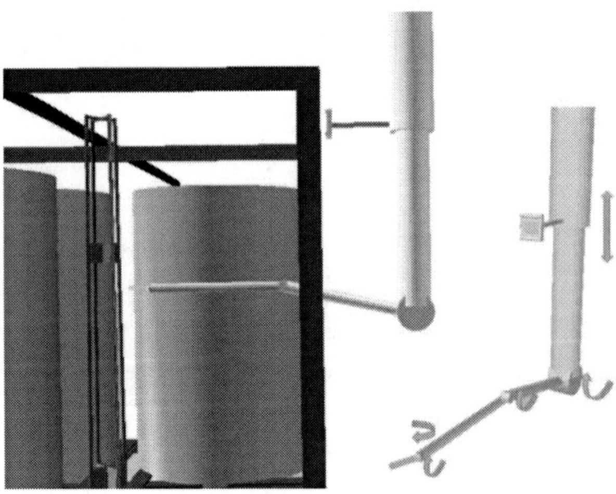

Figure 30. Pictorial view of the robotic arm currently under construction. It features five degrees of freedom: vertical sliding, arm and forearm rotation in the vertical plane, up-down and left-right rotation of the end effector.

---

[1] Acronym for "As Low As Reasonably Achievable." It means making every reasonable effort to maintain exposures to ionizing radiation as far below the dose limits as practical. Be consistent with the purpose for which the licensed activity is undertaken, taking into account the state of technology, the economics of improvements in relation to state of technology, the economics of improvements in relation to benefits to the public health and safety, and other societal and socioeconomic considerations. These means are in relation to utilization of nuclear energy and licensed materials in the public interest. http://hps.org/publicinformation/

As to the radiation detector the best solution would be a germanium semiconductor detector, due to its superlative energy resolution which allows to separate and identify even finely spaced gamma levels in the energy spectrum. Unfortunately such a detector is rather bulky and requires operation at liquid Nitrogen temperature. The best candidate then becomes a scintillator crystal with an associated photodetector. So far the best crystal was believed to be CsI(Tl), but quite recently $LaBr_3$(Ce) has proved to be the closest one can get to a germanium detector performance. The choice will be a tradeoff between the needed performance, the price and the weight: CsI(Tl) is cheap, has quite a reasonable energy resolution and density of 4.5 g/cm^3, whereas $LaBr_3$(Ce) is quite expensive, has a superior resolution and density of 5 g/cm^3. Moreover, CsI(Tl) is easily handled whereas $LaBr_3$(Ce) is highly hygroscopic and must be incapsulated into an air-tight aluminum can. Whichever the choice, a 100 cm^3 scintillator crystal ($\approx$0.5 kg) should provide a reasonable detection efficiency.

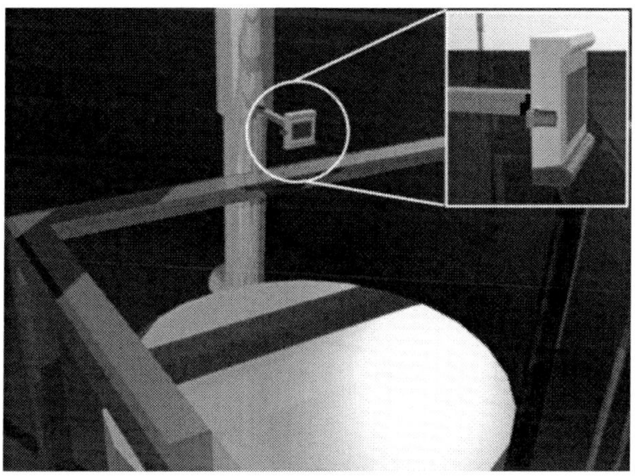

Figure 31. Three-dimensional representation of the robotic arm while being docked to a platform girder by means of its electromagnetic clamp. In the upper right inset a close-up of the magnetic clamp is shown, along with a tiny videocamera helpful during the docking phase.

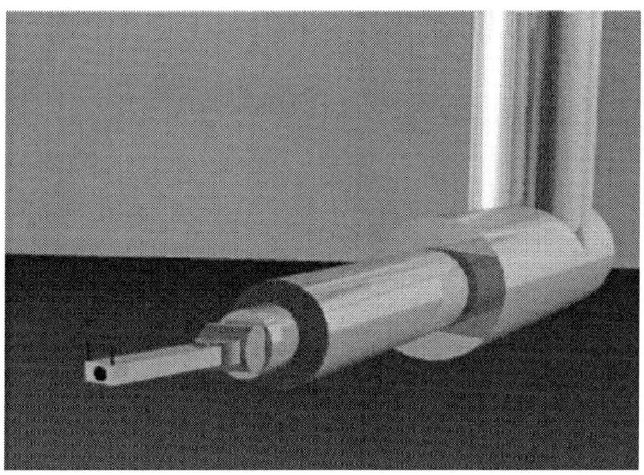

Figure 32. Three-dimensional representation of an alternative version of the robotic arm, where the in/out extension is accomplished by means of telescopic shafts. The end effector still features two degrees of freedom.

Having a video camera on the end effector will allow a thorough visual inspection of the drums, including the barcode reading. Additional features could be a LED illumination of the field of view, digital filtering and analysis of the images, a video-driven counting scan of the drum surface by means of the camera and the gamma detector.

A possible alternative to the swinging arm geometry exploits an in/out extension accomplished by means of telescopic shafts, with the end effector still featuring the same two degrees of freedom (Figure 32). Such a solution, however, is less flexible and requires more leeway between adjacent platform rows. Another alternative under examination is a snake-like arm, even though it is rather complex and so far quite expensive (Figure 33).

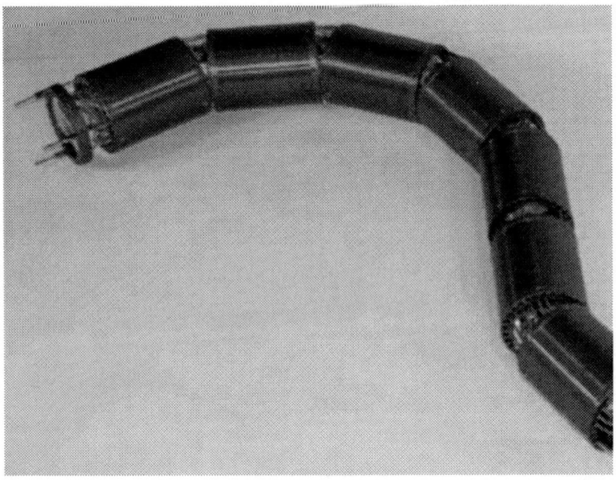

Figure 33. A possible snake-like solution for the robotic arm, even though so far it would be quite expensive.

9. Bench Tests with Prototypes

In order to start performing realistic bench-tests a simple prototype was set up employing three fibers arranged in horizontal coils as in Figure 34. The gamma source used was actually a set of four sources: ^{137}Cs, ^{152}Eu and two ^{60}Co sources, with a global activity of 2.7MBq and covering the energy range ≈0.1÷1.5 MeV. This multisource was put in a box and placed into different positions with respect to the three fibers, as schematically depicted in Figure 35, by means of the toy robot shown in Figure 34. The monitoring system was displaying online data, properly switching into alarm condition whenever the corresponding thresholds were crossed. Moreover, it recorded the data stream onto a logfile for later analysis. By analyzing the logged data the plot shown in Figure 35 was built, where the effect of displacing the source is immediately evident. The time bin for the plot was chosen as 8s, in order to reduce the larger error bars due to the rather low counting rate; however, the actual time bin for counting and data recording was 1s.

The baseline count rate of each fiber without source is different, basically because of the electronic threshold. The system is quite sensitive to counting rate variations, as was immediately understood by displacing the source box around the fibers. Remarkable is the fact that the higher the counting rate, the smaller the relative error will be: this translates immediately into an increased sensitivity. As an example, if one assumes that a fiber in a real waste storage environment is counting 10kHz and uses a 1s time bin, the sigma error bar (that is the square root of the counts) will be 100Hz or equivalently 1%, according to the Poisson statistics. In such a case the alarm thresholds at 3, 4 and 5 sigma units will correspond to 3%, 4% and 5% deviations, with the related tiny probabilities of false alarm reported in section 7. If the condition persists (i.e. the overall statistics increases while integrating the counts over longer and longer time intervals) this will certify that there is an alarm, due to a real increase of radioactivity, to an altered geometry like a displaced or removed drum, to a malfunctioning of the sensor or of the electronics. In any case, the monitor system will be accomplishing its duty.

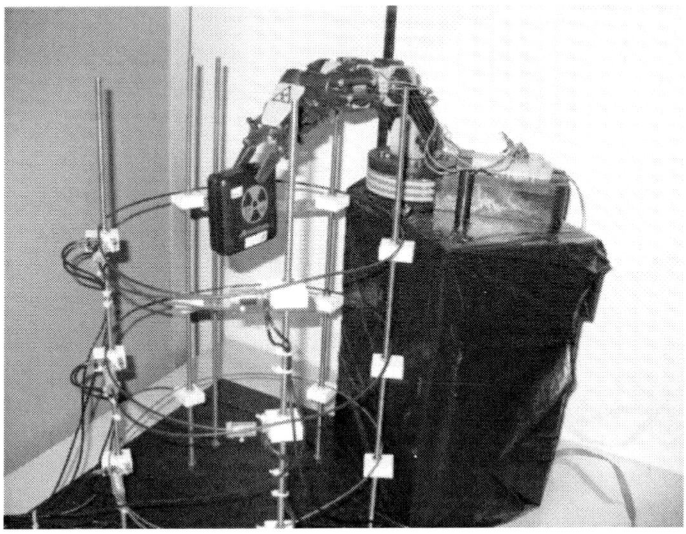

Figure 34. Setup of a preliminary test done with three horizontally coiled fibers and a gamma source.

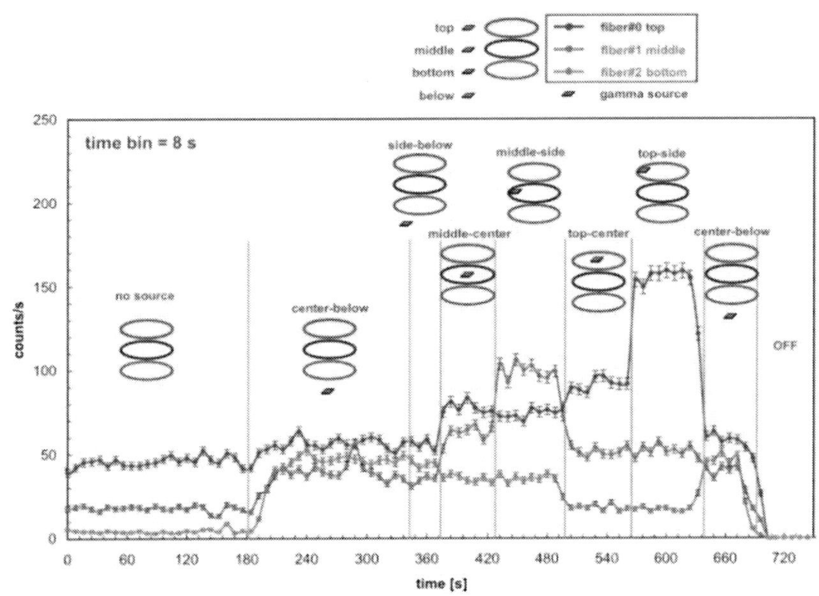

Figure 35. Counting rate versus time for the three fibers of the test done with the setup of Figure 34. The position of the source with respect to the fibers is graphically represented.

A step further with realistic bench-tests was the construction of a second prototype, shown on the lefthand side of Figure 36. The phantom drum was made from an empty big plastic bottle 40 cm high, painted yellow and placed at the center of an array of five fibers. The fibers with their sensors, hosted by the mechanical structures shown in Figure 18 and Figure 19, were arranged three vertical and two horizontal. In Figure 37 counting rate versus time is reported for a short test where the same multisource mentioned in the previous section was placed at the position marked with a red "X" in Figure 36 left. In Figure 36 right a sample 3D image is shown, roughly representing activity levels around the drum, obtained by combining the horizontal and vertical information from the log data shown in Figure 37. This proves that a geometry with crossed fibers can actually provide position information.

Gaining position information could be quite useful under several cases, for instance with liquid waste or in the occurrence of a leak caused by corrosion and gas pressure buildup inside a drum. Indeed the chemical composition and behaviour of the solidified waste are not generally predefined and known. This might happen, for instance, due to radiolysis taking place and producing hydrogen gas on the one side and possibly some corrosive compound on the other. Actually, the terms of such a process are not fully understood, and rough estimates of the time scale are of the order of tens of thousands years. It is clear, however, that too many variables are not under control. It is well known, for instance, that the eventuality of steel embrittlement by hydrogen incorporation during the corrosion process cannot be excluded [23]. Other gases could be produced by radiolysis or decomposition of organic material, and anaerobic corrosion and swelling (e.g. of resins, bitumen) might occur [24]. Due to the above reasons an intermittent leakage type cannot be excluded, as well as a steady one, but the proposed monitoring system should be able to detect both.

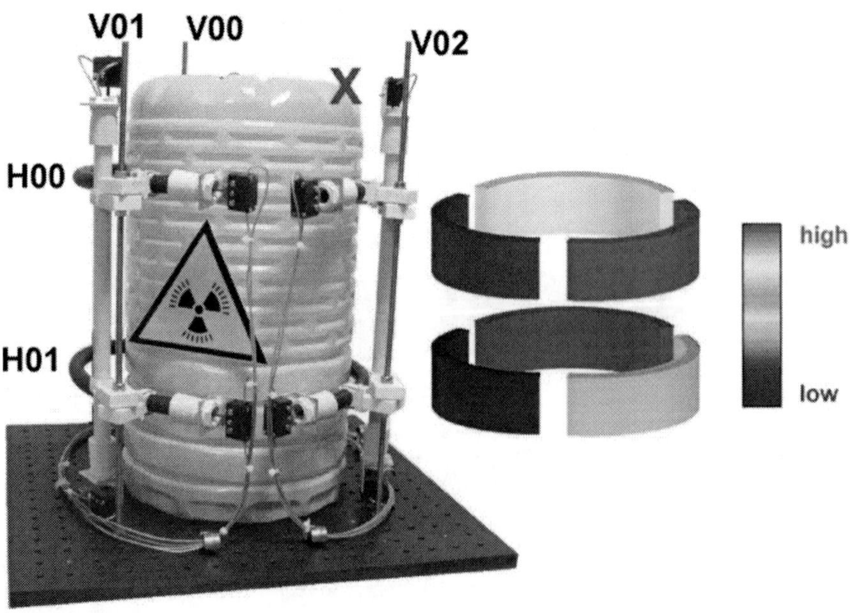

Figure 36. The prototype minidrum with two horizontal and three vertical fibers. The red "X" indicates the position where the gamma source was placed. The graph on the right side represents the 3D rough activity levels obtained by combining the horizontal and vertical information from the log data shown in Figure 37.

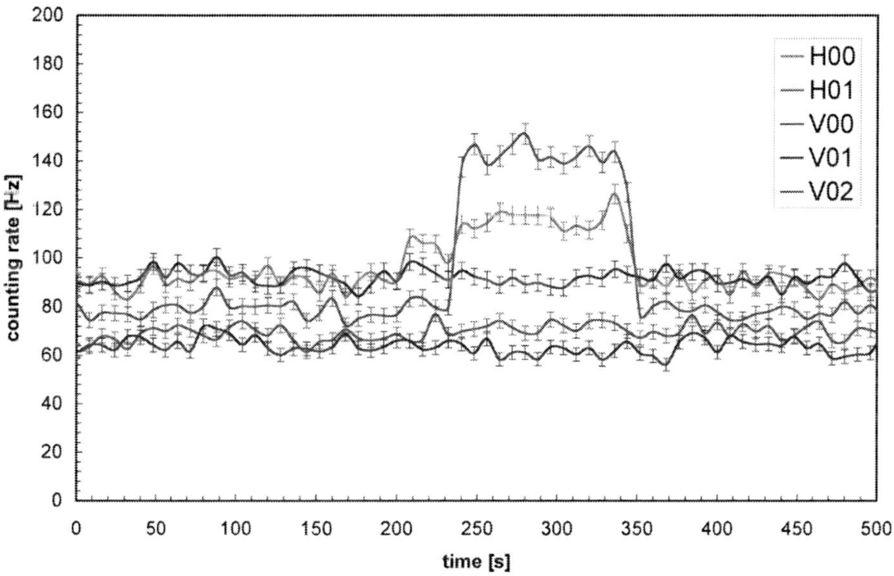

Figure 37. Counting rate versus time for the test done with the five fibers of the minidrum setup shown in Figure 36. H and V stand for Horizontal and Vertical.

10. Open Issues

Obviously the proposed system is not a final one at all, and still many issues are open. First of all is the radiation hardness of the SiPM photosensor: even though the fiber can allegedly stand the expected radiation field for a long time, the same does not apply to the silicon detectors. Recent tests on a family of commercial SiPM sensors have shown that their noise figure degrades considerably after 240 Gy irradiation with ^{60}Co gamma ray source [25]. However, still in ref. [25] a new promising family of SiPM prototypes was tested with encouraging performance in terms of radiation hardness, and this is a clear sign that many improvements are to be expected in the short term. Moreover, the tiny SiPM sensors could be rather easily shielded with a reasonably thick lead cap, thus strongly reducing the damage.

Another issue, still bound to the photosensors, is their thermal stability fluctuations along with their overall stability over a long time span. Tests of these features are currently under way. A radical solution to these problems, if needed, would contemplate some non-dramatic additional electronics.

A nice feature to demand to the system would be the possibility of performing an even rough count-rate vs dose calibration, thus allowing more solid assessments and comparisons with different environments.

An important issue that is already being faced concerns the Monte Carlo simulation of complex geometries for the arrangement of the waste drums in a storage site, in order to be able to estimate the expected activity and corresponding sensitivity levels in a realistic storage environment. This will allow to choose a feasible configuration for the fiber arrays, in terms of number of fibers per drum (or per platform) and their placement.

Still to be studied are the questions concerning the need of additional sensor types to be installed on the platforms and/or on the drums, as for instance to monitor their mechanical stability and the temperature in their vicinity.

As for the robot or robotic arm, in the present version it will basically perform only inspection tasks. However, it would be advisable to develop a more sophisticated version, capable of additional periodical maintenance operations like replacement of batteries, electronic modules, sensors, or also more critical operations involving actions on the drums themselves.

The next step in testing the DMNR system will consist in developing a two-drum sensor array and to install it in a convenient location inside a real storage site around real radwaste drums. Such a step, indeed, is foreseen to be performed quite soon.

CONCLUSION

The experience acquired so far in developing and testing the simple prototypes, along with the related software and hardware, represents a good starting point toward a real study of a technologically advanced monitoring of the short and medium term radioactive waste. Indeed, if no treatment or conditioning facility is readily available, it might be necessary to store LILW for long periods. Also, the need could arise for a long period of storage before disposal. In such cases the DMNR system could ensure the integrity of the waste products, whereas its combination with a suitable database system, as well as the tagging and labelling of waste packages, could guarantee a high-quality identification, maintainance and preservation of the information about the history of each waste package.

In particular one could make sure that important variables do not change unfavourably with respect to the values established when the original safety assessment was carried out, while at the same time strongly reducing the need for direct inspections and measurements to be performed by operators and increasing the frequency and quality of the information records. A facility operated under such conditions could be much more easily assessed as to the safety requirements, contributing to a modern, transparent and responsible handling of the radwaste and to an improved public awareness of the risks and the benefits of the nuclear technology.

ACKNOWLEDGMENTS

This work has been supported so far by INFN within the framework of the INFN-Energy Strategic Project and of the agreement between INFN and Ansaldo Nucleare S.p.A. In particular the latter has also supported five students with one-year fellowships to work at several aspects of the DMNR project. I am indebted with Giovanni Ricco, Marcello Lattuada, Graziano Fortuna, Enrico Botta, who have sustained my idea of online radwaste monitoring since the very beginning.

I have to thank explicitly my collaborators and students, all of them mainly dear friends, without whose contribution this work would not have been possible: Alfio Pappalardo, Luigi

Cosentino, Sergio Scirè, Carlotta Scirè, Giuseppe Greco, Massimo Barbagallo, Giovanni Guardo, Vincenzo Febbraro, Marco Capogni, Vincenzo Bellini.

Finally, I cannot help but mentioning Massimo Piscopo, Pietro Litrico, Claudio Calì, Giuseppe Passaro, Salvatore Marino, Biagio Trovato, Antonino Santonocito: they all provided invaluable technical contributions in terms of mechanical design, mechanical workshop, electronics, in their respective departments at LNS.

REFERENCES

[1] Horsley, D. M. C., Hallington, P., J.; Nuclear power and the management of the radioactive waste legacy, *Chemical Engineering Research and Design*, 83(A7) (2005) 773–776.

[2] Classification of Radioactive Waste, *IAEA Safety Series* No. 111-G-1.1.

[3] IAEA *Nuclear Technology Review* 2010.

[4] Predisposal Management of Low and Intermediate Level Radioactive Waste, *IAEA Safety Guide* WS-G-2.5.

[5] Radiation Protection and Radioactive Waste Management in the Operation of Nuclear Power Plants, *IAEA Safety Guide* NS-G-2.7.

[6] Finocchiaro, P., Pappalardo, A., Bellini, V., Cosentino, L., Febbraro, V., Scirè, S.; Detector Mesh for Nuclear Repositories, *Nuclear Physics B* (Proc. Suppl.) 197 (2009) 35–38.

[7] Finocchiaro, P., Barbagallo, M., Cosentino, L., Greco, G., Guardo, G., Pappalardo, A., Scirè, C., Scirè, S.; Silicon photomultipliers for radioactive waste online monitoring, *Nuclear Instruments and Methods in Physics Research A, in print,* doi:10.1016 /j.nima.2010.07.071.

[8] Papandreou, Z.; *Scintillating Fiber Trapping Efficiency, GlueX-doc-918-v2, GlueX Collaboration, 1*9 November 2007.

[9] Acosta, D., et al.; Effects of radiation damage on scintillating fibre calorimetry, *Nuclear Instruments and Methods in Physics Research* B62 (1991) 116-132.

[10] Botta, E.; Ansaldo Nucleare S.p.A., private communication.

[11] Barbagallo, M., et al.; Development of a thermal neutron detector based on scintillating fibers and silicon photomultipliers, *Review of Scientific Instruments* 81, 093503, 2010.

[12] Finocchiaro, P., et al.; Characterization of a Novel 100-Channel Silicon Photomultiplier – Part I: Noise, *IEEE Transactions on Electron Devices*, vol.55, (2008) 2757.

[13] Finocchiaro, P., et al.; Characterization of a Novel 100-Channel Silicon Photomultiplier – Part II: Charge and Time, *IEEE Transactions on Electron Devices,* vol.55, (2008) 2765.

[14] Giudice, A. C., et al.; A process and deep evaluation tool: afterpulsing in avalanche junctions, *33rd Conference on European Solid-State Device Research*, Vol.1, (2003) 347-350.

[15] Finocchiaro, P., et al.; Test of scintillator readout with single photon avalanche photodiodes, *IEEE Transactions on Nuclear Science*, vol.52, (2005) 3040.

[16] Bonanno, G., et al.; Precision measurements of Photon Detection Efficiency for SiPM detectors, *Nuclear Instruments and Methods in Physics Research* A610 (2009) 93-97.

[17] Finocchiaro, P., et al.; Features of Silicon Photo Multipliers: Precision Measurements of Noise, Cross-Talk, Afterpulsing, Detection Efficiency, *IEEE Transactions on Nuclear Science,* vol.56, (2009) 1033.

[18] Agodi, C., et al.; The HADES time-of-flight wall, *Nuclear Instruments and Methods in Physics Research* A 492 (2002) 14–25.

[19] Bellini, V., Capogni, M., Febbraro, V., Finocchiaro, P.; Implementation of a new low cost detector for low intensity light pulses produced by a radioactive gamma-source, *Applied Radiation and Isotopes* 68 (2010) 1320–1323.

[20] Woods, R.; McAllister, J.; *FPGA-based implementation of signal processing systems,* John Wiley and Sons, 2008.

[21] Cyclone II *FPGA Starter Development Board, Reference Manual,* Altera, http://www.altera.com.

[22] Gislason, D.; Zigbee *Wireless Networking*, Elsevier Science and Technology, August 2008.

[23] Bonin, B., Colin, M., Dutfoy, A.; Pressure building during the early stages of gas production in a radioactive waste repository, *Journal of Nuclear Materials* 281, (2000) 1–14.

[24] Van Dorp, F., Grogan, H., McCombie, C.; Disposal of radioactive waste, *Radiation Physics and Chemistry,* Vol. 34, No. 2, (1989) 337-347.

[25] Matsumura, T.; Effects of Radiation Damage on Multi-Pixel Photon Counter (MPPC), *talk given at the IEEE Nuclear Science Symposium*, Dresden (Germany), 2008.

In: Radioactive Waste: Sources, Types and Management
Editors: Satoshi Yuan and Wenxu Hidaka
ISBN: 978-1-62100-188-1

Chapter 2

RADIOACTIVE WASTE FROM RESEARCH REACTOR OPERATION AND DECOMMISSIONING

J. G. Marques[1,2] *and A. Kling*[1,2]
[1]Instituto Tecnológico e Nuclear, Estrada Nacional 10, Sacavém, Portugal
[2]Centro de Física Nuclear da Universidade de Lisboa, Lisboa, Portugal

ABSTRACT

Research reactors are important tools for radioisotope production, education and training, material structure studies, neutron radiography, silicon doping, neutron activation analysis, and neutron capture therapy. More than 600 research reactors were built worldwide since Fermi's graphite pile in 1942, of which about 200 are currently in operation. Research reactors present a large variety of designs, with different arrangements of moderator, reflector and coolant materials, fuel enrichment up to 93% and a wide power range, from sub-critical assemblies up to 250 MW. In this chapterwe will review the information available on the production of radioactive waste during the operation of research reactors as well as during decommissioning and dismantling of the installations.

1. INTRODUCTION

Research and test reactors comprise a wide range of nuclear fission reactors which are generally not used for power generation. For simplification, we will refer to these two generic types of reactors simply as 'research reactors'. The primary purpose of research reactors is to provide a source of neutrons and the heat released by fission is in most cases simply discarded.

More than 600 research reactors were built worldwide since Fermi's 'Chicago Pile 1' (CP-1) in December 1942 [1], out of which approximately one third is currently in operation [2]. They present a large variety of designs, in contrast with power reactors, where 82% of the world's plants are of just two types [3], Pressurized Water Reactors (PWR) and Boiling Water Reactors (BWR). Research reactors are small relative to power reactors and their power ranges from 'zero' up to 250 MW. In fact, the total power of the currently operating research reactors is only 2.2 GW (thermal), which is less than the thermal power of a typical 1 GW_e nuclear power plant with a PWR or BWR. Research reactors have a simpler design than power reactors and operate at lower temperatures. They need far less fuel, and less fission products build up as the fuel is used. On the other hand, their fuel requires uranium enriched to higher levels, typically just below 20% (thus being 'Low Enriched Fuel', LEU, while fuel enriched to 20% or more is designated 'Highly Enriched Uranium', HEU), although some reactors use uranium enriched to about 93% in ^{235}U. As neutron production is their main function, most research reactors have a compact core surrounded by a reflector to reduce neutron losses from the core.

While some research reactors were built for specific purposes, most are currently used in a wide range of activities, such as radioisotope production, education and training, material structure studies, neutron radiography, silicon doping, neutron activation analysis, and neutron capture therapy [4,5].

In this chapterwe will review the information available on the production of radioactive waste during the operation of research reactors as well as during decommissioning and dismantling of the installations.

2. General Featuresof Research Reactors

The 'Research Reactor Database' (RRDB) [2] maintained by the International Atomic Energy Agency (IAEA) has data pertaining to more than 600 research reactors built since 1942. Table 1 summarizes the different types of research reactors built, along with the number of units built (in percentage of the total) and the power range of the units. The twelve types listed are not strict, e.g., some tank reactors are also heavy-water moderated reactors. Each reactor is counted only once and classified by its most distinctive feature. The classification in the RRDB has been kept in all cases. We will briefly review these types, starting with the oldest reactors and grouping some of the types, whenever applicable.

Table 1. Types of research reactors, as described in the IAEA's Research Reactor Database [2], by order of importance

Type of reactor	Units built (%)	Power range (W)
Pool	23.6	0.01 k – 100 M
Critical assembly	16.1	0.01 k – 20 k
Tank, tank in pool	11.1	0.01 k – 250 M
Homogeneous (solid and liquid)	11.1	0.01 k – 10 k
TRIGA	10.2	20 k – 14 M
Heavy water moderated	5.6	0 – 135 M
ARGONAUT	4.2	0.01 – 300 k
Fast	3.9	0 – 6 k
Graphite moderated	3.3	0.1 k – 120 M
Subcritical	1.7	0
SLOWPOKE	1.5	0 – 20 k
MNSR	1.4	27 k – 33 k
Other	7.3	0 – 200 M

Early research reactors used natural uranium as fuel and, therefore, graphite or heavy water as moderator. A heavy water moderated reactor requires less uranium than a graphite moderated one, so the availability of uranium and of heavy water or graphite were decisive factors in the early days. Amongst early graphite moderated reactors are Fermi's CP-1 in the US, 1942 [1], F-1 in Russia, 1946 [6], GLEEP in the UK, 1947 [7] and BR1 in Belgium, 1956 [8]. Examples of early heavy-water moderated reactors, are ZEEP in Canada, 1945 [9], ZOE in France, 1947 [10] and JEEP in Norway, 1951 [11]. The availability of enriched uranium allowed building higher power reactors, with more compact cores and much higher values of the thermal neutron flux. Overall, only 9% of the research reactors are of these two types.

The oldest homogenous reactors were built at the Los Alamos National Laboratory (LANL) in 1944 for the weapons program, 'LOwPOwer' (LOPO) and its upgrade 'HYghPOwer' (HYPO). LOPO used 565 g of ^{235}U in the form of uranyl nitrate, enriched to 14.5%, dissolved in 13 liters of water [12]. It was the first reactor operating with enriched uranium [13]. LOPO was reflected by beryllium oxide and had no shielding, so its power was limited to 50 mW. HYPO was reflected by graphite and required 808 g of ^{235}U uranium; its maximum power was 6 kW [14]. About thirty liquid homogeneous reactors were built for various uses and most are no longer operational. However, a renewed interest in this type of reactors emerged recently for the production of medical isotopes, namely ^{99}Mo [15]. Solid homogenous reactors were also constructed. Two examples are the 'Aerojet General Nucleonics' [12] and the 'Siemens Training Reactor' (SUR) [16] types. In these reactors the fuel (UO_2 for the AGN and U_3O_8 for the SUR) is mixed with the solid moderator, polyethylene. The first three AGN reactors went critical in US in the same day, January 1st, 1957 [2]. The cost of an AGN-type reactor in the late fifties was USD 100,000 [17]. Overall, 11% of the research reactors are homogeneous ones, of these two types, liquid and solid.

The most common research reactor (ca. 24%) is the pool type, also called swimming pool type. In this type of reactor the core, consisting of the fuel elements and the control rods, is situated in an open water pool. The water acts as neutron moderator, cooling medium and radiation shield. The 'Bulk Shield Reactor' which started operating in 1950 in the Oak Ridge National Laboratory (ORNL) was the first reactor of this type, built to study radiation shield problems for the US Navy propulsion program [18]. An icon of this type of reactors is the one built by ORNL for the First Geneva Conference in 1955. This reactor was later renamed 'SAPHIR' and became the first reactor supplied by the US under the 'Atoms for Peace' program. It operated between 1957 and 1994 in the 'Paul ScherrerInstitut', Switzerland, having undergone several upgrades in power, up to 10 MW. Several US companies supplied pool-type reactors to about twenty countries. Russia also developed pool-type reactors: the IRT reactor built in the Kurchatov Institute in the late fifties [19] is a typical pool-type reactor installed in Belarus, Bulgaria, Georgia, Iraq, Latvia, Libya, North Korea and Ukraine [2,20]. The cost of a 1MW US-made pool-type reactor in the late fifties was approximately USD 1,000,000 [17].

Tank-type reactors are a development of pool-type reactors and allow a better control of the flow path of pumped water. The oldest tank-type reactors in the RRDB are the TVR in Russia, 1949 [21], the 'Low Intensity Test Reactor' (LITR) in the US, 1950 [18], the EL-2 in France, 1952 [10], and the 'Materials Test Reactor' (MTR) in the US, 1954 [18]. The LITR was a low power mock-up of the MTR. The basic design of the fuel elements of the MTR, made by Eugene Wigner [18,22], is used today in a large number of research reactors of several types. Overall, 11% of the built research reactors were of the tank or tank-in-pool type.

The 'Training, Research, Isotope production, General Atomic' (TRIGA) type was developed by General Atomics (GA) [23]. The first TRIGA reactors were put into operation in 1958. Reactors of this type are installed in 24 countries on five continents [24]. The concept that led to the TRIGA design was the one of an 'inherently safe reactor', largely pushed by Edward Teller [25]. In TRIGA reactors much of the moderation is done by the hydrogen that is mixed with the uranium in the fuel rods. In the case of a sudden removal of a control rod the power will increase at first but will return to the previous level within a few thousandths of a second without need for the intervention of engineered safety devices. This feature is unique to TRIGA reactors and enables their operation not only in steady-state but also in pulse mode. For example, the 4.5 MW 'Annular Core Research Reactor' TRIGA in the Sandia National Laboratories reaches a peak power of 30,000 MW in pulse mode [26]. The cost of a typical 1 MW TRIGA reactor in the late fifties was USD 250,000 [2].

The first fast neutron reactor, Clementine, was built in the LANL in 1946 [27]. Although the availability of uranium at reasonably low prices has made less urgent the introduction of fast reactors which can effectively use the more common ^{238}U isotope, a renewed interest emerged meanwhile due to their ability to fission some of the longest lived isotopes in nuclear waste [28]. This category of reactors includes the well known 'Dounreay Fast Reactor' (DFR) in the UK and the Phenix reactor in France. It accounts for 4% of all research reactors.

Subcritical assemblies are mainly intended for neutron physics studies. The first subcritical assembly, the 'Nuclear Test Gauge' (NTG) was built in the Savannah River Site (SRS) in 1958, to assist in the production of plutonium and tritium in the heavy-water moderated reactors of this site [29]. The most recent is the YALINA facility in Minsk-Sosny, Belarus, operational since 2000, which was built to study accelerator driven sub-critical systems [30]. The oldest critical assemblies in the RRDB are COMET and JEZEBEL, built in the LANL in the early fifties for the weapons program [31]. This broad category of research reactors also includes, for example, the RA-8 reactor built to test in the late nineties the fuel for 'Central ARgentina de ElementosModulares' (CAREM), a 100 MW integral PWR design [32], as well as the 'Very High Temperature Reactor Critical Assembly' (VHTRC) built in Japan to carry out nuclear safety evaluations for the 'High Temperature Engineering Test Reactor' (HTTR), a high temperature gas-cooled reactor [33]. Overall, 18% of the research reactors fall into these broad categories of sub-critical or critical facilities.

The 'ARGOnne Nuclear Assembly for University Training' (ARGONAUT) was a low power, low cost (USD 100,000) reactor installed at the Argonne National Laboratory (ANL) in 1957 [34]. The 'Safe LOW-POwerKritical Experiment' (SLOWPOKE) is another low power, pool-type nuclear research reactor, designed by Atomic Energy of Canada Limited (AECL) whose first prototype went critical in 1970. It is beryllium-reflected with a very low critical mass (ca. 1 kg) [35]. The 'Miniature Neutron Source Reactor' (MNSR) built by the China Institute of Atomic Energy shares many features with the SLOWPOKE. These three types account for 7% of all research reactors.

The 'Miscellaneous' category in Table 1 includes, inter alia, prototypes of BWR, PWR and of other reactors.

Figure 1 shows the distribution at the end of 2010 of all research reactors into three categories, 'Operational' (236 reactors), 'Shutdown' (237 reactors, including 12 in temporary shutdown) and 'Decommissioned' (193 reactors). One should emphasize here that less than half of the non-operational reactors have been decommissioned so far.

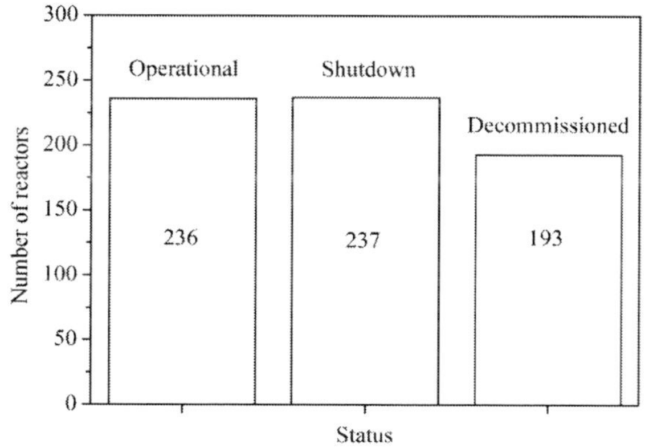

Figure 1.World status of research reactors at the end of 2010, using data from the 'Research Reactor Database' [2].

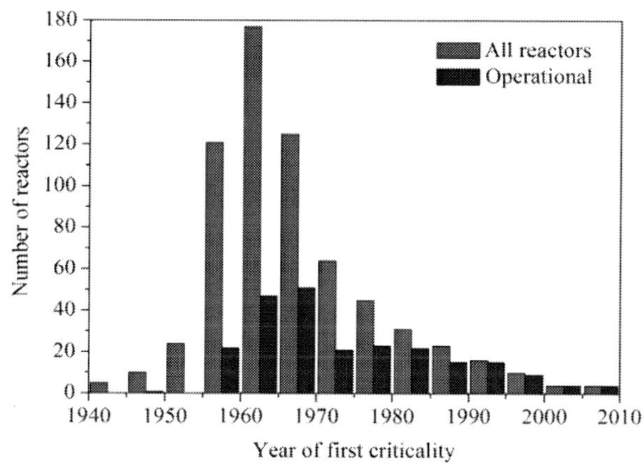

Figure 2.Year of first criticality of all research reactors and of the ones operational at the end of 2010.

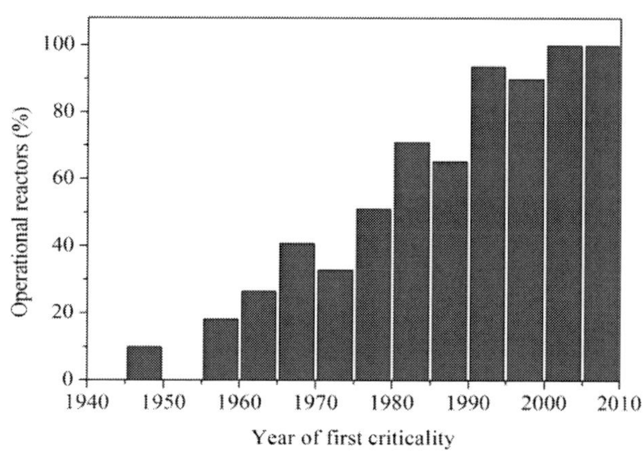

Figure 3.Fraction of operational research reactors at the end of 2010, as function of the year of first criticality.

Figure 2 shows the distribution of the year of first criticality of all research reactors and of the ones which were operational at the end of 2010, in five-year intervals [2]. When looking at the year of first criticality of all built reactors, a peak is clearly visible in the period 1960-1964. In fact, approximately two thirds of all research reactors went critical for the first time between 1955 and 1969, undoubtedly influenced by the ‘Atoms for Peace’ program. This program made it possible for many countries to have access to nuclear technology for peaceful purposes through research reactors. The distribution of the year of first criticality of the reactors which were operational at the end of 2010, also shown in Figure 2, has a different shape. While most of the reactors which went operational in the last two decades are still operational, a significant part of the older ones is shutdown or decommissioned.

Figure 3 shows the fraction (in percentage) of reactors that is still operational as function of the year of their first criticality, in five-year intervals. Interestingly, this fraction is approximately linear, with approximate values of 25% for the reactors which went critical between 1960 and 1964, and 50% for the reactors which went critical between 1970 and 1974. The oldest research reactor still in operation is the graphite-moderated F-1 in Russia, which went critical in 1946 [6].

There is a wide distribution of power of research reactors, as shown by Figure 4, where power is represented in decades. The largest group, with approxi[illegible] of all reactors built since 1942, has power below 1 kW. The second group is the one of reactors with power greater than 1 MW and less than 10 MW. Only 2.4% of the research reactors have a power of 100 MW or higher.

Although in each reactor the thermal neutron flux is proportional to the thermal power, there is no obvious relation between flux and power when a large number of reactors are taken into account, due to the large variety of designs. Figure 5 shows the distribution of maximum thermal neutron flux normalized by reactor power for sixteen reactors – TRIGA Mark II in Austria, BR2 in Belgium, Osiris and RHF in France, BER-II and FRM-2 in Germany, HOR in the Netherlands, JEEP-II in Norway, Maria in Poland, RPI in Portugal, IVV-2M in Russia, ATR, HFIR, MURR, OSTR and UFTR in the US [2]. On a global view, the thermal flux varies approximately between 3×10^{12} n/cm^2/s and 6×10^{13} n/cm^2/s per MW of reactor power. These values are indicative only, as the reactors here considered have different combinations of fuel, moderator, reflector and free volume where this maximum thermal flux is available.

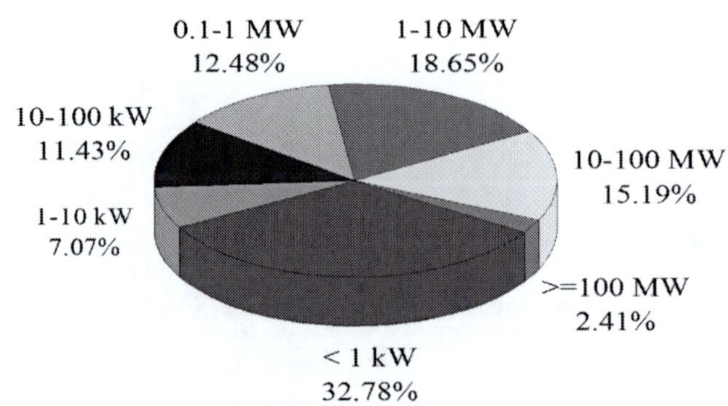

Figure 4. Power distribution of research reactors built since 1942, using data from the ‘Research Reactor Database’ [2].

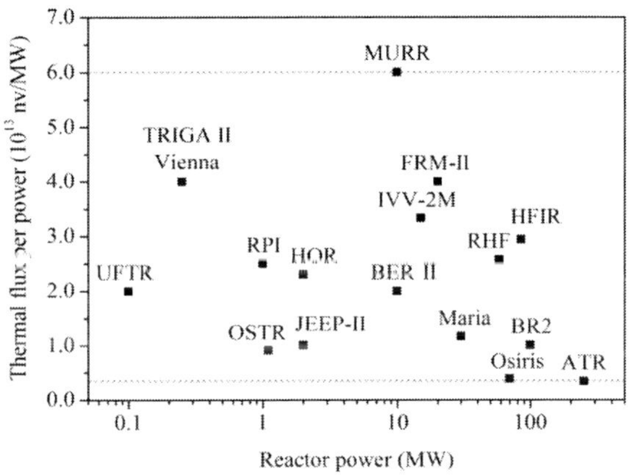

Figure 5. Thermal neutron flux distribution per unit power (in 10^{13}nv/MW=10^{13} n·cm^{-2}·s^{-1}·MW^{-1}) as function of reactor power. The lines were drawn to guide the eye.

3. Radioactive WastefromResearch ReactorOperation

The amounts and types of operational radioactive waste in research reactors depend on many factors, such as reactor type, operational schedule, and main applications. Reactor operation should be done in such way as to minimize the produced radioactive waste in terms of activity, amount and radiotoxicity. Adequate strategies should be already developed in the design stage of a research reactor, considering all pathways for waste production and proper waste management [36]. Improvements in the technical procedures and administrative actions can help to reduce waste generation, especially in installations where such strategies could not be developed at an early stage. Materials that will yield low activation and/or production of short-living isotopes in all locations where substantial neutron fluences occur should be selected and their amounts kept to the amount indispensible. The corrosion of structural materials and components should be minimized in order to reduce propagation of activated material, e.g., by using corrosion-resistant materials and avoiding the use of corrosive and inflammable ones. Surfaces prone to contamination should be easy to decontaminate or covered by material that is easy to remove and results in a minimal waste volume. Further areas in which handling and processing of radioactive materials is permitted should be limited.

3.1. Waste Classification

Radioactive waste can be classified by its activity level, physical form and origin [37]. The IAEA provides a classification system with six levels based on activity level [38]. The lowest level is the 'Exempt Waste' (EW) which is waste that meets the clearance level [39] set by the regulatory body, i.e., it is to be treated as non-radioactive waste. The 'Very Short Lived Waste' (VSLW) is waste that contains only radioisotopes with short-lived activity. This type of waste can be stored up to several years in order to reach the clearance level for EW. A significant part of the waste produced in production and use of radioisotopes belongs to this type. Waste that exceeds slightly the activity levels for EW but does not need a high level of containment and surveillance is classified as 'Very Low Level Waste' (VLLW). On the other hand, 'Low Level Waste' (LLW) comprises waste above the clearance level which contains a limited amount of radioisotopes with long half-life but can contain higher activity concentrations of short-lived radioisotopes. This type of waste requires strong isolation and containment to withstand storage up to a few centuries. Waste that contains a considerable amount of radioisotopes with long half-life (especially α-emitters) is classified as 'Intermediate Level Waste' (ILW) and requires a greater degree of containment and isolation than that provided for LLW. The last category is 'High Level Waste' (HLW). This is waste with levels of activity concentration high enough to generate significant quantities of heat by the radioactive decay process or waste with large amounts of long lived radioisotopes. Research reactors do not produce normally this kind of waste by activation, but spent fuel and the waste from its reprocessing normally fall into this category.

In addition, 'TRansUranic' waste (TRU) encompasses material with alpha-emitters having atomic numbers greater than 92 and half-lives greater than twenty years in concentrations above clearance levels (100 nCi/g in the US [40]), which does not fall under the HLW classification.

Radioactive waste is further divided by its physical form into solid, liquid and gaseous waste. The most important sources of operational waste in research reactors are discussed in the following subsections.

3.2. Solid Radioactive Waste

Common sources of solid VLLW and LLW operational waste are items contaminated during handling of radioactive materials, regular operation and maintenance, such as cleaning materials, personnel protection items and tools. Depending on the utilization of each reactor, various other waste streams occur, e.g., plastic and metallic containers for production of radioisotopes and irradiation of samples for neutron activation analysis, as well as components of installations exposed to neutron beams.

Intermediate-level waste (ILW) can arise from materials used in water cleaning, such as ion exchange resins, and items in the ventilation systems, such as the 'High Efficiency Particulate' (HEPA) and iodine retention filters. Further irradiated components of the reactor monitoring equipment (e.g., ionization and fission chambers, self-powered neutron detectors, thermocouples), control rods, and startup neutron sources can show ILW characteristics. Fission chambers typically contain gram amounts of enriched uranium, which makes them part of the nuclear materials in the facility and therefore subjected to safeguards [41].

A typical procedure for management of the solid waste [42] starts with its separation according to the classification described in subsection 3.1. Contaminated and activated materials containing only radioisotopes with short half-lives are stored under adequate conditions until they can be classified as EW or reused. Decontamination and reuse is the first option to be considered, if feasible, for items contaminated with radioisotopes with long half-lives. The remaining material is segregated according to material classes (e.g., metal, paper, plastic) in order to sort out non-compactable and non-combustible waste that is to be conditioned directly, typically through packaging in steel drums. The most important materials that fall into this category are used ion exchange resins that can be conditioned in various ways [43]. At present the most common approach is immobilization by mixing with cement [44] or with bitumen [45]. The amounts of ion exchange resins to be disposed of depend on the reactor type and power. Periodic regeneration of the resins reduces the amount of resins to be disposed, but increases the amount of liquid waste to be disposed, as the radioactive inventory of the resins is transferred to the washing waters. As the spent resins have a lower volume than the washing waters, less waste is generated by replacing the resins. For example, for the 20 MW OPAL reactor, of the Australian Nuclear Science and Technology Organisation (ANSTO), the expected annual generation of radioactive ion exchange resins amounts to about 1100 kg [46], without performing regenerations.

3.3. Liquid Radioactive Waste

Liquid radioactive waste is normally composed of coolant losses from the reactor pool or vessel, operation of hot chemistry laboratories, wash waters from maintenance and decontamination tasks. Wash water from sinks and showers within controlled areas also contributes to this waste stream.

The radioactive liquid effluents have to be collected at a single or several retention tanks. The further treatment of the liquid waste depends on various factors, namely, the activity concentration, the isotopic composition and the chemical composition. The use of delay tanks can reduce drastically the activity discharged in the case of liquids containing mostly radioisotopes with short half-lives (e.g., ^{24}Na, ^{38}Cl, ^{56}Mn). Other options can be, depending on the requirements of the regulatory authority, the concentration of the liquids, to minimize the volume, or their dilution, to minimize the activity concentration.

Table 2. Examples for annual liquid waste discharges in research reactors (NR=not reported) [50, 51, 52, 53, 54]

Reactor	Thermal power (MW)	Annual liquid waste discharged (Bq)		Year
		βγ-total	^{3}H	
Triga Mark II	0.25	5.6×10^{5}	1.4×10^{7}	2006
RPI	1	1.1×10^{7}	1.5×10^{8}	2009
HOR	2	4.9×10^{6}	NR	2008
JEEP-II [a)]	2	1.4×10^{8}	1.7×10^{12}	2008
FRG-1	5	1.9×10^{7}	1.3×10^{8}	2008
BER-II	10	1.5×10^{5}	4.6×10^{8}	2008
RA-3	10	1.1×10^{8}	NR	2004
FRM II [b)]	20	NR	1.9×10^{10}	2009

a) Heavy water moderated and cooled.
b) Heavy water moderated.

In all cases it is necessary to provide proper sampling and analysis of the retention tank contents according to established guidelines [47, 48,49] in order to decide if the liquid waste can be released. Procedures and facilities for treatment of the liquids or their storage may be required. Table 2 shows examples of the amounts of liquid waste released annually from the reactorsTriga Mark II in Austria [50], RPI in Portugal [51], HOR in the Netherlands [52], JEEP-II in Norway [52], FRG-1 and BER-II in Germany [52], RA-3 in Argentina [53], and FRM-II in Germany [54]. Although the reactors above are ordered by their power, there is no immediate relation between reactor power and the activity released, as many of the higher power reactors have in place measures to decrease the activity released. Tritium is produced via the $^{2}H(n,\gamma)^{3}H$ reaction and is of higher importance in reactors cooled or moderated with heavy water [55].

3.4. Gaseous Radioactive Waste

Gaseous radioactive waste in research reactor is mainly due to activation by neutron capture in the air present in the reactor coolant/moderator and irradiation facilities. The most important isotopes in this respect are ^{41}Ar and ^{14}C. The ^{41}Ar isotope is produced by the $^{40}Ar(n,\gamma)^{41}Ar$ reaction, with argon contributing with about 1% to the air composition. The ^{14}C isotope is produced by the $^{17}O(n,\alpha)^{14}C$ reaction, with ^{17}O contributing with 0.0366% to the natural oxygen. Also ^{14}C is generated in water moderated/cooled reactors by the $^{17}O(n,\alpha)^{14}C$ in the oxygen of the water molecules and the $^{14}N(n,p)$ ^{14}C reaction in the dissolved air (^{14}N is with 99.6% the main isotope of natural N).

In contrast to nuclear power plants, the routine release of radioactive fission products that are noble gases (e.g. ^{85}Kr, ^{131}Xe, ^{133}Xe) and iodine isotopes (mainly ^{131}I and the long-lived ^{129}I) from fuel assemblies has in general a minor importance in research reactors, due to the lower inventories and lower fuel operating temperatures.

Radioactive aerosols can be generated by activated particles stemming mainly from residual corrosion in the fuel cladding and structural materials of the reactor and subsequently release into the air, especially in open pool reactors.

Provisions have to be taken that radioactive gases are properly collected using appropriate ducting. The gaseous waste streams are to be joined at a single point for sampling and control, according to established procedures [48,56], and from there released to the environment. Delay lines for short-living radioisotopes can be used to reduce the release of radioactive gases to the environment.

Table 3. Examples for annual gaseous waste discharges in research reactors (ND=not detected; NR=not reported) [50, 51, 53, 57, 58]

Reactor	Thermal Power (MW_{th})	Annual gaseous waste discharged (Bq)			Year
		Noble gases	Iodine	Aerosol	
Triga Mark II	0.25	4×10^{11}	ND	NR	2006
RPI	1	1.4×10^{12}	7.2×10^{5}	8.5×10^{6}	2009
FRG-1	5	7.5×10^{11}	7.0×10^{5}	1.7×10^{7}	2008
RA-3	10	2.6×10^{13}	9.3×10^{7}	9.3×10^{8}	2004
FRM II	20	2.7×10^{11}	ND	ND	2008
Cabri	25	1.0×10^{12}	ND	7.9×10^{3}	2009
Phébus	38	3.9×10^{12}	1.1×10^{5}	1.6×10^{4}	2009

Table 3 shows examples of the annual amounts of gaseous waste released annually by the reactors Triga Mark II in Austria [50], RPI in Portugal [51], FRG-1 in Germany [57], RA-3 in Argentina [53], FRM-II in Germany [57], Cabri and Phébus in France [58]. The releases of ^{131}I are, in most cases, below the detection limit. A typical monitoring recommendation is that the detection limit for ^{131}I in the gaseous waste stream does not exceed 0.2 Bq/m^3 [48]. With typical stack flow rates between 1000 m^3/h and 10,000 m^3/h, the corresponding annual detection limits amount to 2×10^5Bq and 2×10^6Bq, respectively. For comparison purposes, the average annual release of noble gases by all nuclear power plants in the period 1995-1997 was 2.9×10^{16}Bq while the one of ^{131}I was 9.7×10^{10}Bq [59].

3.5. Management of Operational Waste

Waste generation during operation is unavoidable and a waste management plan should be established [36]. It has to ensure the safety of the research reactor staff and general public as well as reduce the environmental impact of the waste to the minimum reasonably achievable. Important items to be considered in a typical waste management program [47] are the proper classification [37, 38]and separation of the waste types and streams (e.g., separation of high and low activity liquid waste streams, dedicated ventilation systems in locations were gaseous waste is produced). Furthermore, a detailed inventory of solid and liquid waste stored on the premises and control of the amount and type of material removed from the site is required. This includes the control of discharges of gaseous and liquid waste to the environment. Safe handling, transport and discharge procedures for radioactive waste has to be taken into consideration. The plan must include a detailed description of the storage facility for solid and liquid waste, as well as eventual processes for local treatment of radioactive waste.

As in all other operational procedures the radiation protection policy concerning the exposure of the research reactor staff during the handling and processing of radioactive waste has to obey the 'As Low As Reasonably Achievable' (ALARA) principle. The possible exposure pathways in activities concerning waste management have to be identified and appropriate measures, such as shielding or delays before handling, have to be evaluated and put into practice.

4. Radioactive Wastefromthe Fuel Cycle

The fuel cycle of research reactors has unique characteristics. While a large percentage of power reactors use UO_2-based fuel, enriched to less than 5% in ^{235}U, research reactors use metallic uranium or a variety of uranium compounds, e.g., Al-UAl_x (also called aluminide fuel), Al-U_3O_8 (oxide), Al-U_3Si_2 (silicide) and $UZrH_x$(zirconium hidride), with enrichment up to 93%. Several geometries are in use, with the uranium distributed in rods, plates or cylinders, mostly clad with Al or Al alloys (such as AG3NE or Alcoa 6061), although stainless steel or zirconium are used in the cladding of some TRIGA fuel elements [60].

Table 4 shows the most common types of fuels, from the RRDB [2], with typical ^{235}U amounts in LEU and HEU elements. MTR type (also called plate-type) fuel elements are the most common. A plate-type fuel element consists of 12 to 23 flat or curved plates within a total cross-sectional area of 76 × 76 mm (3 × 3 inch). A typical flat-plate element is shown schematically in Figure 6. Each plate contains a fuel core (an alloy or aluminum dispersion), clad with an aluminum alloy with a thickness of 0.25-0.76 mm (10-30 mils). The active core length is usually 610 mm (24 inch). The two most common enrichments are 19.8% and 93%.

Table 4. Most common types of research reactor fuel, by order of importance [2]. The typical amount of ^{235}U per fuel element is also indicated

Research reactor fuel type	Reactors using fuel type (%)	Typical ^{235}U mass in HEU elements (g)	Typical ^{235}U mass in LEU elements (g)
MTR	26	100-400	300-500
TRIGA	21	137	38, 101
EK-10	6	-	8
Rod (non-TRIGA)	6	< 50	< 50
VVR	4	38-54	42, 50
IRT	3	300-550	200-300

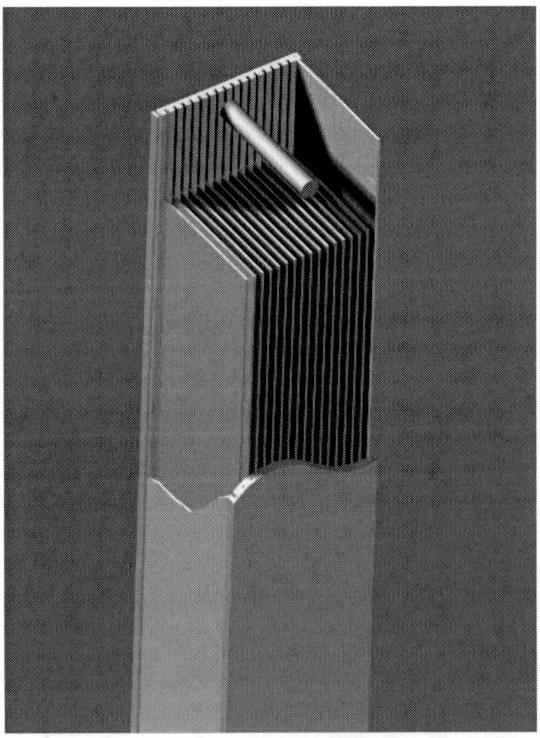

Figure 6. Cut through a MTR-type fuel element with 18 parallel plates. This is the most common fuel type for research reactors, with a number of plates generally between 12 and 23. The plates can also be curved.

TRIGA fuel elements are the second most common. The $UZrH_x$ fuel rods for TRIGA reactors are produced by casting a uranium zirconium alloy into graphite molds, machining to size, hydriding at elevated temperature, and grinding to final size. The 38 mm (1.5 inch) diameter fuel rods contain a 6.4 mm (0.25 inch) drilled central hole to allow faster and more controlled hydriding; the center region is later filled with a zirconium rod. All rods are slip fit into their cladding with a cold gap of 25 to 50 μm (1 to 2 mils). Typical enrichment levels are 20% and 70%. Other types of rod fuels (non-TRIGA) include, e.g., the fuel developed by AECL for the NRU and MAPLE reactors, using initially natural uranium and later UAl_x, U_3Si with HEU and LEU [61'62]. The types EK-10, VVR and IRT are exclusively used in Russian-origin reactors: the EK-10 is a pin-type fuel, with 10% enriched uranium, while the VVR and IRT are tube-type, with 19.8%, 36% or 93% enriched uranium [63]. UO_2-based fuel has only a marginal use in research reactors, e.g., in the PULSTAR-type reactors. The fuel of PULSTAR reactors is essentially the same as used in PWR and BWR reactors, but its enrichment is higher, up to 6% [60].

The burn-up of research reactor fuel is normally expressed in percentage of the initial ^{235}U content. Table 5 shows the burn-up expressed in power reactor equivalent units (MWd/kgU), for LEU fuel (19.75% enriched) and HEU fuel (93% enriched), calculated with the ORIGEN2 code [64]. In contrast with power reactors, the contribution from plutonium fissions is very small and an approximate conversion of units can be done considering a consumption of 1.25 g ^{235}U per MWd [65]. Fuel burn-up is normally much higher in research reactors than in power reactors. The burn-up in HEU aluminide fuels was usually limited to 25% (ca. 180 MWd/kgU), even if tests up to 50% burn-up have shown the fuel to perform satisfactorily [66]. LEU silicide fuel has been tested up to 98% peak burn-up [67]. GA has tested HEU TRIGA fuel elements up to 75% burn-up [68], although burn-up values for most TRIGA reactors are low.

Low power research reactors have little or no need for refueling. For example, a 10 kW research reactor operating 1000 hours per year consumes only 0.5 g of ^{235}U per year. This type of reactors has normally a lifetime core. This picture is significantly different for a high performance research reactor. For example, the 20 MW reactor FRM-II in Munich, Germany, consumes four to five fuel elements per year, each with 8.1 kg of uranium enriched to 93% in ^{235}U [69], with an average burn-up of 20% at the time of discharge [2]. In contrast, a 1 GW_e PWR requires about 27 t of uranium enriched to ca. 3% per year [70].

Table 5. Research reactor burn-up expressed in power reactor equivalent units (MWd/kgU), for LEU fuel (19.75% enriched) and HEU fuel (93% enriched), calculated with the ORIGEN2 code

Research reactor fuel burn-up (%)	Equivalent power reactor fuel burn-up (MWd/kgU) for LEU	Equivalent power reactor fuel burn-up (MWd/kgU) for HEU
10	16	73
20	32	146
30	48	219
40	65	292
50	82	366
60	99	440

At the start of the 'Atoms for Peace' program, both the US and the Soviet Union supplied foreign research reactors with LEU fuels, although at different enrichment levels, 19.8% in the case of the US and 10% in the case of the Soviet Union. In many US-supplied research reactors, power upgrades required an increased ^{235}U loading per element, as well as a reduction in fabrication costs. These factors pushed HEU fuel at 93% enrichment to become the standard. In Soviet-supplied research reactors, power upgrades required changing from pin to tubular fuel element geometries, as well as increasing the ^{235}U element loading and HEU at 80-90% enrichment became the standard enrichment in these research reactors.

The US Department of Energy (DOE) established in 1978 the 'Reduced Enrichment for Research and Test Reactors' (RERTR) program to develop and qualify LEU fuels that can replace HEU fuels in research reactors. Russia's Ministry of Atomic Energy also initiated around the same time a program to convert foreign research reactors to 36% enriched uranium [71] as a first step before conversion to LEU fuel [72]. After 1989, Russia exported only fuels containing 36% enriched uranium to foreign research reactors [73].

To complement the RERTR program, DOE established the 'Foreign Research Reactor Spent Nuclear Fuel Acceptance Program' (FRRSNF) in 1996. Under this program, US-supplied spent and unused fresh fuel elements are accepted for management and final disposition in the US [74]. The Russian Federation established in 2004 an acceptance program for fuels supplied by Russia and the former Soviet Union [75].

The US and Russian spent fuel acceptance programs provide an end-of-cycle solution for the fuel of many facilities. The FRRSNF program was extended until May 2019, for eligible fuels in use at research reactor facilities until May 2016, without significant changes to the number of assemblies to be returned. The end of this program may lead to the closure of research reactors which otherwise would have no alternative for the disposal of their spent fuel, namely in countries without nuclear power plants. Even in countries with large commercial nuclear installations, research reactor operators will need custom packages for their spent fuel or waste from reprocessing that meet the requirements of these facilities [76].

Table 6. Estimated number of Aluminum-based and TRIGA spent nuclear fuel elements as of January 2006 in the 'Foreign Research Reactor Spent Nuclear Fuel Acceptance Program' of the US Department of Energy [74, 77]

Type of fuel	Number of elements	Initial U mass (kg)	Volume (m^3)	Elements per year [a)]
Al-based, HEU	10,109	4,528	105	194
Al-based, LEU	7,672	13,654		559
TRIGA, HEU	1,106	82	5	6
TRIGA, LEU	3,834	950		79
Total	22,721	19,214	110	838

[a)] Average yearly increase from 1996 to 2006 [77].

In the early nineties, DOE estimated that the foreign research reactors would have in January 2006 a total of over 22,000 fuel elements eligible to be returned to the US, corresponding to an initial uranium mass of 19 t, as detailed in Table 6 [74,77]. The number of elements per year is the average yearly increase from 1996 to 2006, from reference [77]. The estimated total amount of spent fuel of US-origin, including the fuel from US research reactors, is 28 t [78]. As of 2010, 9,175 fuel elements had been returned to the US under this program [79] and 67 reactors had been converted to LEU or shutdown prior to conversion

[80]. The RERTR program is also actively promoting the reduction from HEU to LEU in the targets for production of isotopes from fission [80].

About 2 t of HEU and 2.5 t of LEU spent fuel will be shipped to the Mayak reprocessing complex near Chelyabinsk, under the Russian acceptance program [81]. At the end of 2010 no Russian research reactors had been converted to LEU, but the RERTR program had supported the conversion to LEU of Russian-supplied reactors in the Czech Republic, Hungary, Libya, Vietnam, Ukraine and Uzbekistan [82], with LEU fuels made in Russia [72]. RERTR-supported conversion projects are underway in Belarus, Bulgaria, the CzechRepublic, Kazakhstan and Poland. A 2002 study estimated that Russian research reactors required 180 kg of HEU per year [83], while a 2008 study indicated 350 kg of HEU per year, but included reactors outside Russia and some test reactors in Russia which may be used for non-civil purposes [73].

The US and Russia are not the only sources of enriched research reactor fuel – the UK, China, France, India and South Africa also produced enriched fuel for their reactors, with limited exports in the cases of UK, China and France [84]. Some countries have significant stocks of irradiated natural uranium. For example, Norway has 16 t of spent fuel of which 6 t is stored at Kjeller and 10 t in Halden. This includes approximately 12 t of Al-clad fuel, of which 10 t is metallic uranium fuel and the remainder is UO_2 [85,86].

The estimates of global amounts of research reactor fuel have necessarily significant uncertainties, being affected by the annual variations in the fuel consumption by the facilities, the conversion from HEU to LEU fuels, and the shipments to the country of origin. The RRDB has records for the number of fuel elements in storage at the facilities in 2010 and for the number of assemblies required per year of operation [2], as communicated by the operators. These values are detailed in Table 7. There is a reasonable agreement between the values shown in Tables 6 and 7. The current consumption of HEU assemblies of US-origin is close to the value previously estimated by DOE, while the one of LEU assemblies is somewhat lower, probably due to a slower rate of reactor conversions. Considering the number of assemblies received by the US and the number of on-site assemblies of US origin reported by the operators, the total is close to the 22,000 assemblies estimated by DOE [77].

Table 7. LEU and HEU fuel elements in storage in research reactors in 2010 and forecasted use of elements per year for the two major suppliers [2]

Enrichment	US-origin elements in storage	US-origin elements use per year	Russian-origin elements in storage	Russian-origin elements use per year
LEU	6,351	474	15,990	75
HEU	4,461	228	8,833	267

In the US, the currently 104 operating nuclear power reactors discharge annually an average of 2,100 t (heavy metal) of fuel, totaling approximately 60,000 t (heavy metal), or about 200,000 assemblies at the end of 2008 [87]. If all these reactors receive a 20-year life extension the total spent fuel amount will double by 2055. Thus spent fuel from research reactors is only a small part of the whole inventory of spent fuel. The limit of the Yucca Mountain repository in the US is 70,000 t (heavy metal).

DOE does not currently plan to reprocess the research reactor spent fuel returned to the US. Although aluminum based spent nuclear fuel has been chemically reprocessed at SRS for decades [88], currently only fuels considered as 'at risk' are reprocessed. All other aluminum based fuels are stored in a pool until a final disposition path is implemented. Some fuel elements have been in water basins for over 50 years, considering the time they were in a reactor core and the storage time. Corrosion of Al-clad fuels is prevented by having water conductivity in the 1-3 μS/cm range, with chlorides at less than 100 ppb [89]. TRIGA fuel elements are handled at the INL [90].

Although reprocessing in the US was halted to avoid plutonium separation, this is not a significant problem with conventional reprocessing of enriched fuel from research reactors. The higher ^{235}U enrichment of these fuels leads to significantly less plutonium than in the fuel of nuclear power plants and the separation of plutonium is not a required part of the reprocessing treatment, as it can be left in the liquid waste stream along with the fission products for later vitrification [91].

DOE developed the 'melt and dilute' process applicable to aluminide, oxide and silicide fuels [78]. This process allows a decrease in volume by about 70%, when compared with direct disposal [92], producing at the same time a robust waste form for long term storage. In this process an intact fuel element is placed in a furnace and melted. Depleted uranium is added, as necessary to dilute the ^{235}U enrichment to less than 20%, in order to eliminate criticality and proliferation concerns during storage and transportation. Aluminum is also added to the melt in order to keep the mixture near the eutectic point for the UAl_x alloy and consequently near its lowest melting point (646 °C). The process is expected to operate at 850 ± 50 °C, minimizing the release of volatile radioisotopes from the fuel [93,94]. Neutron absorber materials, such as gadolinium and hafnium, can also be added to the melt to minimize the potential for long-term criticality in the repository [95]. The melt-dilute process takes approximately six hours, including one hour to completely melt and dilute an MTR element. Once solidified the ingot is placed in a steel canister, with other ingots. The canisters will be back-filled with helium, sealed, and temporarily stored at SRS in concrete storage modules before final disposal.

AREVA-NC (former COGEMA) in France has a long experience in recycling spent fuel, including the one from research reactors [96, 97,98, 99, 100,101]. About 20 t of aluminide research reactor fuel were recycled in the plant at La Hague [101]. The treatment is based on a modification of the classical 'Plutonium - URaniumEXtraction' (PUREX) process [88,102], with the dissolution of the research reactor fuel in nitric acid [101]. To avoid criticality problems research reactor fuel is diluted into power reactors UO_2 fuel to blend down the ^{235}U content to a maximum of 2%. The plant can process a few tons of research reactor fuel per year, in function of the throughput for power reactor fuel (maximum capacity of 1600 t/year). Research reactor fuel enters directly the continuous dissolver, bypassing the shearing process necessary to reduce the dimensions of the fuel rods from power reactors before dissolution. The chemical operations after dissolution are basically performed following the reference process. In addition, the integrated waste conditioning technologies lead to a drastic reduction of the volume of residues that require disposal. A volume reduction by a factor 30 to 50 is achieved: 0.4 m^3 per ton of spent fuel, compared with 12 to 18 m^3 per ton of spent fuel for the direct disposal option [98]. The volume here indicated for the direct disposal is 2-3 times higher than the one considered for the non-TRIGA fuel in Table 6. However, the volume indicated in Table 6 is only the one of the fuel elements (on average ca. 6000 cm^3 for 1 kg uranium, per element) and not the effective volume for their direct disposal, which is necessarily higher, for criticality control and waste isolation.

Table 8 shows an estimate of the volumes of radioactive waste generated by the different options for disposal of the US-origin non-TRIGA fuel of Table 6, with a total volume of 105 m^3. For the direct disposal, an effective volume 2-3 times higher than the one of the elements has been considered, as per reference [97]. For the waste generated from the melt and dilute process a reduction of 70% of the estimated volume for direct disposal has been considered [92], while for reprocessing a factor of 0.4 m^3 per ton of fuel has been used [97]. The option that generates less waste is reprocessing, but melt and dilute is also an effective option.

Table 8. Estimated waste volume of US-origin Al-based research reactor fuel under various disposal options. The total volume of the fuel elements is 105 m^3

Mode	Volume (m^3)
Direct disposal	210 - 315
Melt and dilute	63 - 95
Vitrified waste after reprocessing	42

TRIGA and silicide fuels are not readily reprocessed in conventional facilities. The reprocessing of silicide fuel was initially demonstrated at SRS [103]. This type of fuel was reprocessed in Dounreay [104], but this facility in the UK is no longer operational. Modifications in the PUREX at La Hague process are required for silicide fuel or a dilution with other fuels, such as aluminide [105]. Although the new UMo fuel, still under qualification, can be reprocessed [106], the Si additions necessary to improve its radiation behavior [107] may introduce limitations in this process.

5. Radioactive WastefromRadioisotope Production

Radioisotopes have innumerous applications in medicine, agriculture as well as industrial and basic research. The production of radioisotopes in research reactors is of high economic relevance due to its impact on health care and industrial process control. Waste is produced in the target fabrication, as well as during handling and processing of the irradiated material.

About fifty radioisotopes are produced in larger scale at research reactors, using nuclear fission of ^{235}U or neutron capture in targets [108]. Although most of the existing research reactors produce radioisotopes, at least for research purposes [109], the bulk of industrial scale production has been concentrated in only five reactors worldwide, NRU in Canada, HFR in The Netherlands, SAFARI in South Africa, BR2 in Belgium, and Osiris in France [110].

5.1. Radioisotope Production by Neutron Capture Reactions

Target encapsulation is an important source of solid radioactive waste in the case of radioisotope production by neutron capture. Aluminum, zircaloy and stainless steel are commonly used for encapsulation [108]. Aluminum has a low cross section for neutron capture and its main activation occurs through the $^{27}Al(n,\alpha)^{24}Na$ reaction leading to the short-lived ^{24}Na ($T_{1/2}$ = 15.02 h) isotope and thus not hampering the reuse of the containers. The use of zircaloy for encapsulation results in the production of ^{123m}Sn with short half-life ($T_{1/2}$ = 40.08 m) and ^{93}Zr with long half-life ($T_{1/2} = 1.53\times10^6$ y) from neutron capture in the main constituents of the alloy. The short half-life of ^{123m}Sn and the low radiological risk of ^{93}Zr (pure β-emitter) allow in general reusing the containers. The use of stainless steel for encapsulation yields various radioisotopes with longer half-lives (e.g., ^{59}Fe, ^{51}Cr, ^{60}Co) which may inhibit the reuse of containers.

While in some cases, e.g., ^{192}Ir seed for brachytherapy, the radioisotopes are in their final physical state, chemical processing is required in most cases. This process can generate considerable amounts of liquid radioactive waste which requires proper treatment, as previously discussed. The laboratory equipment used in the chemical processing can be highly contaminated. In most cases its reuse is not economically feasible, resulting in considerable amounts of solid waste. However, a large part of this waste can be classified as VSLW, since many important radioisotopes for medical applications have short half-lives.

5.2. Radioisotope Production by Fission of Uranium

The production of radioisotopes by nuclear fission of ^{235}U is the economically more important technique. Nuclear fission of ^{235}U produces several hundred radioisotopes of which ^{90}Sr, ^{99}Mo, ^{131}I, ^{133}Xe and ^{137}Cs are of major interest for applications. In particular, the ^{99m}Tc obtained from the ^{99}Mo/ ^{99m}Tc generator system is used in 80-85% of all diagnostic nuclear medicine procedures in the world [111]. In the US only, ^{99}Mo is used daily in over 50,000 procedures [80]. The weekly worldwide demand for ^{99}Mo amounts to about 12,000 Ci, with almost 50% for the US market [112, 113]. The estimated selling prices of EUR 45 per Ci of ^{99}Mo for the raw product from the reactor and EUR 1,810 per Ci of ^{99}Mo for the final radiopharmaceutical [114] make this radioisotope also economically the most important. Therefore the waste generation and management of the production process of this isotope will be reviewed in the following in more detail. An exhaustive discussion can be found in reference [115].

Targets for the production of ^{99}Mocan consist of uranium or uranium alloys with cladding in various shapes, e.g., plates, pins or cylinders, or without cladding as encapsulated metal foils. The fabrication of these targets can lead to the generation of low solid LLW in the form of metal scraps containing or being contaminated with uranium. The recovery of the uranium from this waste has been successfully tested [116].

The irradiation of the targets produces the full set of fission products and a small amount of actinides. For a typical target with 35 g of ^{235}U, irradiated for 104 h at a thermal neutron flux of 1.5×10^{14} n $cm^{-2}s^{-1}$, the activity inventory due to fission products is 4×10^{16}Bq after irradiation and 7×10^{12}Bq after one year. The activity due to the actinides is 4×10^{12}Bq after irradiation and 1×10^{8}Bq after one year[115].

The irradiated targets have to be processed chemically for the extraction of the ^{99}Mo. Two different dissolution processes are currently used: alkaline dissolution in NaOH for cladded or aluminide targets and acid dissolution in NHO_3 for uranium metal or UO_2 targets. An overview on the currently used processes is given in references [117, 118]. In both cases more than 85% of the ^{99}Mo can be extracted from the solution. While the chemical characteristics of the liquid waste streams are different, their properties in terms of activity concentration and radioisotope content are similar. Both processes also yield gaseous waste in almost identical amounts.

The gaseous waste consists mainly of radioactive noble gases, e.g., ^{85}Kr, ^{133}Xe, ^{135}Xe. The short-lived noble gas radioisotopes ^{133}Xe and ^{135}Xe can be retained by cryogenic methods and stored until they are sold or decay. Since other methods for ^{133}Xe production yield higher radioanalytical purities this isotope is normally not commercialized. The long-lived ^{85}Kr ($T_{1/2}$ = 10.72 y) is released into the atmosphere. In addition, a fraction of the radioactive iodine enters the gaseous waste stream due to its volatility and has to be retained by activated charcoal filters. Finally, considerable amounts of tritium are released in the chemical process and have to be removed from the gaseous waste stream by oxidation and subsequent filtering [115].

The liquid waste stream contains the remaining uranium, all non-volatile fission products (except molybdenum) and the major fraction of radioactive iodine isotopes, tritium and actinides generated by neutron capture in uranium. Furthermore, if a target cladding and/or uranium alloys are used the respective activation products enter this waste stream. For the most used Al cladding, mainly short-lived isotopes, e.g., ^{28}Al, ^{56}Mn and ^{64}Cu are present.

The liquid waste from the ^{99}Mo extraction process is classified as HLW. Typical values for the volume produced are 5-15 liters per kCi of ^{99}Mo extracted [115]. The liquids are stored in appropriate tanks for six to twelve months in order to reduce the activity of the fission products (especially ^{131}I) and actinides. After this cooling period the activity is dominated by a few main fission radioisotopes (in decreasing order of activity): $^{144}Ce/^{144}Pr$, ^{137}Cs, $^{90}Sr/^{90}Y$, and $^{106}Ru/^{106}Rh$. While in some installations the liquid waste is conditioned and immobilized by cementation [115], most installations opt for long-term storage awaiting a final solution.

Solid waste classified as ILW stems mainly from the chemical extraction process itself (e.g., extraction columns, small laboratory equipment) and the air filtration process (HEPA and charcoal filters). Typically, the amount of solid waste produced is in the order of a few tens of kg per kCi of ^{99}Mo produced [115].

The solid, liquid and gaseous waste identified above is not necessarily produced at the reactor where the uranium targets are irradiated, as in many cases these are shipped to processors where the dissolution and separation of the isotopes is performed.

HEU is still used in the majority of the fission targets. As the waste resulting from processing the targets contains about 97% of the initial uranium, about 45 kg of HEU end up in the liquid waste stream per year, posing a proliferation risk [119].Although the uranium from the HEU containing waste could be recovered by chemical processing, this option has not been chosen by any producer [119]. An alternative method to reduce the proliferation risk is the downblending of the HEU waste by mixing with natural or depleted uranium to LEU levels. However, the most promising option to overcome proliferation risks is to eliminate the use of HEU by using LEU targets or alternative production routes [80,120]. The SAFARI reactor was the first large producer to use LEU targets [121], joining the smaller producers OPAL in Australia and RA-3 in Argentina [110]. In addition, it is planned that ^{99}Mo will be produced using LEU targets at the reactors MURR in the USA, RECH-1 in the Czech Republic, MARIA in Poland, and TRIGA II in Romania [110].

Although the use of LEU would mitigate the proliferation risk, its application has also some drawbacks in terms of waste generation. In the first place, about six times more uranium has to be dissolved in the extraction process. Since the concentration of metals in the solutions used in the present processes is limited, this leads also to an increase of the volume of the liquid waste stream by a factor of about five. On the other hand, it has been pointed out that advanced chemical processes such as the 'modified Cintichem' [122] may fully compensate this disadvantage [123]. Second, the amount of ^{239}Pu produced in the irradiation of the targets increases by a factor of about 30 [124], which may turn part of the waste into TRU with higher demands regarding its handling and disposal. However, like in the case of HEU, uranium and other actinides can be extracted chemically from the liquid waste stream, reducing significantly the amount of potentially TRU waste.

6. Radioactive Waste from Decommissioning and Dismantling of Research Reactors

Although less than half of the non-operational reactors have been decommissioned so far, there is a significant experience in this field. Several of the earlier decommissioning projects in the US are summarized in reference [125] and in Russia in reference [126]. Research and development activities were performed in several projects, e.g., in the decommissioning of the CP-5 reactor [127] in the US and in the decommissioning of the BR3 [128] and JEN-1 [129] reactors in Europe. The IAEA collected much data and operator experience, including research and developmentefforts, on the decommissioning of research reactors [130, 131, 132, 133,134].

Decommissioning strategies range from immediate dismantling and removal of all radioactive materials from the site, allowing unrestricted release, to an option of in situ disposal involving encapsulation of the reactor and subsequent restriction of access [135].

Immediate dismantling involves the decontamination, dismantling and removal of all equipment, structures and other parts of the facility that are contaminated, typically within three years after permanent shutdown. This option normally has the fewest uncertainties, eliminates the risk associated with the facility in a small time scale, costs less than deferred dismantling, and allows the use of operational staff who knows the facility and its history. However it may lead to higher worker doses, due to the relatively short decay time of the radioactive isotopes present. An obstacle to this option may be the absence of a waste disposal route, namely in the case of the spent fuel.

The deferred dismantling decommissioning strategy requires that the facility is placed and maintained in a safe, stable, and monitored condition for an extended period of time (some decades) until it is decontaminated or dismantled. A certain number of activities still need to be done immediately after shutdown. This strategy minimizes the initial commitments of time, funds, radiation exposure, waste disposal capacity, and may reduce the quantities of radioactive waste produced. It is limited to facilities where the appropriate decay of the nuclides in the nuclide vector is shorter than the expected life structure. However, there are additional costs associated with providing long term surveillance and maintenance, and the operator has to ensure that sufficient expertise and knowledge will be available at the time of decommissioning. Funding and legislative uncertainties may also be higher. This option puts an undue burden on future generations, which may be socially unacceptable.

The entombment is a decommissioning strategy similar to immediate dismantling, except that not all of the radioactive material is removed from the site and the radioactive contaminants are encased in a structurally long-lived material, such as concrete. The entombment structure must be appropriately maintained and surveyed until the radioactivity decays to a level permitting an unrestricted release of the facility. In principle, entombed facilities become a near surface waste repository and must comply with radiological requirements for waste disposal facilities. This may be an option for countries with a very limited number of facilities, e.g., just one research reactor, and not having the resources to develop or obtain the infrastructure needed for dismantling and waste disposal [135].

The Nuclear Regulatory Commission (NRC) evaluated these strategies in the early eighties [136]. Table 9 presents the total costs of the different strategies together with the collective dose from the decommissioning operations for two reference reactors. The first reference reactor is the 1 MW Oregon State University TRIGA Reactor (OSTR), at Corvallis, while the second is the 60 MW Plum Brook Reactor Facility (PBRF), of the National Aeronautics and Space Administration (NASA) at Sandusky. The OSTR was recently converted to LEU fuel within the RERTR program [137], while the PBRF is shutdown since 1973 and its decommissioning is expected to be complete at the end of 2011 [138].

Table 9. NRC estimated costs and collective doses for different decommissioning strategies for two types of research reactors [136]

Strategy	OSTR		PBRF	
	Cost (MUSD) [a)]	Collective dose (man·Sv)	Cost (MUSD) [a)]	Collective dose (man·Sv)
Immediate	0.85	0.18	15.6	3.22
Deferred – 10 y	1.64	0.15	17.6	1.98
Deferred – 30 y	2.24	0.13	20.0	1.18
Deferred – 100 y	4.50	0.13	27.2	1.12
Entombment	0.56	0.17	14.6	4.25

[a)] 1981 values.

Deferred dismantling has been considered after 10, 30 and 100 years in the NRC study. The collective doses in the case of deferred dismantling include the operations that need to be performed immediately as well as the operations to be performed after the waiting period. The prices are given in 1983 USD, which can be converted to 2010 values multiplying by 2.40, as determined using the Consumer Price Index (CPI) inflation calculator of the US Bureau of Labor Statistics [139]. The most cost-effective strategy is immediate dismantling. Deferred dismantling is more costly than immediate dismantling, although a 50% reduction of the collective dose is achievable after a delay of 10-15 years in the case of a large research reactor. The reduction of the collective dose is not significant for the deferred dismantling of a less complex, lower power reactor, where the amounts and activation of materials are necessarily lower. Entombment has no obvious advantages, from the cost and collective dose perspectives.

Selected examples of decommissioning and dismantling of facilities are given in the sections below, addressing small and medium-sized facilities, as well as some problems with specific materials.

6.1. Decommissioning and Dismantling of Small Facilities

DR1 was a homogeneous reactor with a maximum power of 2 kW, at the Risø National Laboratory (RNL), Denmark [140]. Uranium in the form of UO_2SO_4 was dissolved in water, with a total core volume of 14 liters. It contained 5 kg of uranium enriched to 19.9% in ^{235}U, with a very low burn-up (0.5%) when it was shutdown in 2001. The core vessel was a spherical stainless steel vessel with a diameter of 0.32 m. The reactor had a graphite reflector with 1.50 m diameter and 1.30 m height, placed in a steel tank. A recombiner served to recombine the hydrogen and oxygen that were formed by radiolysis when the reactor was in operation. The biological shield consisted of magnetite concrete with a thickness of 1.20 m. In the case of DR1, RNL wished to re-use the building for other purposes, so the reactor block was removed and the building cleaned to meet clearance levels. The whole project took 13 months to execute [141]. The major part of the activity in DR1 was concentrated in the reactor vessel, the recombiner and the pipe connecting the two, all made of stainless steel. Most of the biological shielding was cut with a wire saw; other methods were only used close to the experimental channels. This allowed a better control of the origin of the concrete waste for segregation. Table 10 shows the materials balance from the decommissioning of the reactor, totaling 212 t [141]. About two thirds of the total waste is due to the dismantling of the biological shielding. Although 80% of the materials from the biological shielding could be released without restrictions, this block still contributed with about 80% of the total radioactive waste. The collective dose during the operations was approximately 1 man· mSv [141]. The total cost of the decommissioning of the DR1 reactor was DKK 5,400,000 (approximately USD 970,000) [142].

Table 10. Materials balance from the decommissioning of the DR1 homogenous reactor in Denmark [141]

Type of waste	Biological shield (t)	Whole facility (t)
Exempt waste	111	174
Radioactive waste (LLW/ILW)	31	38
Total	142	212

The ‘Moata’ reactor was an Argonaut type 100 kW reactor operated by ANSTO from 1961 to 1995. The HEU fuel was removed from the reactor in 1995 and sent to the US in 2006. Once the nuclear materials were removed from the reactor, the demineralised water was drained in 1996. The control system was removed in 1998 and the control room was converted into office and laboratory space [143]. Some internal components were removed in 2009. The concrete biological shield was demolished and removed during 2010. The anticipated amount of waste generated is 136 m^3 (55% EW, 45% LLW, <0.1% ILW). The collective dose was 10 man·mSv [144]. The budget for the operation was AUD 3,800,000 (approximately USD 3,800,000) [145].

Jason was another Argonaut-type reactor, with 10 kW power, located in the KingWilliamBuilding, within the RoyalNavalCollege in Greenwich, London. It was stopped in 1996 and fully dismantled by 1999, with the complete preservation of the historical building [146]. The HEU fuel was sent to Sellafield for long-term storage. About 270 t of radioactive waste were removed. The collective dose during the operations was below 2 man·mSv [132]. The cost of this operation was GBP 7,000,000 (approximately USD 11,000,000).

6.2. Decommissioning and Dismantling of Medium-Sized Facilities

The 'Ames Laboratory Research Reactor' (ALRR) was a 5 MW, tank-type research reactor, which operated in the IowaStateUniversity campus between 1965 and 1977. The ALRR was moderated and cooled with heavy water and used MTR-type fuel assemblies. The reactor core was housed in a 1.52 m diameter aluminum core tank. Six 2.54 cm thick curved stainless steel plates (i.e., the thermal shield) surrounded the core tank on all sides except at the thermal column. These plates were cooled with light water, which also served as a neutron reflector and circulated through its own heat exchanger in the pump room. The core tank and thermal shield were contained in a 2.44 m diameter aluminum tank, which in turn was surrounded by a concrete biological shield. Decommissioning of the reactor started in 1978 and was completed in 1981. The total amount of waste generated was 1224 t. The collective dose during decommissioning was 0.69 man·Sv. The total cost was USD 4,300,000 [125].

The DR2 is the second of three research reactors on the RNL site that has been decommissioned. The DR2 was a 5 MW water cooled and moderated pool-type reactor, designed by the Foster Wheeler Corporation [13]. The pool comprised an open aluminum tank, with 2 m diameter and 8 m height. The biological shield had an octagonal shape, with a thickness of ca. 2 m in the lower half of the tank (barite concrete) and ca. 1 m in the upper half (ordinary concrete). The DR2 first reached criticality in late 1958 and was stopped in 1975. The reactor used MTR-type fuel, which was returned to the US. The decommissioning of the reactor block started in May 2006, after a decay period of ca. 30 years, and was completed in mid-2008. The building of DR2 was kept and the pool was demolished. Table 11 shows the materials balance from the decommissioning of the reactor, totaling 616 t [147]. The collective dose during the operations was below 10 man·mSv. The estimated cost for this operation was DKK 28,000,000 (approximately USD 5,000,000) [148].

Table 11. Materials balance from the decommissioning of the DR2 pool-type reactor in Denmark [147]

Type of waste	Biological shield (t)	Whole facility (t)
Exempt waste	375	421
Radioactive waste (LLW/ILW)	161	195
Total	536	616

The 'AdaptierterSchwimmbecken-Typ-Reaktor Austria' (ASTRA) was a pool-type reactor built by AMF Atomics, which operated between 1960 and 1999 in the Austrian

Research Center Seibersdorf. ASTRA had several upgrades in power, having reached 10 MW in the last decade of operation. The integrated power per year was 500-900 MWd/year since 1966 [149]. The decommissioning was started in 2003 and finished in 2006 [150, 151,152]. The Research Center Seibersdorf decided to keep the reactor building for other uses. Figure 7 shows the layers defined for cutting the biological shield (pool) of the ASTRA reactor, using wire saw technology [151]. The most active part of the biological shield was between cuts 4 and 5, which includes the zone closest to the core. Table 12 shows the materials balance from the decommissioning of the reactor, totaling 2,174 t [153]. More than 80% of the total waste is due to the dismantling of the pool, although this block represents only one third of the total radioactive waste, as over 70% of the materials from the pool could be released without restrictions. The pool of ASTRA had 400 m^3 of reinforced barite concrete, totaling approximately 1,500 t [151]. The graphite from the inner part of the thermal column was annealed in a hot-cell prior to interim storage, to release its Wigner energy [154,155]. The collective dose during the operations was below 0.1 man·Sv [156]. The total cost of the decommissioning of the ASTRA reactor was EUR 15,000,000 (approximately USD 18,000,000) [153].

Table 12. Materials balance from the decommissioning of the 10 MW ASTRA pool-type reactor in Austria [153]

Type of waste	Biological shield (t)	Whole facility (t)
Exempt waste	1,553	2,030
Radioactive waste (LLW/ILW)	27	83
Total	1,580	2,174

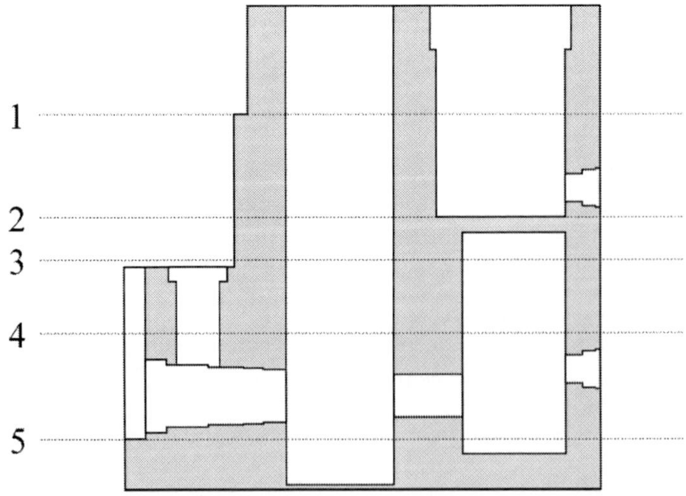

Figure 7. Simplified view of the biological shield of the ASTRA reactor, showing the layers used for cutting. Adapted from references [150, 151].

These examples show that the main source of waste comes from the dismantling of the biological shield. Nevertheless, with proper segregation it is possible to clear a large part of the waste. Notably, in the case of the decommissioning of the ASTRA reactor 96% of the total amount of waste could be exempted. The decommissioning costs are not far from the initial cost of the installation, once inflation is taken into account. As reference, the installation costs in 1960-1965 would have to be multiplied by 7 to obtain 2010 prices, taking inflation into account according to the CPI inflation calculator of the US Bureau of Labor Statistics [139].

6.3. Issues with Specific Materials

Graphite is used in research reactors as a moderator and reflector. Early graphite reactors contained considerable amounts of this material, e.g., CP-1 had 385 t, GLEEP had 550 t, while the 'Brookhaven Graphite Research Reactor' (BGRR) had 700 t [12].

The IAEA estimates that the total amount of graphite used in moderators and reflectors of fission reactors is 250,000 t [157], or roughly 150,000 m^3. Graphite moderated plutonium-producing reactors, as well as graphite moderated power reactors, are the major contributors to this value. In the UK only, the total amount of graphite used in MAGNOX and 'Advanced Gas Reactor' (AGR) units is 81,000 t [158,159].

The RRDB lists 22 graphite reactors [2], which we estimate use 6,000 m^3 of graphite. Additionally, a large number of research reactors have a stacking of graphite in one of their irradiation facilities, the 'thermal column'. Other research reactors have graphite reflectors, of varied sizes. As a rough estimate, if two thirds of all research reactors had 4-5 m^3 of graphite, about 2,000 m^3 of this material would be in use in facilities around the world. This rough estimate results in a total amount of 8,000 m^3graphite used in research reactors.

The disposal options for graphite are, generically, (i) placement in repository, (ii) incineration, with release of gases, sequestration, or partial recycling, and (iii) recycling of solid graphite for further use within the nuclear industry [157]. Annealing is normally performed before long-term storage to release the Wigner energy [155].

The ^{14}C isotope ($T_{1/2}$=5740 y) is produced by neutron activation in the graphite, through the $^{13}C(n,\gamma)^{14}C$ reaction. The amount of accumulated ^{14}C in the graphite may make its incineration unacceptable, triggering other disposal options [55]. The graphite from BGRR, in the form of approximately 60,000 blocks, has a total activity of 760 Ci, 71% of which due to ^{14}C. Other significant (> 1%) radionuclides identified included ^{63}Ni, ^{3}H and ^{153}Eu. This graphite will be sent to the Nevada Test Site for disposal as LLW. The estimated volume is 886 m^3, at a packing efficiency of 50% [160].

Beryllium is used in research reactors as a source of neutrons, moderator and reflector. The material itself is toxic and its inhalation can lead to chronic beryllium disease, an incurable lung disease [161]. The neutron irradiation induces the formation of helium atoms through various nuclear reactions [162] which cause dimensional changes such as swelling and radiation growth [163]. The swelling caused by helium leads to the need to replace the beryllium periodically. This need is more obvious in high-flux reactors – e.g., the 250 MW 'Advanced Test Reactor' (ATR) at the INL replaces its beryllium reflector every 10 years [164]. In addition to the currently operating reactors, several decommissioned reactors have beryllium reflectors that require disposal. Beryllium was disposed in the US as LLW until the early nineties, when it was found that typical impurity levels of uranium could result in this material being classified as TRU and thus requiring different disposal provisions [164].

Conclusions

The 'Atoms for Peace' program made it possible for many countries to have access to nuclear technology for peaceful purposes through research reactors. More than 600 research reactors were built since 1942, out of which one third are currently in operation. The amounts of radioactive waste produced during normal operation of the facilities are very small.

Decommissioning has been underway in particular in the last two decades but a large number of shutdown reactors are not yet decommissioned. Recentdecommissioning projects show that this can be performed with reasonably low collective doses, at values close to the initial cost of the installation, once inflation is taken into account. Compared with the spent fuel from nuclear power plants, the one from research reactors has some specific issues, namely due to different materials and enrichment, but the amount is much smaller, and the issues can be addressed in several ways.

References

[1] Fermi, E. (1952). Experimental Production of a Divergent Chain Reaction. *American Journal of Physics* 20, 536-558.

[2] International Atomic Energy Agency. *Research Reactor Database*. <http://www.iaea.org/worldatom/rrdb/>.

[3] Marques, J.G. (2011). *Review of Generation III/III+ Fission Reactors*. Nuclear Energy Encyclopedia: Science, Technology, and Applications, Lehr, J., Krivit, S.B., Kingery, T.B.(Eds.), John Wiley and Sons, New York, 231-254. ISBN 978-0470894392.

[4] International Atomic Energy Agency (2001). *The Applications of Research Reactors*. TECDOC-1234, IAEA, Vienna.

[5] International Atomic Energy Agency (2007). Utilization Related Design Features of Research Reactors: A Compendium. *Technical Reports Series N*o. 455, IAEA, Vienna.

[6] Kruglov, A. (2002). *The History of the Soviet Atomic Industry*. Taylor and Francis, London, 189-191. ISBN 978-0415269709.

[7] Graham, I.P., Fowler, J.P. (1996). Decommissioning of Western Europe's Oldest Reactor The Graphite Low Energy Experimental Pile (GLEEP) at UKAEA Harwell. *Proceedings 9th Congress of the International Radiation Protection Association*, Vienna, 1996, Vol 3, 335-337.

[8] Wagemans, J., Vittiglio, G., Malambu, E., Aït Abderrahim, H. (2009). The BR1 Reactor: a Versatile Irradiation Facility for Fundamental Research and Industrial Applications. *Proceedings 1st International Conference on Advancements in Nuclear Instrumentation Measurement Methods and their Applications*, Marseille, 2009, doi 10.1109/ANIMMA.2009.5503727.

[9] Atomic Energy of Canada Limited (1997). *Canada Enters the Nuclear Age: A History of Atomic Energy of Canada Limited.* McGill-Queen's University Press, Montreal and Kingston, 215-232. ISBN 0-773516018.

[10] Goldschmidt, B. (1982). *The Atomic Complex: A Worldwide Political History of Nuclear Energy*. ANS, La GrangePark, 121-125. ISBN 0-894485504.

[11] Barendregt, T.J., Hidle, N., Lundby, A., Sceland, E., Trumpy, G (1956). Practical Experiences with the JEEP Reactor. *Proceedings of the International Conference on the Peaceful Uses of Atomic Energy,* Geneva, 1955, United Nations, Geneva, Vol. 2, 259-265.

[12] Glasstone, S. (1955). *Principles of Nuclear Reactor Engineering*. D. Van Nostrand Company, Princeton, 788-838.

[13] International Atomic Energy Agency (1959). *Directory of Nuclear Reactors*. Vol. II, IAEA, Vienna.

[14] Los Alamos Scientific Laboratory of the University of California (1951). An Enriched Homogeneous Nuclear Reactor. *Review of Scientific Instruments* 22, 489-499.

[15] International Atomic Energy Agency (2008). *Homogeneous Aqueous Solution Nuclear Reactors for the Production of Mo-99 and other Short Lived Radioisotopes*. IAEA, Vienna. ISBN 978-9201097088.

[16] Klein, F. (2009). Konditionierung von Brennstoffplatten aus Siemens-Unterrichtsreaktoren Verfahrensentwicklung und-anwendung. *Atw. Internationale Zeitschrift für Kernenergie* 54, 242-247.

[17] Chastain, J.W. (1958). *US Research Reactor Operation and Use.*Addison-Wesley, Reading, 316-341.

[18] Rosenthal, M.W. (2010). *An Account of Oak Ridge National Laboratory's Thirteen Nuclear Reactors.*ORNL/TM-2009/181, Rev. 1, ORNL, Oak Ridge, 45-46.

[19] Ryazantsev, E.P. (2008). 50th Anniversary of the IRT Reactor: History of Creation and Development. *Atomic Energy* 104, 470-482.

[20] Aerospace Information Division of the Library of Congress (1983). Soviet Nuclear Research Reactors of the IRT-2000 Type: Review or Soviet Literature. *AID Report P-63-53*, Library of Congress, Washington.

[21] Volkov, E.B., Shvedov, O.V., Gavrilin, N.D., Konev, V.N. (2003). Decommissioning of the TVR Research Heavy-Water Reactor at the Institute of Theoretical and Experimental Physics. *Atomic Energy* 95, 735-740.

[22] Weinberg, A.M. (1994). *The First Nuclear Era: The Life and Times of a Technological Fixer.*AIP, New York, 57-58. ISBN 978-1563963582.

[23] Taylor, T.B., Mc-Reynolds, A.W., Dyson, F.J. (1964). *Reactor with Prompt Negative Temperature Coefficient and Fuel Element Therefo*r. US Patent 3127325, issued March 31, 1964.

[24] Fouquet, D.M., Razvi, J., Whittemore, W.L. (2003). TRIGA Research Reactors: A Pathway to the Peaceful Applications of Nuclear Energy. *Nuclear News* 46:12, 46-56.

[25] Dyson, F.J. (1981). *Disturbing the Universe*. Basic Books, New York, 94-106. ISBN 978-0465016778.

[26] Schmidt, T.R., Reuscher, J.A. (1994). Overview of Sandia National Laboratories Pulse *Nuclear Reactors*. SAND-94-2466C, SNL, Albuquerque.

[27] Glasstone, S. (1955). *Principles of Nuclear Reactor Engineering*. D. Van Nostrand Company, Princeton, 832-834.

[28] Marques, J.G. (2010). Evolution of Nuclear Fission Reactors: Third Generation and Beyond. *Energy Conversion and Management* 51, 1774-1780.

[29] Parkinson, T.F., Baumann, N.P. (2000). The Nuclear Test Gauge. *Proceedings Symposium 50 Years of Excellence in Science and Engineering at the Savannah River Site, Savannah River Site,* Aiken, WSRC-MS-2000-00061, WSRC, Aiken, 57-66.

[30] Gohar, Y., Smith, D.L. (2010). *YALINA Facility: A Sub-Critical Accelerator-Driven System (ADS) for Nuclear-Energy Research Facility Description and an Overview of the Research Program.* ANL-10/05, Argonne National Laboratory, Argonne.

[31] Paxton, H.C. (1976). *Safety Analysis of the Los Alamos Critical Experiments Facility.* LA-6206, Vol. I, LANL, Los Alamos.

[32] Fukami, M.V.I., Santecchia, A. (2000). CAREM Project: Innovative Small PWR. *Progress in Nuclear Energy* 37, 265-270.

[33] International Atomic Energy Agency (2001). Critical Experiments and Reactor Physics Calculations for Low Enriched High Temperature Gas Cooled Reactors. *TECDOC-*1249, IAEA, Vienna, 27-43. ISSN 1011–4289.

[34] Holl, J.M., Hewlett, R.G., Harris, R.R. (1997). *Argonne National Laboratory*: 1946-96. University of Illinois Press, Champaign, 175-206. ISBN 978-0252023415.

[35] Kennedy, G., St. Pierre, J., Bennett, L.G.I., Nielsen, K.S. (2002). LEU-Fuelled Slowpoke-2 Research Reactors: Operational Experience and Utilisation. *Proceedings 24nd International Meeting on Reduced Enrichment for Research and Test Reactors,* Bariloche, 2002, ANL, Argonne. <http://www.rertr.anl.gov>.

[36] International Atomic Energy Agency (2008). Radiation Protection and Radioactive Waste Management in the Design and Operation of Research Reactors. *IAEA Safety Standards Series No.* NS-G-4.6, IAEA, Vienna.

[37] International Atomic Energy Agency (2007). Categorizing Operational Radioactive Wastes, *TECDOC*-1538, IAEA, Vienna.

[38] International Atomic Energy Agency (2009). Classification of Radioactive Waste. *Safety Standards Series No.* GSG-1. IAEA, Vienna.

[39] International Atomic Energy Agency (2004). Application of the Concepts of Exclusion, Exemption and Clearance. *Safety Standards Series No.* RS-G-1.7. IAEA, Vienna.

[40] US Department of Energy (2008). Transuranic Waste Acceptance Criteria for the Waste Isolation Pilot Plant. *DOE/WIPP*-02-3122, Carlsbad.

[41] International Atomic Energy Agency (2002). IAEA Safeguards Glossary 2001 Edition. *International Nuclear Verification Series No.* 3, IAEA, Vienna.

[42] International Atomic Energy Agency (2001). Handling and Processing of Radioactive Waste from Nuclear Applications. *IAEA Technical Reports Series No.* 402, IAEA, Vienna.

[43] Speranzini, R.A., Buckley, L.P. (1983). A Summary of Methods for Conditioning and Immobilizing Ion Exchange Resins. AECL-7976, *Chalk River Nuclear Laboratories,* Chalk River.

[44] Li, J.F., Wang, J.L. (2006). Advances in Cement Solidification Technology for Waste Radioactive Ion Exchange Resins: *A Review. Journal Hazardous Materials* 135, 443-448.

[45] Roth, A., Centner, B., Lemmens, A. (2008). Radioactive Spent Resins Conditioning by the Hot Super-Compaction Process, *Proceedings of the 11th International Conference on Environmental Remediation and Radioactive Waste Management,* Bruges, 2007, ASME, New York, 687-693. ISBN 978-0791843390.

[46] Australian Nuclear Science and Technology Organization (2004). *ANSTO Replacement Research Reactor Project, SAR Chapter 12 - Operational Radiological Safety.* RRRP-7225-EBEAN-002, ANSTO, Lucas Heights, Australia.

[47] German Nuclear Safety Standards Commission (1994). Monitoring and Assessing of the Discharge of Radioactive Substances in Liquid Effluents, *Safety Standard KTA* 1504 (6/94), Salzgitter.

[48] German Nuclear Safety Standards Commission (1998). Monitoring the Discharge of Radioactive Substances from Research Reactors, *Safety Standards KTA* 1507 (6/98), Salzgitter.

[49] US Department of Energy (1991). *Environmental Regulatory Guide for Radiological Effluent Monitoring and Environmental Surveillance*. DOE/EH-0173T, DOE, Washington.

[50] Austrian Federal Ministry for Health and Austrian Federal Ministry of Agriculture, Forestry, Environment and Water Management (2009). *Radioaktivität und Strahlung in Österreich 2005 und 2006 – Daten und Bewertung*. BMG and BMLFUW, Vienna.

[51] Kling, A. (2010). *Radiological Control at the Portuguese Research Reactor (RPI): Report January 1 - December 31*, 2009. ITN/RPI-R-10/111, ITN, Sacavém.

[52] OSPAR Commission (2010). *Liquid Discharges from Nuclear Installations in 2008.*OSPAR, London.

[53] Argentinean National Commission for Atomic Energy (2005). *Joint Convention on the Safety of Spent Fuel Management and on the Safety of Radioactive Waste Management – Second National Report,* Comisión Nacional de Energía Atómica, Buenos Aires.

[54] Bavarian State Agency for Environment (2010). Strahlenhygienischer *Jahresbericht 2009–Allgemeine Umweltradioaktivität und Umgebungsüberwachung der kerntechnischen Anlagen in Bayern*, LfU, Augsburg.

[55] International Atomic Energy Agency (2004). Management of Waste Containing Tritium and Carbon-14. *IAEA Technical Reports Series No*. 421, IAEA, Vienna.

[56] German Nuclear Safety Standards Commission (2002). Surveilling the Release of Gaseous and Aerosol-bound Radioactive Substances, Part 1: Surveilling the Release of Radioactive Substances with the Stack Exhaust Air during Specified Normal Operation, *Safety Standards KTA* 1503.1 (06/02), Salzgitter.

[57] German Federal Ministry of Environment, Natural Protection, and Reactor Safety (2009). *Umweltradioaktivität und Strahlenbelastung – Jahresbericht* 2008. BMU, Bonn.

[58] Commission for Atomic and Alternative Energies (2010). *Rapport Transparence et Sécurité Nucléaire – 2009*. CEA Cadarache - Tome 1, CEA, Cadarache.

[59] United Nations Scientific Committee on the Effects of Atomic Radiation (2000). Sources and Effects of Ionizing Radiation: *United Nations Scientific Committee on the Effects of Atomic Radiation (UNSCEAR) 2000 Report to the General Assembly.*Vol. 1, Bernan Associates, Blue Ridge Summit. ISBN 978-9211422429.

[60] Stahl, D. (1983). *Fuels for Research and Test Reactors: Status Review* July 1982. ANL-83-5, ANL, Argonne.

[61] Chalder, G.H., Bourns, W.T., Feraday, M.A., Veeder, J. (1967). U3Si as a *Nuclear Fuel*. AECL-2874, AECL, Chalk River.

[62] Sears, D.F., Conlon, K.T. (2006). Development of LEU Fuel to Convert Research Reactors: NRU, MAPLE and SLOWPOKE. *International Symposium on Minimization of Highly Enriched Uranium in the Civilian Nuclear Sector,* Oslo, 2006.< http://www.nrpa.no/>.

[63] Rozhikov, V.V., Enin, A.A., Alexandrov, A.B., Tkachev, A.A. (2007). Design and Manufacture of Fuel Assemblies for Russian Research Reactors. *Proceedings of the NATO Advanced Research Workshop on Safety Related Issues of Spent Nuclear Fuel Storage,* Almaty, 2005, Lambert, J.D.B., Kadyrzhanov, K.K. (Eds.), Springer, Dordrecht, 95-105. ISBN 978-1402059025.

[64] Croff, A.G. (1980). *A User's Manual for the ORIGEN2 Computer Code*. ORNL/TM-7175, ORNL, Oak Ridge.

[65] Pond, R.B., Matos, J.E. (1996). *Photon Dose Rates from Spent Fuel Assemblies with Relation to Self-Protection (Rev. 1).*ANL/RERTR/TM-25, ANL, Argonne.

[66] Miller, L.G., Brown, K.R., Beeston, J.M., McGinty, D.M. (1986). *Extended Life Aluminide Fuel:Final Report*. EGG-2441, EGandG Idaho, Idaho Falls.

[67] Nuclear Regulatory Commission (1988). *Safety Evaluation Report Related to the Evaluation of Low-Enriched Uranium Silicide-Aluminum Dispersion Fuel for Use in Non-Power Reactors*. NUREG-1313, NRC, Washington.

[68] Simnad, M.T., Foushee, F.C., West, G.B. (1976). Fuel Elements for Pulsed TRIGA Research Reactors. *Nuclear Technology* 28, 31-56.

[69] Roehrmoser, A. (2010). Core Model of New German Neutron Source FRM II. *Nuclear Engineering and Design* 240, 1417-1432.

[70] Lamarsh, J.R., Baratta, A.J. (2001). *Introduction to Nuclear Engineering*. Prentice Hall, UpperSaddleRiver, 185-187. ISBN 978-0201824988.

[71] Nasonov, V.A., Ryazantsev, E.P., Taliev, A.V., Egorenkov, P.M. (2010). Work on the Program for Lowering the Fuel Enrichment in Research Reactors to 36% at the Russian Science Center Kurchatov Institute. *Atomic Energy* 108, 207-213.

[72] Nasonov, V.A., Ryazantsev, E.P., Taliev, A.V., Egorenkov, P.M. (2010). Work on the Program for Lowering the Fuel Enrichment in Research Reactors to 19.7% at the Russian Science Center Kurchatov Institute. *Atomic Energy* 108, 364-369.

[73] Sokova, E.K. (2008). Phasing Out Civilian HEU in Russia: Opportunities and Challenges. *Nonproliferation Review* 15, 209-236.

[74] US Department of Energy (1996). *Final Environmental Impact Statement on a Proposed Nuclear Weapon's Nonproliferation Policy Concerning Foreign Research Reactor Spent Nuclear Fuel.* DOE/EIS-0218F, Vol. 1, DOE, Washington.

[75] Bolshinsky, I., Tyacke, M., Allen, K., Moses, S., Thomas, J., Dewes, J. (2005). Russian Research Reactor Fuel Return to the Russian Federation. *Proceedings of 27th International Meeting on Reduced Enrichment for Research and Test Reactors,* Boston, 2005, ANL, Argonne. <http://www.rertr.anl.gov>.

[76] International Atomic Energy Agency (2010). Cost Aspects of the Research Reactor Fuel Cycle. *IAEA Nuclear Energy Series* No. NG-T-4.3. ISBN 978-9201019103.

[77] Matos, J. (1994). *Foreign Research Reactor Irradiated Nuclear Fuel Inventories Containing HEU and LEU of US-Origin*. ANL/RERTR/TM-22, ANL, Argonne.

[78] Peacock, H.B., Adams, T.M., Duncan, A.J., Iyer, N.C. (1999). Melt-Dilute Treatment of Spent Nuclear Fuel Assemblies from Research and Test Reactors, *Proceedings of 22nd International Meeting on Reduced Enrichment for Research and Test Reactors*, Budapest, 1999, ANL, Argonne. <http://www.rertr.anl.gov>.

[79] Messick, C.E., Taylor, J.L., Niehus, M.T., Landers, C. (2010). The United States Foreign Research Reactor Spent Nuclear Fuel Acceptance Program: 2010 Update.

Transactions 14th International Topic Meeting on Research Reactor Fuel Management, Marrakech, 2010, ENS, Brussels. <http://www.euronuclear.org/>.

[80] Staples, P. (2010). The NNSA Global Threat Reduction Initiative's Efforts to Minimize the Use of Highly Enriched Uranium for Medical Isotope Production. *Transactions 14th International Topic Meeting on Research Reactor Fuel Management,* Marrakech, 2010, ENS, Brussels. <http://www.euronuclear.org/>.

[81] Choi, J-S., Isaacs, T.H. (2005). Changing Perspectives on Nonproliferation and Nuclear *Fuel Cycles*. UCRL-CONF-211009, LLNL, Livermore.

[82] National Nuclear Security Administration (2009). GTRI: Reducing Nuclear Threats. *Fact Sheet*, January 2, 2009. <http://nnsa.energy.gov>.

[83] Bukharin, O., Ficek, C., Roston, M. (2002). *U.S.-Russian Reduced Enrichment for Research and Test Reactors (RERTR) Cooperation*. Russian American Nuclear Security Advisory Council (RANSAC), Washington.

[84] World Nuclear Association (2010). Research Reactors. *Information Paper No.* 61, WNA. <http://www.world-nuclear.org/>.

[85] Oberländer, B.C., Wethe, P.I., Bennett, P. (2009). Storage of Spent Research Reactor Fuel in Norway. *IAEA Technical Meeting on Guidelines of Good Practice for the Management and Storage of Research Reactor Spent Fuel,* Thurso, 2009.

[86] Bennett, P., Oberländer, B., Adelfang, P., Eriksson, E. (2010). *Recommendations for the Conditioning of Spent Metallic Uranium Fuel and Aluminium Clad Fuel for Interim Storage and Disposal*. IFE, Halden.

[87] Nuclear Energy Agency (2009). *Radioactive Waste Management and Decommissioning in the United States of America,* NEA, Paris. <http://www.oecd-nea.org/rwm/profiles/>.

[88] Gray, L.W. (1986). *Reprocessing of Nuclear Fuels at the Savannah River Plant.*DP-MS-86-157, SRL, Aiken.

[89] International Atomic Energy Agency (2003). Corrosion of Research Reactor Aluminium Clad Spent Fuel in Water. *Technical Reports Series No.* 418, IAEA, Vienna, 52-53. ISBN 978-9201137036.

[90] Hill, T.J., Fillmore, D.L. (2007). Managing Spent Nuclear Fuel at the Idaho National Laboratory. *Proceedings of the NATO Advanced Research Workshop on Safety Related Issues of Spent Nuclear Fuel Storage, Almaty*, 2005, Lambert, J.D.B., Kadyrzhanov, K.K. (Eds.), Springer, Dordrecht, 71–84. ISBN 978-1402059025.

[91] National Research Council (1998). *Research Reactor Aluminum Spent Fuel: Treatment Options for Disposal*. NationalAcademy Press, Washington, 16. ISBN 0-309-06049-4.

[92] Iyer, N.C., Vinson, D.W., Sindelar, R.L., Thomas, J.E., Adams, T.M. (2007). Interim Storage and Long-Term Disposal of Research Reactor Spent Fuel in the United States. *Proceedings of the NATO Advanced Research Workshop on Safety Related Issues of Spent Nuclear Fuel Storage,* Almaty, 2005, J.D.B. Lambert, K.K. Kadyrzhanov (Eds.), Springer, Dordrecht, 225-241. ISBN 978-1402059025.

[93] Shibata, T., Tamai, T., Hayashi, M., Posey, J.C., Snelgrove, J.L. (1984). Release of Fission Products from Irradiated Aluminide Fuel at High Temperatures, *Nuclear Science and Engineering* 87, 405-417.

[94] Parker, G.W., Creek, G.E., Barton, C.J., Martin, W.J., Lorenz, R.A. (1967). *Out-of-Pile Studies of Fission-Product Release from Overheated Reactor Fuels at ORNL*, 1955-1965. ORNL-3981, ORNL, Oak Ridge, 59-65.

[95] Vinson, D.W., Adams, T.M., Sindelar, R.L. (2000). *Evaluation of Neutron Absorbers for the Melt-Dilute Treatment of Aluminum-Based Spent Fuel.* WSRC-MS-2000-00181, WSRC, Aiken.

[96] Thomasson, J. (1998). MTR Spent Fuel Back-End COGEMA Long Term Commitment. *Transactions 2nd International Topical Meeting on Research Reactor Fuel Management,* Bruges, 1998, ENS, Berne, 129-135. <http://www.euronuclear.org/>.

[97] Thomasson, J. (1999). A Durable and Dependable Solution for RTR Spent Fuel Management. *Transactions 3rd International Topical Meeting on Research Reactor Fuel Management,* Bruges, 1999, ENS, Berne, 144-149. <http://www.euronuclear.org/>.

[98] Thomasson, J. (2000). RTR Spent Fuel Treatment and Final Waste Storage. *Transactions 4th International Topical Meeting on Research Reactor Fuel Management,* Colmar, 2000, ENS, Berne, 67-74. <http://www.euronuclear.org/>.

[99] Thomasson, J., Drain, F., David, A. (2001). Reprocessing RTR Fuel in the La Hague Plants. *Transactions 5th International Topical Meeting on Research Reactor Fuel Management,* Aachen, 2001, ENS, Berne, 112-116. <http://www.euronuclear.org/>.

[100] Auziere, P., Thomasson, J. (2002). *Management of the Acceptance Process of RTR Aluminide Type Spent Fuel. Transactions 6th International Topical Meeting on Research Reactor Fuel Management*, Ghent, 2002, ENS, Berne, 123-131. <http://www.euronuclear.org/>.

[101] Auziere, P. (2004). UMo Spent Fuel Acceptance for Treatment at La Hague Plant. *Transactions 8th International Topical Meeting on Research Reactor Fuel Management*, Munich, 2004, ENS, Berne, 131-133. <http://www.euronuclear.org/>.

[102] Anderson, H.H., Asprey, L.B. (1960). Solvent Extraction Process for Plutonium. US patent 2924506, issued 9 February, 1960.

[103] Rodrigues, G.C., Gouge, A.P. (1983). *Reprocessing RERTR Silicide Fuels*. DP-1657, SRL, Aiken.

[104] Cartwright, P. (1996). Reprocessing of LEU Silicide Fuel at Dounreay. *Proceedings 19th International Meeting on Reduced Enrichment for Research and Test Reactors,* Seoul, 1996, ANL, Argonne. <http://www.rertr.anl.gov>.

[105] Touron, E., Cheroux, L. (2001). Silicon Behavior during Reprocessing of Uranium Silicide Fuel by PUREX Process. *Transactions 5th International Topical Meeting on Research Reactor Fuel Management,* Aachen, 2001, ENS, Berne, 117-121.

[106] Durand, J.P., Maugard, B., Gay, A. (1998). *Technical Ability of New MTR High Density Fuel Alloys Regarding the Whole Fuel Cycle. Transactions 2nd International Topical Meeting on Research Reactor Fuel Management,* Bruges, 1998, ENS, Berne, 55-71. <http://www.euronuclear.org/>.

[107] Garcés, J.E., Bozzoloa, G., Hofman, G., Rest, J. (2007). Role of Si on the Al Behavior in the Reaction Layer of Al/UMo Diffusion Couples. *Applied Physics Letters* 90, 074104.

[108] International Atomic Energy Agency (2003). *Manual for Reactor Produced Isotopes*. TECDOC-1340, IAEA, Vienna.

[109] International Atomic Energy Agency (1999). *The Applications of Research Reactors.* TECDOC-1234, IAEA, Vienna.

[110] International Atomic Energy Agency (2010). Production and Supply of Molybdenum-99. *Nuclear Technology Review* 2010 Supplement, IAEA, Vienna.

[111] Mushtaq, A. (2010). Reactors are Indispensable for Radioisotope Production. *Annals of Nuclear Medicine* 24, 759–760.

[112] European Commission (2009). *Preliminary Report on Supply of Radioisotopes for Medical Use and Current Developments in Nuclear Medicine Established by an Internal European Commission ad-hoc Interservice Group.* SANCO/C/3/HW D(2009), Luxemburg.

[113] Ramamoorthy, N (2009). Commentary: Supplies of Molybdenum-99 – Need for Sustainable Strategies and Enhanced International Cooperation. *Nuclear Medicine* Communications 30, 899-905.

[114] Nuclear Energy Agency (2010). *The Supply of Medical Radioisotopes - An Economic Study of the Molybdenum-99 Supply Chain, Report* 6967, NEA, Paris. ISBN 978-9264991491.

[115] International Atomic Energy Agency (1998). *Management of Radioactive Waste from 99Mo Production.* TECDOC-1051, IAEA, Vienna.

[116] Blaumann, H., Bonini, A., Cristini, P., de Lio, L., Dell'Occhio, L., Gil, D., Gonzalez, A.G., López, M., Mirandou, M., Taboada, H. (2010). *2010 Progress Report on RERTR Activities in Argentina. 32nd International Meeting on Reduced Enrichment for Research and Test Reactors*, Lisboa.

[117] Ali, S.A., Ache, H.J. (1987). Production Techniques of Fission Mo-99. *Radiochimica Acta* 41, 65-72.

[118] van der Walt, T.N., Coetzee, P.P. (2004). The Isolation of Mo-99 from Fission Material for Use in the Mo-99/Tc-99m Generator, *Radiochimica Acta* 92, 251-257.

[119] National Research Council, Committee on Medical Isotope Production without Highly Enriched Uranium (2009). *Medical Isotope Production without Highly Enriched Uranium.* The National Academies Press, Washington. ISBN 0-309130409.

[120] Nuclear Energy Agency (2010). *The Supply of Medical Isotopes – Review of Potential Molybdenum-99/Technetium-99m Production Technologies.* NEA, Paris.

[121] World Nuclear News (2010). South African Radioisotope Production on Target. WNN, 17 September. <http://www.world-nuclear-news.org/>.

[122] Wu, D., Landsberger, S., Vandegrift, G.F. (1996). Progress in Chemical Treatment of LEU Targets by the Modified Cintichem Process. *Proceedings 19th International Meeting on Reduced Enrichment for Research and Test Reactors, Seoul,* 1996, ANL, Argonne. <http://www.rertr.anl.gov>.

[123] Vandegrift, G. (2006). HEU vs. LEU Targets for 99Mo Production - Facts and Myths. *International Symposium on Minimization of Highly Enriched Uranium in the Civilian Nuclear Sector*, Oslo, 2006. < http://www.nrpa.no/>.

[124] Vandegrift, G.F., Conner, C., Hofman, G.L., Leonard, R.A., Mutalib, A., Sedlet, J., Walker, D.E., Wiencek, T.C., Snelgrove, J.L. (2000). *Modification of Targets andProcesses for Conversion of 99Mo Production from High- to Low-Enriched Uranium.*Industrial and Engineering Chemistry Research 39, 3140-3145.

[125] Konzek, G.J. (1983). *Technology, Safety and Costs of Decommissioning Reference Nuclear Research and Test Reactors.*NUREG/CR-1756 Addendum, Nuclear Regulatory Commission, Washington.

[126] Morozov, S.I. (2005). Decommissioning Russian Research Facilities. *Proceedings of International Conference on Research Reactor Utilization, Safety, Decommis-*

sioning,Fuel and Waste Management, Chile, 2003, IAEA, Vienna, 573-578. ISBN 92-01139047.

[127] Noakes, M.W. (1998). CP-5 *Reactor Remote Dismantlement Activities: Lessons Learned in the Integration of New Technology in an Operations Environment.* ORNL/CP-97762, ORNL, Oak Ridge.

[128] European Commission (1995). *Studies in the Frame of the EC Pilot Dismantling Project* BR3 Mol. EUR-16645, EC, Brussels.

[129] European Commission (1996). *Decommissioning of the JEN-1 Experimental Reactor.* EUR-16899, EC, Brussels.

[130] International Atomic Energy Agency (1999). State-of-the-art Technology for Decontamination and Dismantling of Nuclear Facilities. *Technical Reports Series No.* 395, IAEA, Vienna.

[131] International Atomic Energy Agency (2002). *Decommissioning Techniques for Research Reactors.*TECDOC-1273, IAEA, Vienna.

[132] International Atomic Energy Agency (2006). Decommissioning of Research Reactors: Evolution, State of the Art, Open Issues. *Technical Reports Series No.* 446, IAEA, Vienna.

[133] International Atomic Energy Agency (2008). Decommissioning of Research Reactors and Other Small Facilities by Making Optimal Use of Available Resources. *Technical Reports Series No.* 463, IAEA, Vienna.

[134] International Atomic Energy Agency (2008). *Innovative and Adaptive Technologies in Decommissioning of Nuclear Facilities.*TECDOC-1602, IAEA, Vienna. ISBN 978-9201100085.

[135] Sed, L.J., Reisenweaver, D. (2005). International Decommissioning Strategies. Proceedings of International Conference on Research Reactor Utilization, Safety, *Decommissioning, Fuel and Waste Management*, Chile, 2003, IAEA-CN-100/83, IAEA, Vienna, 593-602. ISBN 92-01139047.

[136] Konzek, G.J., Ludwick, J.D., Kennedy, W.E. Jr., Smith, R.I. (1982). *Technology, Safety, and Costs of Decommissioning Reference Nuclear Research and Test Reactors.*NUREG/CR-1756. NRC, Washington.

[137] Keller, S.T., Reese, S.R. (2009). Going From HEU to LEU: Conversion of the Oregon State TRIGA Reactor. *Proceedings of 31st International Meeting on Reduced Enrichment for Research and Test Reactors,* Beijing, 2009, ANL, Argonne. <http://www.rertr.anl.gov>.

[138] National Aeronautics and Space Administration (2009). *Plum Brook Reactor Facility Decommissioning Project: 10 Years of Progress.* NASA Glenn Research Center, Cleveland. <http://www.grc.nasa.gov/>.

[139] US Department of Labor (2011). Consumer Price Index. <http://www.bls.gov/cpi/>.

[140] Nielsen, H., Nielsen, K., Petersen, F., Jensen, H.S. (1998). Risø National Laboratory – *Forty Years of Research in a Changing Society*. Risø National Laboratory, Roskilde. ISBN 87-55023924.

[141] Lauridsen, K. (2006). *Decommissioning of DR 1: Final report, DD-18(EN) rev.1*, Danish Decommissioning, Roskilde. <http://www.dekom.dk>.

[142] Andersson, I., Backe, S., Iversen, K., Lindskog, S., Salmenhaara, S., Sjöblom, R. (2006). Cost Calculations for Decommissioning and Dismantling of Nuclear Research

Facilities: Phase 1. NKS-146, *Nordic Nuclear Safety Research (NKS),* Roskilde, 48-62. ISBN 87-78932092.

[143] Kim, S. (2003). Plan for Moata Reactor Decommissioning. *Proceeding 9th International Group on Research Reactors Conference (IGORR 9), Sydney*, 2003. <http://www.igorr.com/>.

[144] Maharaj, P. (2010). *Decommissioning and Dismantling of the Moata Reactor-A Radiation Protection Perspective. 35th Annual Conference of the Australasian Radiation Protection Society, Adelaide,* 2010. <http://leishman-associates.com.au /arps2010/>.

[145] Malone, G., Kim, S. (2007). Moata Decommissioning and Dismantling, *IAEA R2D2P Workshop on Transition Phase*, Sydney, 2007.

[146] Lockwood, R.J.S., Beeley, P.A. (2001). Decommissioning the JASON Argonaut Research Reactor at a World Heritage Site. *Transactions 5th International Topical Meeting on Research Reactor Fuel Management,* Aachen, 2001, ENS, Berne, 195-199.

[147] Lauridsen, K. (2009). *Decommissioning of DR 2: Final report, DD-38(EN) rev.1,* Danish Decommissioning, Roskilde. <http://www.dekom.dk>.

[148] Lauridsen, K. (2001). *Decommissioning of the Nuclear Facilities at Risø National Laboratory: Descriptions and Cost Assessment.* R-1250(EN), Danish Decommissioning ,Roskilde. <http://www.dekom.dk>.

[149] Mück, K., Casta, J. (2000). Planning for the Decommissioning of the ASTRA-Reactor. *Proceedings 10th International Congress of The International Radiation ProtectionAssociation*, Hiroshima, 2000, Paper P-5-307.

[150] Meyer, F. (2003). Decommissioning of the ASTRA Research Reactor – Review and Status on July 2003. *Nuclear Technology and Radiation Protection* 18, 61-64.

[151] Meyer, F., Steger, F. (2006). Decommissioning of the ASTRA Research Reactor – Dismantling of the Biological Shield, *Nuclear Technology and Radiation Protection* 21, 79-91.

[152] Meyer, F. Steger, F., Steininger, R. (2008). Decommissioning of the ASTRA Research Reactor – Dismantling the Auxiliary Systems and Clearance and Reuse of the Buildings. *Nuclear Technology and Radiation Protection* 23, 54-62.

[153] Meyer, F. (2007). *Nuclear Engineering Seibersdorf GmbH. Technical Meeting on Establishment of an International Decommissioning Network*, Vienna, 2007. <http://www.iaea.org/>.

[154] Wittels, M.C. (1964). Stored Energy in Reactor Materials. *Nuclear Safety* 4, 40-44.

[155] Lexa, D., Kropf, A.J. (2006). Thermal, Structural, and Radiological Properties of Irradiated Graphite from the ASTRA Research Reactor – Implications for Disposal. *Journal of Nuclear Materials* 348, 122-132.

[156] Steger, F., Kurz, H., Urbanich, E., Meyer, F. (2009). Strahlenschutz bei Abbau und Dekommissionierung des ASTRA-Forschungsreaktors Seibersdorf. *Strahlenschutz aktuell* 43:1, 8-23.

[157] International Atomic Energy Agency (2010), *Progress in Radioactive Graphite Waste Management,* TECDOC-1647, IAEA, Vienna.

[158] Holt, G. (2001). Radioactive Graphite Waste Management at UK Magnox Nuclear Power Stations. *Technical Committee Meeting on Nuclear Graphite Waste Management,* Manchester, 1999, NGWM/CD01-00120, IAEA, Vienna.

[159] Hagos, B., Jones, A.N., Marsden, B.J., Marrow, T.J. (2009). Characterisation and Treatment of Irradiated Graphite Waste. *Proceedings Conference on Decommissioning, Immobilisation and Management of Nuclear Waste for Disposal,* York, 2009. < http://www.diamondconsortium.org/>.

[160] Brookhaven National Laboratory (2007). Brookhaven Graphite Research Reactor Decommissioning Project: *Final Remedial Design / Remedial Action Work Plan for the Graphite Pile Removal.*BNL, Upton.

[161] Laul, J.C., Norman, R. (2008). Beryllium: Hazards Evaluation, Facility Classification, Consequence Analysis of Releases during Potential Accidents, and Protection of Public and Workers. *Journal of Chemical Health and Safety* 15:4, 13-25.

[162] Tomberlin, T.A. (2004). Beryllium: A Unique Material In Nuclear Applications. *Proceedings 36th International SAMPE Technical Conference,* San Diego, 2004, Society for the Advancement of Material and Process Engineering, Covina.

[163] Sannen, L., De Raedt, Ch., Moons, F., Yao, Y. (1994). Helium Content and Induced Swelling of Neutron Irradiated Beryllium. *Fusion Engineering and Design* 29, 470-474.

[164] Longhurst, G.R. (2007). *The Idaho National Laboratory Beryllium Technology Update.* INL/CON-07-13516, INL, Idaho Falls.

In: Radioactive Waste: Sources, Types and Management ISBN: 978-1-62100-188-1
Editors: Satoshi Yuan and Wenxu Hidaka

Chapter 3

PROMISING ANALYTICAL TECHNIQUES FOR HLW ANALYSIS

A. V. Legin,[1] D. O. Kirsanov,[1] V. A. Babain,[2] L. N. Gall,[3] and N. R. Gall[4]

[1]St. Petersburg State University, Chemistry Department, St. Petersburg, Russia
[2]Khlopin Radium Institute, St. Petersburg, Russia
[3]Institute for Analytical Instrumentation RAS, St. Petersburg, Russia
[4]N.R., Ioffe Physical Technical Institute RAS, St. Petersburg, Russia

ABSTRACT

Reprocessing of spent nuclear fuel requires thorough analytical control at every stage of the process to avoid alarm situations, environmental failures and economical losses. There are only limited possibilities for this monitoring nowadays – the standard in the area are de facto expensive and time-consuming techniques like ICP-AAS and ICP-MS. There is a growing interest to the developing of alternative approaches for HLW analysis in the last several years.

One of the very attractive fields is the use of MS ERIAD – a version of the well-known electrospray-mass-spectroscopy. This method seems being a very promising one not only for the analysis of complexes in solution, but for elemental analysis at the same time. The advantages of the method are extremely small sample volumes (several microliters), high sensitivity and very simple sample preparation. Furthermore, it is possible to design MS ERIAD interface for standard mass- spectrometers.

An extensive growth of the use of chemometric techniques in analytical chemistry through the last decade has lead to the appearance of several other promising analytical alternatives. One of them is the employment of arrays of non-specific, cross-sensitive chemical sensors in combination with chemometric tools. Although not competing in preciseness with ICP-based methods this approach allows for fast and simple quantification of rare earth elements in complex mixtures in the presence of major components like uranium and thorium. Another approach brought into life by chemometrics is the use of classical optical spectroscopy methods in UV, Vis and IR ranges in combination with modern multivariate regression algorithms.

This chapter is devoted to the detailed description of these promising analytical techniques for HLW analysis.

1. Introduction

Nuclear energy seems to be an attractive alternative for organic fuel in the near future. However, there is a strong need for the development of more efficient and safer processes in the nuclear industry. Current situation with spent nuclear fuel (SNF) and high level waste (HLW) processing must be improved and modified to be adopted for modern ecological and safety requirements. There is a number of reprocessing technologies under study at the moment, but aqueous processing methods are most popular now and this will be the case in the near future. The general scheme of the aqueous processing is as follows: spent fuel is dissolved and uranium and plutonium are recovered for use in a new fuel fabrication, the HLW containing long-living radionuclides can be vitrified directly. On the other hand, HLW can be processed to obtain special fractions of radionuclides. For example, americium can be irradiated for transmutation or it can be used as an additional fuel for fast neutron reactors. Many various ways of partitioning are being studied in France, USA, Russia, China, India, etc. The main goal of all the studies is to reduce the volume of HLW and, therefore, to decrease perceptibly the cost of storage. It is important to mention that thorough analytical control is the basis for any technological process management and ideally one should know the concentration of all the components in each technological solution at any moment and at every process stage to avoid abnormal process run and to prevent economical losses and ecological dangers. In real life this is far from being so. When talking about e.g. SNF reprocessing it is difficult to determine on-line metal concentration in solutions. A standard scheme of analysis includes sample preparation, sample transportation to the analytical lab and the analysis itself. Modern analytical methods such as ICP MS or ICP OES seem to be preferable for metal detection in solutions. These methods, particularly ICP MS, are very sensitive and not so time-consuming compared to "classical" analytical methods. However, the equipment for such analysis is very expensive, well skilled operators are needed to run it and there is a number of limitations for automation of ICP MS. Extremely high sensitivity of ICP MS is even excessive for routine technological applications when metal concentrations are rather high. This sensitivity becomes a certain disadvantage of the method since it assumes the use of highly purified chemicals, air quality control in the labs, etc., thus leading to a high price per analysis. It is also important that such a routine and simple steps as sample preparation, dilution and transportation are turning into rather complex procedures when dealing with radioactive solutions. All preliminary stages in this case should be performed in "hot cells" or boxes, sample preparation with radioactive solutions leads to the secondary wastes and radiation exposure risk of the operator should be also kept in mind. These issues are the reason why the whole preliminary procedure becomes very expensive and time consuming. There are several reports on the trials to fit serial ICP equipment for handling radioactive solutions [1, 2]. These modifications are fully possible, however the price of such equipment is high and the speed of analysis is still not fast enough.

The considerations above imply the necessity of the development of new approaches for HLW analysis. The main direction for this development should be the elaboration of techniques with simple (e.g. just dilution) sample preparation or direct on-line techniques.

The purposes of technological monitoring do not assume extreme selectivity and sensitivity but the fast analysis is a must, because fast analytical result means fast technological correction and provides for a stable process run.

Obviously, analytical methods based on radioactivity detection are fully available for on-line implementation but the problem in many cases is that metals of interest are not so highly radioactive and their detection is not possible on the background of highly active fission products.

This chapter is devoted to the review of several analytical techniques that, in our opinion, can be developed and used in the near future for the fast and simple HLW monitoring. The authors had no intention to give a broad and excessive review of all suitable methods for HLW analysis, which are quite numerous and diverse.

2. MS ERIAD Method

Mass spectrometry is used widely for measuring the isotopic content, and recently it was also used for quantitative analysis of metals in solutions. Below we are discussing ERIAD historically the forerunner of well-known ESI MS (electrospray ionization) for elemental, isotopic and chemical analysis of inorganic matter.

2.1. The Method Description

A sensitive method based on ion extraction from solutions at the atmospheric pressure was proposed in the beginning of 80-s of the last century for non-volatile matter like bio-polymers and metal salts. It was developed at the Institute for Analytical instrumentation of USSR Academy of Science in Leningrad and was called ERIAD [3, 4]; and about the same time in the USA under the name "electrospray" ESI [5], which is now generally accepted in scientific literature. Later ESI MS was developed mainly for thermally unstable polymers and biopolymers [3, 5] while it is also used widely for metal complexes in solutions (see e.g. [4]).

The main difference between ERIAD and classical ESI is that the gas dynamic ion input is combined with controlled in-source fragmentation of molecular ions in ERIAD. If the fragmentation conditions are rigid enough, the atomization takes place, and ERIAD can be used for isotopic and elemental analysis of metal solutions [1, 2]. This approach is more informative than the application of ESI described in [6-10].

2.2. Registration of Metal Cations by ERIAD MS

Earlier we have shown that most part of a metal can be registered as single-charged ions M^+, MO^+, and MO_2^+ in ERIAD MS [3, 11]. The simplest cases are the mass-spectra of the individual univalent metals solutions, such as Li, Na, K, Rb, Cs, Ag, Tl(I), Cu(I). Spraying of such solutions results in the spectra with a limited number of bands attributed to a cluster ions like $Me^+(H_2O)_n(Sol)_m$, where n, m = 0, 1, 2, and Sol is the solvent molecule. The increase of the fragmentation voltage ΔU will lead to the production of a single ion in mass-spectra, namely the metal ion M^+. The relation between the metal concentration and ion current depends both on the cation and on the nature of the solvent. Figure 1 represents the mass-spectrum of the equimolar mixture of alkaline cations in water.

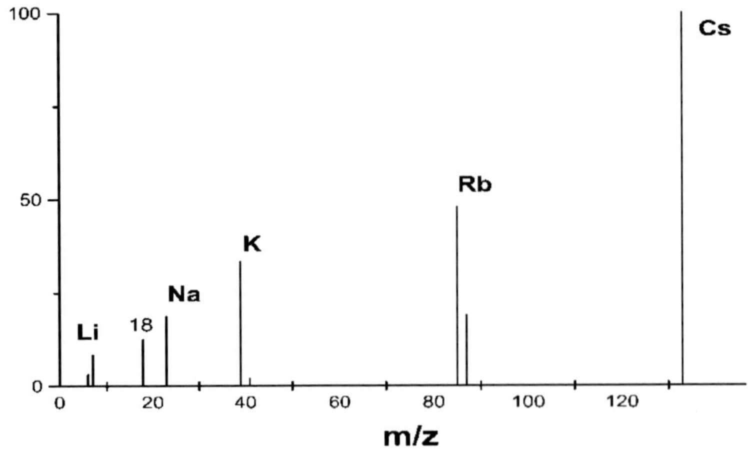

Figure 1. The mass-spectrum of the equimolar mixture of alkaline metals in water. Each component concentration is 2*10^{-4} M, Uk = 3.4 keV, ΔU = 400 V.

One can see that the ion current ratio for Li^+: Na^+: K^+: Rb^+: Cs^+ is 0,06 : 0,2 : 0,5 : 0,8 : 1,0, respectively, and it becomes more uniform when the solvent is changed from water to methanol solution.

Figure 2 shows the dependence of ion current ratio Li^+/Cs^+ on the increase of methanol content in the solution. The observed tendency is typical of all alkaline metals. The ion current ratio for pure methanol solution samples is as follows: Li^+ : Na^+: K^+ : Rb^+ : Cs^+ = 0,3 : 0,5 : 0,8 : 1,0 : 1,0. It is important that the ion current ratio is constant for the constant solution composition and this can be used for quantitative analysis [3].

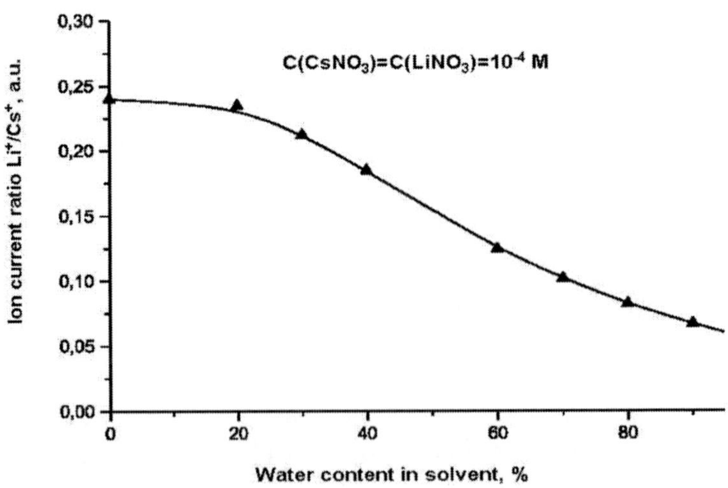

Figure 2. Relative sensitivity to Cs and Li vs. water content in water-methanol solution. Uk = 3.4 keV, ΔU = 400 V.

More complex dependencies can be registered for double-charged metal ions like Ba, Sr, Ca, Mg, Be, Zn, Cd, Hg(II), Mn(II), Fe(II), Co, Ni, and Pb. One can expect to get multi-linear spectra taking into account the diversity of possible ways of interaction between charged and neutral particles both in the liquid and in the gas phase. The ions like $Me^{2+}(H_2O)_n(Sol)_m$, $(Me^{2+}An^-)^+(H_2O)_n(Sol)_m$, $MeOH^+(H_2O)_n(Sol)_m$, were observed at low fragmentation voltages ΔU, where Sol is the molecular solvent An^- is a univalent anion and n, m = 0, 1, 2. However, such ions as Me^+, MeO^+ and MeH^+ were also detected while they are not present in the initial sample solution. The increase of ΔU leads to the destruction of the cluster and molecular complexes and only such ions as Me^+, $MeAn^+$, MeO^+, $MeOH^+$ might be observed.

Polyvalent cations of transition and rare earth metals such as La, Ce, Pr, Nd, Sm, Eu, Gd, Tb, Dy, Ho, Er, Tm, Yb, Lu; U, Th, Fe, Cr, Hf. Zr, In, Y, and Sc are typical for the samples, which are relevant to radiochemistry and geology. Metal ions M^+ of these metals appear at about ~250 V but the simplest spectra can be observed at < 350 V when only Me^+, MeO^+, MeO_2^+ are present. The doubly charged ions M^{++} were registered only for europium and lanthanum. Ion current of M^+ grows with the increase of ΔU while the intensity of other ions have a tendency to decrease, however, MO^+ ion is important for Er, Lu, and Tb even at the highest ΔU. At high ΔU = 400–800 V the mass-spectra shape have no dependency on the anion either of sulfate, nitrate, chloride or perchlorate.

Mass-spectra of the metals with higher valence (Th, Zr, Hf, U (IV, VI)) typically show a higher content of the ion MO^+ being compared with the rare earth metal spectra. The spectra are independent of the initial chemical form of the element for ΔU>400V. The spectra of U(IV) and U(VI), Fe(II) and Fe(III), Ce(III) and Ce (VI) are pretty similar for these conditions.

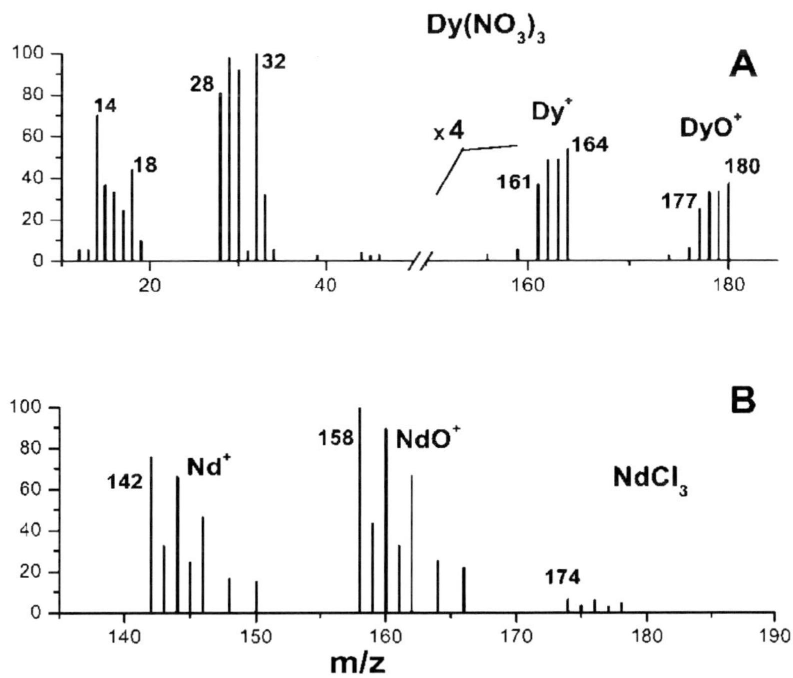

Figure 3. The mass-spectra of neodymium and dysprosium solutions in 90% CH_3OH - 10% H_2O, ΔU = 450 V. A – $Dy(NO_3)_3$, C = 10^{-4} M; B – $NdCl_3$, C = $5*10^{-5}$ M.

Isotopic content for all elements fits well with the reference data. The mass-spectra of the rare earth metals are shown in the Figure 3 and isotopic lines can be easily identified both for M^+ and MO^+ ions. The measured isotopic content is in a good agreement with the natural content with accuracy of 1% even for a single registration.

2.3. Origination of Ions MeO^+ and MeO_2^+ in ERIAD Mass-Spectra

The chemical elements from the groups 2-7 of the Periodic System do not yield ions like Me^+, MeO^+, MeO_2^+ in solution, neither do they in mass-spectra registered at low values of ΔU when a molecular ion is observed. However, these ions are important for quantitative analysis and the mechanism of oxy-ion formation is of high interest. We believe that these ions are formed due to the gas phase reaction for which the oxygen can be supplied from corresponding anions from the solvent and from the counter gas.

One should see the differences in the spectral shape for oxygen-containing and oxygen-free anions, e.g. between nitrates and chlorides, if the oxygen from the anion works. However the spectra of Nd, La, and Y are closely similar for nitrates, perchlorates and chlorides so this is not the way for oxygen to participate in the reaction.

We have used water enriched with ^{18}O as a solvent to choose between two other possibilities. For water with 82% of ^{18}O and counter gas with ^{16}O only ^{16}O was observed in oxy-ions for La and Pr nitrates at sample concentration 10^{-4} - $5*10^{-5}$ mol/l, and ΔU=500V. Thus, the oxygen from the solvent does not take part in oxy-ion formation either.

When the air with 21% of oxygen was replaced by dry nitrogen with less then 0.1% of ^{16}O, mass-spectra has changed significantly. For example, the metal to oxy-ion current ratio is 1:6 for yttrium when the air is used but is 1:2 for dry nitrogen. The molecular oxygen peak at m/z = 32 is also diminished strongly. Yttrium yields three peaks, Y^+ (m/e=89), $Y^{16}O^+$ (m/e=105), and $Y^{18}O^+$ (m/e=107) with the ratio Y^+: $Y^{16}O^+$: $Y^{18}O^+$ = 1 : 2 :0,1 when sprayed in the pure nitrogen atmosphere. If the ratio $^{16}O^+/^{18}O^+$ in the initial solvent are known it is possible estimate the role of the oxygen for liquid and gas phase for YO^+ origination. Such estimation shows that at least 95% of oxygen in oxy-ions is from the gas phase. It provides for the main effect for oxy-ion formation even at the level of 0.1% of the counter gas. The purification of the counter-gas from oxygen to a high degree is needed to diminish the oxy-ion content.

2.4. Analytical Features of ERIAD MS with Respect to Radiochemical Applications

Development of the nuclear fuel cycle was running in parallel with the development of isotopic and elemental mass-spectrometry. Nowadays, mass-spectrometer devices are the main analytical control option for isotopic enrichment industry at all stages of the cycle, both for laboratory testing and for automatic control systems. Mass spectrometry with surface thermo ionization is used for isotopic measurements for more than 50 years [12]. At present time, high and low resolution ICP MS is widely used to measure elemental content of solutions including radiochemical industry. All three methods employ different mechanisms of ionization; however they have an important common feature - the sample is prepared and introduced into the source in a form of a solution.

ERIAD MS possesses very high sensitivity to the metals. The volume of 1-10 microliters of the sample with concentration 10^{-7} M (which is equivalent to 10^{-13} mole (10^{-11}-10^{-12} gram) is enough for a measurement. Obviously, there are certain problems with representative sampling of 10 microliter from the technological process stream, but this is manageable with modern sampling techniques. The dynamic range is about five orders of magnitude in one experiment. Higher concentration determination is hindered by the instrument limit of registration, and lower concentration is limited by the stability of the spraying system.

It is interesting to compare ERIAD MS with its closest analogues such as ICP MS for elemental analysis and surface thermo ionization MS for isotopic measurements. In favorable conditions, the detection limit of ICP MS is about 10^{-15} mol/liter; this exceeds radiometric sensitivity for most of the isotopes [11]. However this value grows strongly, by 5-6 orders of magnitude for mixtures; the total amount of sample should be about 1 ml or even more and rather complex sample preparation does not allow realizing the measurements in a rapid way. ICP MS instruments with high resolution intended to work with artificial isotopic content are very expensive and require ultra pure reagents and argon for analysis. These problems result from the nature of the ionization process such as desolvatation, atomization and ionization occuring in ICP plasma at the atmospheric pressure and temperature of 6-8 thousand degrees. Switching on and maintaining of plasma at atmospheric pressure requires high power and rather complex, expensive, and unreliable control systems [13]. Plasma stability depends strongly on the sample composition its total amount and especially on the presence of easy-to-ionize matter. Plasma chemical reactions in ICP torch produce large amount of unusual chemical substances, which interfere with all peaks in a resulting mass-spectra. Summarizing, one can say that ICP MS is too sensitive, too huge, and too expensive to be used widely in routine analysis.

Surface thermo ionization provides for very high accuracy in isotopic measurements, however, it is very difficult to automate the method, and even now all commercial instruments are exploiting only manual sample input.

ERIAD MS combines possibilities of measuring the isotopic and elemental content for nano-amounts of the sample with high accuracy and low detection limits. When compared with currently used methods, ERIAD MS ensures the following:

- allows for direct input of radioactive samples after minimal sample preparation using an automated input systems and, thus, diminishes staff irradiation,
- the sample volume can be of 1^{-10} microliters with the salt concentration as low as 10^{-7} M, which corresponds to 10^{-13} mole (10^{-11}-10^{-12} gram) per analysis,
- the dynamic range is at least 10^5, so both major and minor components can be registered,
- both elemental and chemical content can be measured in one experiment only by varying the sampler-skimmer voltage,
- ERIAD interface is much simpler and cheaper than that of ICP or thermo ionization ion sources,
- ERIAD is a cost-effective method as it does not require the use of expensive ultra pure reagents. Water, methanol or ethanol can be used as solvents, and power

consumption for ionization step is less than 1 Watt (compared to ICP where plasma maintaining requires 1.5-2 kWt).

ERIAD MS is especially convenient for the determination of stable or β-emitting nuclides in the presence of excessive amounts of γ-nuclides. Radiometric determination in these cases is possible only after chemical separation of elements, which is time-consuming and associated with staff irradiation. Small sample volume and the absence of chemical separation steps make ERIAD MS very attractive for these purposes.

HLW analysis by ERIAD MS is also very attractive due to the small sample volume and special design of ion source. The final contamination of the instrument is 2-3 orders of magnitude lower comparing to that of standard ICP MS.

ERIAD MS looks very promising instrument for radiochemical and nuclear fuel technology, it is much simpler, more reliable and cheaper comparing to ICP instruments and it is much more universal than surface thermo ionization MS. An important issue is that ERIAD MS might be fully automated.

3. Potentiometric Multisensor Method

As mentioned above most of the analytical methods in use are highly complicated ones with limited possibilities for automation and on-line implementation. The use of some simple, fast and cost-effective techniques for the monitoring purposes seems to be very attractive. Chemical sensors and namely ion-selective electrodes could be a good alternative. This type of sensors is widely known for their cheap and robust construction which allows for easy and fast quantification of various ions in water solutions with a very simple measuring protocol [14]. Nowadays the most popular ion-selective sensors are those based on plasticized polymeric membranes. Sensor membranes (sensing elements) of this type usually consist of polymer (poly(vinylchloride) is most common) plasticized by lipophilic organic solvent and containing small amounts (1-5 wt %) of neutral or charged ligand (ionophore) and ion-exchanger. The weight ratio between PVC and plasticizer of ½, correspondingly, leads to the formation of mechanically stable membranes with reasonable electric resistance and electrochemical mobility of active substances and ions inside the material. Ionophore is the component responsible for selective complex formation with targeted ion in the membrane phase [14]. Right from the time of the invention of potentiometric sensors based on plasticized polymeric membranes a lot of attention was being paid to the development of ionophores and sensors for transition metals. Hundreds of research papers were devoted to this topic. Main activities during the end of 20th century are reviewed in [15, 16]. Ion-selective sensors based on PVC-plasticized membranes allows for sensor properties control in a wide range. This is achieved by the variation of chemical structure of membrane active substance – ionophore (neutral ligand); by the selection of ion-exchanger/ionophores ratio and by use of different solvent-plasticizers. The complex formation reaction between the ligand and the primary ion at the phase boundary between water solution and polymeric membrane leads to the change in electromotive force value in the galvanic cell consisting of membrane electrode and reference electrode:

Ag| AgCl, KCl(sat)| sample solution| sensor membrane| inner solution| AgCl| Ag.

The EMF value of this galvanic cell is an analytical signal and it is connected to the activity of the target ion according to Nikolsky-Eisenman equation:

$$E = E^0 + (RT/z_iF)\ln(a_i + \sum_j K_{ij}^{pot} a_j^{z_i/z_j})$$

where E^0 – standard electrode potential; z_i – charge of the primary ion; a_i – activity of the primary ion; K_{ij}^{pot} – selectivity coefficient of the electrode; a_j – activity of the interfering ion; z_j – charge of the interfering ion, R – universal gas constant, T – temperature, F – Faraday constant.

The main application field of these sensors is diluted water solutions which are far from what the reprocessing medium is.

Literature survey reveals that there is a number of reports on the development of ion-selective sensors for determination of rare earth elements. PVC-plasticized electrochemical membrane sensors were reported for lanthanum [17], cerium [18], praseodymium [19], neodymium [20], samarium [21], gadolinium [22], etc. The selectivity coefficients, reported in these papers are usually very low (which means that selectivity is high) even when interfering ion is the nearest neighbour of the target ion in the lanthanide series. This is somewhat surprising taking into account chemistry of the lanthanides, their similar ionic radii and electronic structure. We were not able to reproduce the reported data following the description in the publications for samarium and ytterbium [23, 24]. Another important issue is the limited pH working range of the reported sensors, most of them are functional at pH values starting from 3-4 and the values of the sensor sensitivities reported in these papers are usually obtained at pH 6. This is not quite clear to what kind of species in solutions the observed sensitivities can be assigned, since it is widely known that significant amounts of Me^{3+} lanthanide ions can only present in acidic solutions with pH around 2.

In general since the lanthanides are quite similar in nature, the development of the selective sensors for each particular ion appears to be not very promising approach for the quantification of these metals in a complex mixtures because it is very hard to develop the ligands with pronounced sensitivity towards one particular lanthanide.

However, there is another way to fit sensor performance to various analytical needs – in recent years there is a rapid development of so called "electronic tongue" multisensor approach [25]. The main idea is the employment of the array of the sensors without sharp selectivity towards particular ions, but with rather high cross-sensitivity to the variety of analytes. Each sensor of the array has its own particular sensitivity pattern and the number of such sensors being combined together yields a kind of "unresolved spectrum" signal in a sample solution. Yet this response contains information about all spices (to which the sensors of the array possess sensitivity) in a sample and can be processed by modern chemometric techniques (application of mathematical and statistical methods for chemical data analysis) to extract qualitative and quantitative data about analyzed media. Chemometric approach for the processing of multidimensional data of various natures is gaining wide use and acceptance in science and technology through the last 30 years [26-29]. It is widely used for atomic spectroscopy [30], UV-Vis spectroscopy [31, 32], potentiometry [33, 34], chromatography [35] and for many other analytical methods.

Multisensor technology looks highly attractive for different monitoring tasks, as it allows for rapid control of alarm situations with rather simple measuring protocol. To be properly used multisensor approach requires proper sensor development, e.g. when talking about rare earth metal (RE) monitoring one should have a variety of sensors with high cross-sensitivity to these metals. This task is substantially different from developing highly selective sensors towards each particular contaminant, but some general approaches of course can be transferred.

Figure 4. General view of the potentiometric multisensor system.

Interaction between a ligand and an ion is an elementary act which is responsible both for liquid extraction processes and polymeric sensor membrane sensitivity. Comparison of extractive ability of a ligand with a corresponding sensor performance reveals a lot of common trends for these two areas. Thus employment of the data on liquid extraction of RE metals by various neutral ligands gives a good basis for development of novel sensors. This approach proved to be rather fruitful [33, 36, 37]. Of course in some cases still there are noticeable deviations from conformity between extraction and sensors which cannot be satisfactory explained yet due to the lack of experimental material.

One of the first attempts to develop cross-sensitive sensors with RE sensitivity was performed with a series of three bidentate organophosphorous compounds [38]. Organic ligands such as octylphenyl-*N,N*-diisobutylcarbamoylmethyl phoshineoxide (CMPO), diphenyl-*N,N*-di-*n*-butylcarbamoylmethylphoshine oxide (DPCMPO), and tetraphenylmethylenediphosphine dioxide (TPDO) and their mixtures with chlorinated cobalt dicarbollide (CCD) were employed for sensor preparation. Chemical structures of the active substances are shown in the Figure 5. These substances are reported to be useful extracting agents for RE and actinides from acidic HLW [39, 40]. Sensors contained 50 mmol/kg ligand (3-4% wt) and 10 mmol/kg CCD (around 2%).

Polymeric membranes were produced according to the following standard procedure. Weighed amounts of membrane components were dissolved in freshly distilled tetrahydrofuran (THF) and magnetically stirred for 20 min. Once the components were dissolved in the THF, the membrane cocktail was poured into a flat bottom Teflon beaker and allowed to stand overnight at room temperature to evaporate the solvent. Disks 8mm in diameter and 0.5mm thick were cut from the parent membranes and glued with PVC-cyclohexanone mixture to the end of PVC tubes (10mm in diameter) for use as electrode bodies. At least three replicate sensors of each respective composition were prepared. The electrodes were filled with 0.01M NaCl solution and then immersed and stored in the same solution for 24 h prior to measurements.

Tetraphenylmethylenediphosphine dioxide

(TPDO)

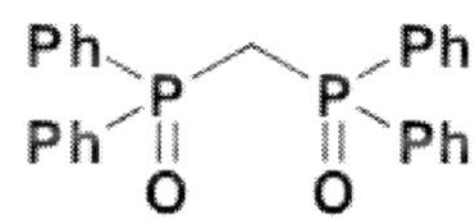

Octylphenyl-*N,N*-diisobutylcarbamoylmethyl

phoshine oxide

(CMPO)

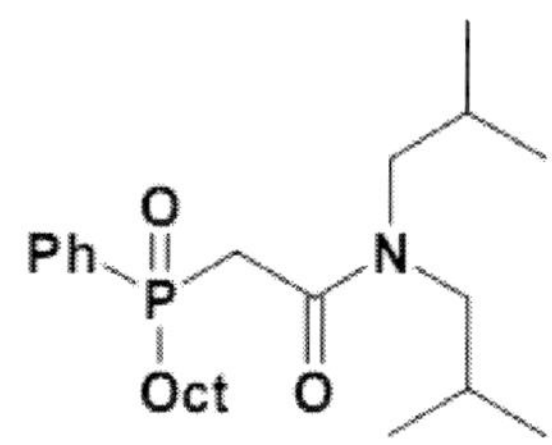

Diphenyl-*N,N*-di-*n*-

butylcarbamoylmethylphoshine

oxide

(DPCMPO)

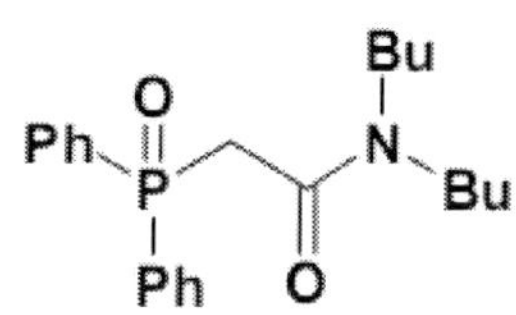

Figure 5. Chemical structures of the bidentate organophosphorous ligands.

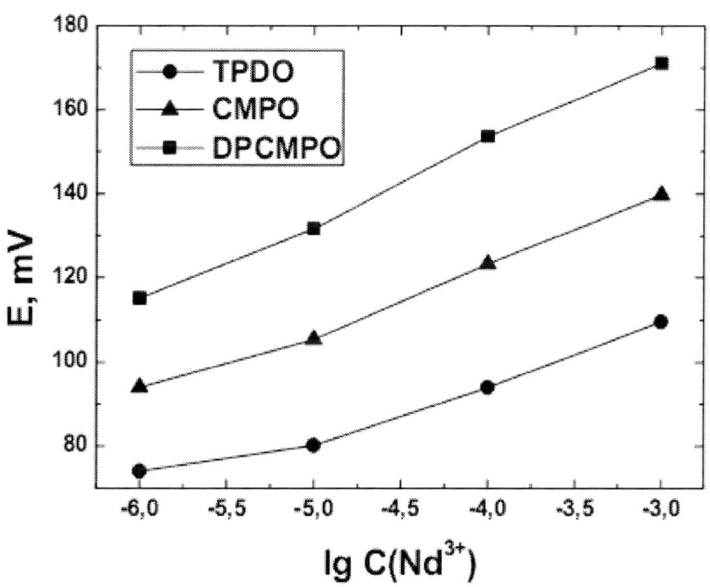

Figure 6. Typical view of the sensor response in solutions of $Nd(NO_3)_3$. pH 2 fixed with nitric acid addition.

The response of these sensors was tested in nitric acid solutions (pH 2) of various RE metals. It was found that the developed sensors possess high sensitivity towards La^{3+}, Pr^{3+}, Nd^{3+} and Eu^{3+} in the acidic media. The slopes of the linear parts of the electrode calibration functions were in the range 11-30 mV/pMe^{3+}. In the Figure 6 typical view of the sensor response curve in neodymium solutions is shown.

Selectivities of the sensors measured by fixed interference method [14] revealed that there are no sharp preferences for some particular RE ion and selectivity patterns are rather smooth. These results indicate that different phosphine oxides with chlorinated cobalt dicarbollide can be successfully used to develop a variety of novel polymeric electrochemical sensors. Excellent chemical and radiation stability of the bidentate organophosphorous compounds that has been previously reported in relation to solvent extraction studies combined with the known chemical and radiolytical stability of plasticized PVC indicates the potential utility of these sensor materials in "electronic tongue" sensor arrays for analysis of radioactive solutions.

Another example of successful use of the substances suggested for the liquid extraction for sensor preparation is given in [37]. Diamides, phosphine oxides and modified calixarenes were employed as active membrane compounds for sensor preparation. Chemical structures of some of these substances are shown in the Figure 7.

5,11,17,23-tetrakis-dibutylphosphinylmethyl-25,26,27,28-tetrapropoxycalix[4]arene (cone conformer)
(C1)

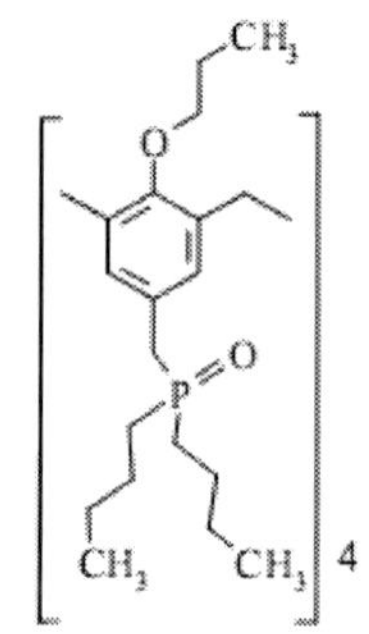

5,11,17,23-tetra(diethylcarbamoylethoxymethyl-carboxamido)-25,26,27,28-tetrapropoxycalix[4]arene
(C2)

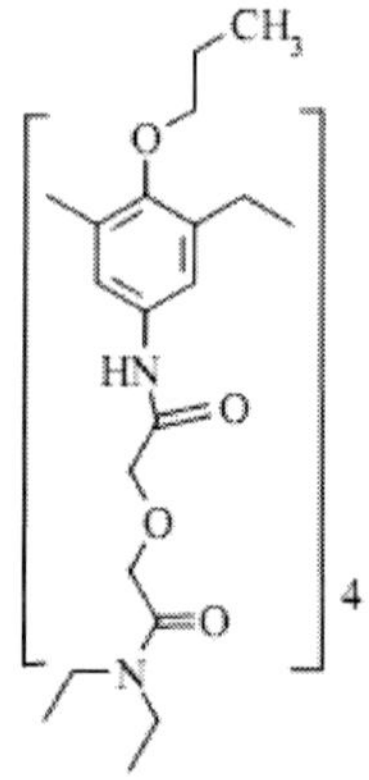

Figure 7. Chemical structures of two calixaren ligands.

Calixarene ligands were compared with corresponding phosphine oxide (tri-*n*-octylphosphine oxide (TOPO)) and diamide ligands which are not attached to the calixarene platform. Extraction and sensor properties of the ligands were studied in different metal ion solution with special attention paid to RE metals. Attachment of phosphine oxide groups to the calixarene platform leads to the sharp increase of both extraction and sensing ability of the corresponding systems comparing to non-bonded phosphine oxide. In case of the diamide derivatives attached to the calixarene performance of corresponding ligands was similar to those of nonbonded diamides. Comparison of the sensor sensitivities and extraction data on distribution coefficients revealed a very good agreement between these techniques. Data on the distribution coefficients of the ligand in extraction system can be used to predict further electrochemical properties of the sensor (such as sensitivity, selectivity) based on the same ligand.

Sensitivity of the sensors was studied in the nitric acid solutions of La, Ce, Pr, Nd, Sm, Eu, Gd, Lu at pH=2. It was found that the sensors based on tetraoctyl-3-oxapentanediamide (TODGA) and C2 have sensitivity values of 12-25 mV/pMe^{3+} in these conditions. Detection limits for all of the studied metals by these sensors were around 10^{-6} mol/liter of the RE. TODGA sensors exhibit growing selectivity with the growth of the lanthanide atomic number and the most preferable ion is Lu^{3+}. It is noteworthy that in spite of rather similar chemical properties of the lanthanides the $\log K_{La,RE}$ pot values differs quite strong—almost three orders of magnitude between La and Lu. The attachment of the diamide groups to the calixarene cycle leads to the sharp smoothing of selectivity and C2 sensors do not discriminate between different RE ions. Probably the sensors with this type of selectivity pattern may be practically applicable for the determination of the sum of the lanthanides in the sample solution. These observations are in a good agreement with the data on liquid extraction of the RE metals from nitric acid solutions by TODGA and C2 – distribution coefficients grow from La to Lu for TODGA and have slightly changing values for C2. As for the sensors based on calixarene with phosphine oxide functions (C1), selectivity pattern was rather smooth, but taking into account their low (<15 mV/dec) sensitivities these electrodes could hardly be applied for any practical task. The response time for all of the studied sensors does not depend on the membrane composition and t_{95} was 10–15 seconds in RE solutions at pH 2.

It is noteworthy that sometimes this approach based on the use of extracting agents for sensor preparation allows for construction of a sensors with high selectivity towards particular ions like e.g. cadmium [41] and lead [42].

Further development of these ideas towards more practical things like quantifying several RE in a mixture simultaneously by the sensor array was shown in [33]. In this study the array of the potentiometric cross-sensitive sensors was employed for the quantitative analysis of the double and triple mixtures of cerium, neodymium, samarium and gadolinium. These elements were chosen on the basis of the following considerations: Ce^{3+} and Nd^{3+} are typically present in the highest quantities among the other RE in spent nuclear fuel. Sm^{3+} and Gd^{3+} are neutron poisons and must be completely removed from the reprocessing medium before it can be used repeatedly as a nuclear fuel. The concentration range 10^{-5} – 10^{-3} mol/liter of RE was chosen for these mixtures taking into account typical RE contents in the nuclear waste. The pH of all mixtures was fixed at 2 by nitric acid. Content of each RE in the mixture was varied over 5 different concentration levels for double mixtures and over 3 levels for triple mixtures. 9 different solutions both for double and triple mixtures were prepared and 5 replicated measurements were done in each of them with the sensor array. PLS1 (partial least squares regression) algorithm was then applied for the analysis of multidimensional data from the sensor array. This regression technique is widely known and well established for multivariate calibration, see e.g. [43].

The PLS regression model can be presented in a matrix form:

$$\mathbf{Y} = \mathbf{XB} + \mathbf{F} \tag{1}$$

where **X** matrix of size *I* x *J* is a result of e.g. potentiometric measurements of *I* samples with *J* sensors, thus each element represents the EMF value of this particular sensor for this particular sample. **Y** is a matrix (vector) of *I* x *K* size, each element represents the reference value of the parameter of interest which is going to be measured afterwards by means of this calibration model. Usually, each element of **Y** is a concentration of a substance of interest in a given sample (e.g. rare earth metal content).

B is a matrix of regression coefficients and **F** is residual modeling error matrix.

During the PLS modeling process both **X** and **Y** are decomposed according to the following equations:

$$\mathbf{X} = \mathbf{TP}^{\mathrm{T}} + \mathbf{E} \tag{2}$$

$$\mathbf{Y} = \mathbf{UQ}^{\mathrm{T}} + \mathbf{F}_{\mathbf{y}} \tag{3}$$

and then with a condition of maximum covariation between **T** and **U**

$$\mathbf{W} = \max(\mathrm{cov}(\mathbf{T},\mathbf{U})) \tag{4}$$

matrix regression coefficients B is calculated:

$$\mathbf{B} = \mathbf{W}(\mathbf{P}^{\mathrm{T}}\mathbf{W})^{-1}\mathbf{Q}^{\mathrm{T}} \tag{5}$$

After that calibration model can be used for prediction of the parameters of interest ($\mathbf{Y}_{\mathbf{new}}$) from the measurements in a new samples ($\mathbf{X}_{\mathbf{new}}$):

$$\mathbf{Y}_{\mathbf{new}} = \mathbf{X}_{\mathbf{new}}\mathbf{B} \tag{6}$$

Potentiometric data processing was performed in The Unscrambler 9.7 (CAMO, Norway) software package with PLS1 algorithm. All variables were mean centered before building the models.

The first four replicas were used to build the multivariate regression model. Test set validation method was employed for its verification. The fifth replica was used as an “unknown” sample for prediction purposes.

The results of simultaneous determination of several RE metals in double and triple mixtures are presented in Table 1.

Table 1. The results of simultaneous quantitative determination of several RE cations in a mixtures at pH=2

Mixture	Cation	Average error of prediction of RE content, ± logC(RE)
Cerium and neodymium	Ce	0.3
	Nd	0.2
Samarium and gadolinium	Sm	0.3
	Gd	0.2
Cerium, neodymium and samarium	Ce	≤0.4
	Nd	≤0.4
	Sm	0.3

The developed sensor system has demonstrated promising performance and was capable of simultaneous quantitative determination of three rare earth metals with acceptable analytical errors.

Noteworthy, the suggested method of RE mixture analysis is a highly attractive alternative to the existing techniques, since it is not associated with time-consuming sample preparation, any consumables, it is fast (typical sensor response time in RE mixtures was not higher than 2 minutes) and comparatively cheap. There is usually no need for high precision determination of RE content during technological process monitoring – a technologist would rather need an estimate of normal or abnormal run of the process of RE partitioning. An effective fulfillment of this task with the electronic tongue analytical methodology seems to be realistic and promising.

Analysis of the triple RE mixture in a presence of uranium and thorium as a further gradual sophistication of the samples is now in progress. The multisensor system based on previously discussed sensors is applied for simultaneous quantification of neodymium, samarium and gadolinium in a presence of 10^{-5} – 10^{-4} mol/liter of uranium and thorium. In Table 2 the parameters of the PLS1 regression models validated by independent test set for prediction of RE metals in acidic solutions with 10^{-4} mol/liter content of uranyl ion are shown. Reasonable precision of simultaneous RE quantification is achievable even in the presence of uranium.

Table 2. Parameters of the calibration and validation lines of PLS1 models for simultaneous determination of Nd, Sm and Gd cations in a mixtures with 10^{-4} mol/liter UO_2^{2+}

RE metal		Slope	Offset, lgC_{RE}	RMSE, lgC_{RE}	R^2
Nd^{3+}	calibration	0,899	-0,382	0,234	0,899
	validation	0,847	-0,571	0,319	0,823
Sm^{3+}	calibration	0,821	-0,666	0,296	0,821
	validation	0,779	-0,802	0,345	0,771
Gd^{3+}	calibration	0,829	-0,659	0,322	0,829
	validation	0,790	-0,817	0,409	0,738

As a summary of the multisensor method we would like to outline its advantages. The main issue is that the measurements are comparatively cheap and very fast. A possible limitation which could hinder its use is the fact that with HLW reprocessing solutions sample dilution is required before the measurements. Obviously, the method is in need for further development both from the point of view of sensors for the arrays and from the data processing perspective. For example, one of the promising ways of further chemometric improvement of multisensor methodology could be an employment of 3wayPLS regression where the third dimension in the data is time, i.e. for each of the samples one will get not the static time cut sensor reading numbers after 2-3 minutes of measurements, but the whole kinetic curve of the sensor response. This kinetic curve and its shape can contain further useful chemical information on the species in a sample, since different ions can have different kinetics of interaction with sensor membrane. This dynamic information can be extracted by means of multiway regression methods.

4. Optical Spectroscopy in Combination with Chemometrics

Spectrophotometric determination of various ions in visible and near UV spectral range is a classical analytical method. It can be successfully employed for determination of various oxidation states of actinides in solutions. However, when several actinides with significantly different concentrations are simultaneously present in the sample, classical spectral analysis of the minor components can become difficult or even impossible without preliminary chemical separation of the elements.

Chemometric processing of multidimensional data of various natures is nowadays a common practice in numerous fields of analytical chemistry [26-35].

Application of multivariate spectral data processing for UV-Vis spectral data of actinides is very limited. One of the first papers on the chemometric approach was published in 1991 [44], it was dealing with qualitative analysis of plutonium using PLS regression in the nitric acid media. Analysis of the visible spectra of Pu(IV) influenced by various concentrations of nitric acid, fluoride and oxalate ions was also performed in [44]. This approach was not developed further towards the possibility of on-line monitoring. Spectrophotometric determination of uranium and nitric acid in mixtures with chemometric data processing (PLS regression) was suggested in [50]. Quantification of both components was possible with reasonable precision: RMSEP for U was 0.19 g/L, RMSEP for HNO_3 0.098 mol/L.

In [45] there is a report on the study of uranyl complexes structures using chemometric approach to spectroscopic experimental results.

Combination of UV-Vis spectroscopic analysis chemometric data processing seems to be promising for on-line determination of various constituents in SNF reprocessing solutions without preliminary chemical separation of the elements. The use of the modern optical fibers eliminates the limitations of the detector location and all control units of such spectroscopic system can be placed in the clean zone. Moreover, one can simultaneously apply not only UV-Vis detectors, but also the detectors of IR and other types of radiation.

The use of UV-Vis spectroscopy for on-line monitoring of uranium in SNF reprocessing was proposed in [46]. The system studied there contained two components (uranium and nitric acid) and the experimental spectral data were studied from the point of view of possibility to discriminate the changes in uranyl and nitric acid concentrations by UV-Vis spectroscopy. No multivariate processing of the data was performed.

It seems very promising to treat optical spectra of SNF reprocessing solutions in a multivariate way. This could yield more chemical information about the species in the media and in general is more robust way to treat the data since the input for this data processing is a whole spectrum but not a single spectral variable which can be noisy, overlapping, etc.

In [47] we report on simultaneous determination of neptunium and plutonium in the model mixtures simulating solutions obtained in the course of the PUREX spent nuclear fuel reprocessing. UV-Vis spectroscopic measurements were performed in the mixtures contained constant concentration of nitric acid and uranium and varied content of neptunium and plutonium. Spectral data were processed with classical chemometric technique – PLS regression. This approach has shown significant improvement in analytical precision of actinides determination comparing to standard univariate data processing.

For this study we have prepared 31 solutions containing various concentrations of Np and Pu and constant content of uranium (50 g/L) and HNO_3 (2 mol/L) were prepared. The concentrations of both Np and Pu in the mixtures were varied from 0.05 to 1.55 g/L with 0.05 g/L step. Spectroscopic measurements were performed with AvaSpec spectrometer (Avantes B.V., Holland) in the range 184 – 1100 nm with 0.53 nm step. The spectra were recorded in the cell 10x10 mm. All measurements were performed at room temperature. In the Figure 8 spectra of two solutions with the most different content of the elements are shown as an example.

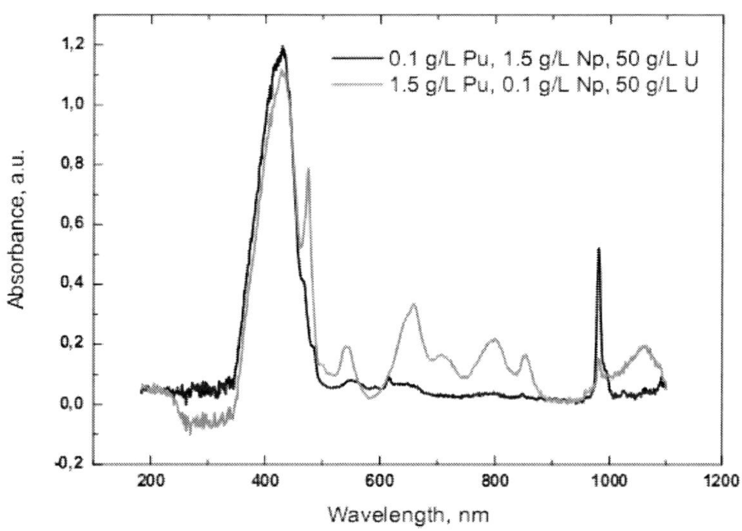

Figure 8. Typical spectra of Np and Pu in solutions with uranium in 2 M nitric acid.

The broad peak at 430 nm corresponds to uranium (VI), and the sharp signal at 980 nm and small peak at 613 nm belong to neptunium (V). Plutonium (IV) gives a sharp signal at 475 nm right at the uranium peak slope, and the broader peaks at 540 and 660 nm belong to plutonium (III) [48, 49]. However, it should be noted that in the case of multivariate approach to the spectral data processing one could even have no information on the peak assignment.

All the spectral data for 31 mixtures were arranged in a matrix form to produce PLS1 calibration models. Sample set was split into two subsets: first one (21 sample) for calibration and second one (10 samples) for assessing the predictive performance of the models. Obtained calibration models were employed for prediction of Np and Pu content in 10 unknown samples. Mean relative error for 10 test samples was 4.4 and 7.1 % for plutonium and neptunium, respectively. The same spectral data were also subjected to the traditional univariate calibration procedure. Just like in the case of PLS processing, the data on the same 21 samples were used for calculating univariate linear regression and the rest 10 samples were employed for estimation of this linear regression predictive ability. Absorbance values at 980 nm were used for calibration against neptunium content, signal intensities at 475 nm – for plutonium. Linear fit of the experimental data points yielded regressions with R^2 0,98 for plutonium and 0,95 for neptunium. In spite of the high correlation, predictive ability of these regressions was much poorer than those of PLS models and averaged relative error of prediction for 10 unknown test samples was 12,7 and 21% for plutonium and neptunium, respectively.

Another example of the multivariate regression application for SNF relevant task is simultaneous determination of uranium and nitric acid in mixtures. This study was to some extent similar to that reported in [46] but with more stress on a quantitative determination of uranium and nitric acid. 25 sample mixtures containing uranium in the concentration range 0.01-1.3 mol/L and nitric acid in the concentration range 0.01-8 mol/L were prepared and UV-Vis measurements in the range 184 – 1100 nm were performed. Spectral data from 17 samples were employed for PLS calibration and 8 samples were used as an independent test set to assess the predictive ability of the models. Mean relative error for prediction of uranium content was 7.7%, while determination of nitric acid was possible only in a semi-quantitative way with mean relative error 25.1%. Further experimental studies are now in progress to improve these results with more complex model solutions as well as by use of more sophisticated chemometric techniques.

While preparing this chapter for publication two papers of American authors have appeared [51, 52]. They are devoted to the analysis of actinides in SNF solutions with combination of Raman, UV-Vis and NIR spectroscopy. The idea of combination of several modes for measurements is very nice and it confirms wide perspectives of various optical spectroscopic methods for on-line control in radioactive waste technology.

CONCLUSION

We have described three methods which can be perspective for HLW analysis in our opinion. They may need some further research and development efforts to be adopted by radiochemical and nuclear industry and each of them has its own limitations and advantages. MS ERIAD and potentiometric multisensor technologies can offer fast and simple measuring protocol but both of them would likely require preliminary dilution of the sample. The application of optical spectroscopy methods in combination with chemometric tools seems to be a feasible option for the on-line technological monitoring though analytical procedures for more complex solutions and possible methodology of long-term measurements are to be developed and verified.

REFERENCES

[1] Todolí, J. L.; Mermet J. M. *Spectrochim Acta B* 2006, 61, 239-283.
[2] Lee, C. H.; Suh, M. Y.; Choi, K. S.; Kim, J. S.; Song, B. C.; Jee, K. Y.; Kim, W. H. *Anal. Chim. Acta* 2001, 428, 133-142.
[3] Aleksandrov, M. L.; Gall, L.N.; Krasnov, N.V.; Nikolaev, V.I.; Pavlenko, V.A.; Shkurov, V.A. *Dokl. Akad. Nauk. SSSR* 1984, 277, 379-383.
[4] Aleksandrov, M. L.; Gall, L.N.; Krasnov, N.V.; Shkurov, V.A.; Nikolaev, V.I.; *Tech. Phys.+* 1984, 54, 1559-1571.
[5] Yamashita, M.; Fenn, J.B. *J. Chem. Phys.* 1984, 88, 4451-4459.
[6] Hirabayashi, A.; Sakairi, M.; Takada, Y.; Koizumi, H. TrAC *Trend Anal. Chem.* 1997, 16, 45-52.
[7] Lamouroux, C.; Rateau, S.; Moulin C. *Rapid Commun. Mass Sp* 2006, 20, 2041 – 2052.
[8] Moulin, C.; Amekraz, B.; Colette, S. ; Doizi, D.; Jacopin, C.; Lamouroux, C.; Plancque, G. *J. Alloy Compd.* 2006, 408–412, 1242–1245.
[9] Di Marco, V. B.; Bombi, G. G. *Mass Spectrom. Rev.* 2006, 25, 347– 379.
[10] Pasilis, S.; Somogyi, Á.; Herrmann, K.; Pemberton, J. E. *J. Am. Soc. Mass Spectrom.* 2006, 17, 230–240.
[11] Gall, L.N.; Shkurov, V.A.; Babain, V.A.; Gall N.R. *Mass-spektrometria* (in Russian) 2007, 4, 11-18.
[12] Zandberg, E.Ya.; Ionov, N.I. Poverhnostnaya ionizaciya (in Russian), *Nauka* publishers, Moscow, 1969.
[13] *Inductively coupled plasma mass spectrometry*; Montaser, A.; Ed.; Wiley VCH Inc., L, 1998.
[14] Bakker, E.; Bühlmann, P.; Pretsch, E. *Chem. Rev.* 1997, 97, 3083-3132.
[15] Bühlmann, P.; Pretsch, E.; Bakker, E. *Chem. Rev.* 1998, 98, 1593-1687.
[16] Umezawa, Y.; Bühlmann, P.; Umezawa, K.; Tohda, K.; Amemiya, S. *Pure Appl. Chem.* 2000, 72, 1851-2082.
[17] Miital, S.; Ashok Kumar, S.K.; Sharma, H.K. *Talanta* 2004, 62, 801–805.
[18] Shamsipur, M.; Yousefi, M.; Ganjali, M.R. *Anal. Chem.* 2000, 72, 2391-2394.
[19] Gupta, V.K.; Goyal, R.N.; Pal, M.K.; Sharma, R.A. *Anal. Chim. Acta* 2009, 653, 161–166.
[20] Ashok Kumar, S.K.; Singh, J.; Jain, A.K. *Electroanal.* 2010, 22, 2443 – 2452.
[21] Miital, S.; Sharma, H.K.; Ashok Kumar, S.K.; *Sensors* 2004, 4, 125-135.
[22] Ganjali, M.R.; Emamia, M.; Rezapour, M.; Shamsipur, M.; Maddah, B.; Salavati-Niazari, M., Hosseini, M.; Talebpoui, Z. *Anal. Chim. Acta* 2003, 495, 51–59.
[23] Ganjali, M.R.; Pourjavid, M.R.; Rezapour, M.; Haghgoo, S. *Sens. Actuat. B-Chem.* 2003, 89, 21-27.
[24] Ganjali, M.R.; Norouzi, P.; Tamaddonb, A.; Adib, M. *Sens. Actuat. B-Chem.* 2006, 114, 855-861.
[25] Legin, A.; Rudnitskaya, A.; Vlasov Y. In *Comprehensive Analytical Chemistry;* Alegret, S.; Ed.; Elsevier BV, Amsterdam, 2003, Vol. 39, 2003, pp 437-486.
[26] Meher, L.C.; Vidya Sagar D.; Naik, S.N. *Renew Sust. Energ. Rev.* 2006, 10, 248-268.
[27] Nyström, J.; Dahlquist, E. *Fuel* 2004, 83, 773-779.
[28] Shepherd, K.D.; Walsh, M.G. *J. Near Infrared. Spec.* 2007, 15, 1-19.

[29] Karoui, R.; De Baerdemaeker, *J. Food Chem.* 2007, 102, 621-640.
[30] Basic Chemometric Techniques in Atomic Spectroscopy; Andrade-Garda, J. M.; Ed.; RSC Publishing, 2009.
[31] *Ren, S.; Ling Gao, L. Anal.* Bioanal. Chem. 2007, 388, 215-225.
[32] Kandelbauer, A.; Kessler, W.; Kessler, R. W. *Anal. Bioanal. Chem.* 2008, 390, 1303-1315.
[33] Babain, V.; Kirsanov, D.; Legin, A.; Rudnitskaya, A.; Tatuev, Y.; Baulin, V. *Radiochim. Acta*, 2009, 97, 479-484.
[34] Polshin, E.; Rudnitskaya, A.; Kirsanov, D.; Legin, A.; Saison, D.; Delvaux, F.; Delvaux, F.; Nicolaï, B.; Lammertyn, J. *Talanta* 2010, 81, 88-94.
[35] Daszykowski, M.; Walczak, B. TrAC *Trend Anal. Chem.* 2006, 25, 1081-1096.
[36] Legin, A.V.; Babain, V.A.; Kirsanov, D.O.; Mednova, O.V. *Sens. Actuat. B-Chem.* 2008, 131, 29–36
[37] Alyapyshev, M. Yu.; Babain, V. A.; Boyko, V. I.; Eliseev, I. I.; Kirsanov, D. O.; Klimchuk, O. V.; Legin, A. V.; Mikhailina, E. S.; Rodik, R. V.; Smirnov, I. V. *J. Incl. Phenom. Macrocycl. Chem.* 2010, 67, 117–126.
[38] Legin, A.V.; Kirsanov, D.O.; Babain, V.A.; Borovoy, A.V.; Herbst R.S. *Anal. Chim. Acta* 2006, 572, 243–247.
[39] Myasoedov, B.F.; Chmutova, M.K.; Kochetkova, N.E.; Koiro, O.E.; Pribylova, G.A.; Nesterova, N.P.; Medved, T.Ya.; Kabachnik, M.I. *Solvent Extr. Ion Exc.* 1986, 4, 61-68.
[40] Rozen, A.M.; Volk, V.I.; Vakhrushin, A.Y.; Zakharkin, B.S. *Proceedings of GLOBAL'97*, Yokohama, Japan, 1997, p. 1523.
[41] Alyapyshev, M.; Babain, V.; Borisova, N.; Eliseev, I.; Kirsanov, D.; Kostin, A.; Legin, A.; Reshetova, M.; Smirnova Z. *Polyhedron*. 2010, 29, 1998–2005.
[42] Kirsanov, D. O.; Mednova, O. V.; Pol'shin, E. N.; Legin, A. V.; Alyapyshev, M. Yu.; Eliseev, I. I.; Babain, V. A.; Vlasov, Yu. G. *Russ. J. Appl. Chem.* 2009, 82, 247−254.
[43] Wold, S.; Sjöström, M.; Eriksson, L. *Chemometr. Intell. Lab.* 2001, 58, 109-130.
[44] Carey, W.P.; Wangen, L. E. *Chemometr. Intell. Lab.* 1991, 10, 245-257.
[45] Quilès, F.; Burneau A. *Vib. Spectrosc.* 2000, 23, 231-241.
[46] Warburton, J.; Smith, N.; Czerwinski, K. *Separ. Sci. Technol.* 2010, 45, 1763-1768.
[47] Kirsanov, D.; Babain, V.; Agafonova-Moroz, M.; Lumpov, A.; Legin, A. *Radiochim. Acta* 2011, accepted.
[48] Vdovenko, V.M. *Sovremennaya Radiohimia* (in Russian), Moscow, Atomizdat, 1969.
[49] Milyukova, M.S.; Gusev, N.I.; Sentyurin, I.G.; Sklyarenko, I.S. *Analiticheskaya himiya plutoniya* (in Russian), Moscow, USSR Academy of Science, 1965.
[50] Burck, J. *Anal. Chim. Acta* 1991, 254, 159-165.
[51] Bryan, S.A.; Levitskaia, T.G.; Johnsen, A.M.; Orton, C.R.; Peterson, J. M. *Radiochim Acta* 2011, 99, 563-572.
[52] Bryan, S.A.; Levitskaia, T.G.; Casella, A.J.; Peterson, J.M.; Johnsen, A.M.; Lines, A.M.; Thomas, E.M. In *Advanced separation techniques for nuclear fuel reprocessing and radioactive waste treatment;* Nash K.L.; Lumetta G.J.; Eds.; Woodhead publishing, Philadelphia, 2011, pp 95-119.

In: Radioactive Waste: Sources, Types and Management ISBN: 978-1-62100-188-1
Editors: Satoshi Yuan and Wenxu Hidaka

Chapter 4

TREATMENT OF HIGH-LEVEL RADIOACTIVE WASTE ARISING FROM PYROPROCESSING AND ITS WASTE FORM

Ippei Amamoto
Japan Atomic Energy Agency,Tokai-mura, Naka-gun, Ibaraki,Japan

ABSTRACT

The molten salt, 3LiCl(59.5mol%)-2KCl(40.5mol%), as the medium used in the pyroreprocessing by the electrorefining method will be contaminated by the accumulation of various actinoid elements (An) and fission products (FP) due to prolonged electrolytic operation. Though the An can beremoved at the pyrocontactor step by extraction using cadmium melt, the FP remain in the eutectic medium, which is regarded as a spent electrolyte at certain stage, can cause the rising of the melting point of the electrolytic bath and /or lowering the current efficiency. Some measures e.g., its replacement by a virgin medium, etc. should be taken to maintain its stable condition.

The constant replacement of the electrolyte, however, could lead to the generation of enormous volume of high-level radioactive waste (HLW). In terms of environmental load reduction and economical improvement, it is desirable to have the spent electrolyte purified for recycling by removing its FP.

Some technical developments on spent electrolyte treatments have been carried out in several countries. One of them is the zeolite sorption of FP following An removal which is being developed in the United States, Japan, etc.. Meanwhile, the contaminants precipitation methods by converting contaminants to insoluble compounds are undertaken in Russia and a few other countries and are anticipated to reduce the total volume of HLW. In the case of the Russian process, its purpose is to remove contaminants such as the minor actinoids (MA) and rare earth elements (RE) from the medium before its disposal. In our case, the main objective is to recycle the purified medium, delaying its disposal for as long as possible.

It is with this objective in mind that this study was undertaken. Author et al. have introduced a simple filtration method to remove RE particles which were formed due to the conversion of RE chlorides to phosphates. Here, the iron phosphate glass system is used as a filtration medium for the removal of FP particles. However, some soluble FP such as compounds of alkali metals, alkaline earth metals, etc. still remain in the medium.

Next, we attempted to apply the iron phosphate glass system as a sorbent instead, to remove the soluble FP. We have obtained some positive results and have intention to incorporate it into the spent electrolyte recycle process as a part of the FP separation and immobilization system. Although this method is still on a laboratory scale level, it is a potential rational method to use the iron phosphate glass form which possesses a larger loading capacity for FP than the conventional glass form.

In this manuscript, the author intends to describe the progress hitherto of the iron phosphate glass system functioning as a FP filter medium and sorbent as well as being the waste form itself.

1. INTRODUCTION

The Fast Reactor (FR) fuel cycle system (see Figure1-1) is composed of the FR and two other processes: the fuel fabrication and reprocessing. In general, the aqueous process is operated for reprocessing to separate and purify the uranium, etc. from the spent oxide fuel after dissolving it in nitric acid while pyroprocessing which does not use water, is applied to reprocess metallic fuel. Among pyroprocessing, the electrorefining method which applies electrochemistry, is one of the most advanced techniques. It uses the melting alkali metals salt (molten salt) as a medium to dissolve the spent fuel. In this paper, the 'medium' here is defined as a kind of electrolyte by itself but is not contaminated by the constituents of the spent fuel. Figure1-2 showspyroprossesing by the electrorefining method called the ANL Process[1] which was developed at Argonne National Laboratory (ANL) in the United States. The ANL Process uses 3LiCl-2KCl (molar ratio) eutectic salt as a medium to reprocess the spent oxide fuel by converting oxides to metals at the reduction step.

Besides the ANL Process, Research Institute of Atomic Reactor (RIAR) in Russia has developed the RIAR Process [2][3] by the electrowinning method for reprocessing of the spent oxide fuel using NaCl-KCl or NaCl-2CsCl (molar ratio) as a medium.

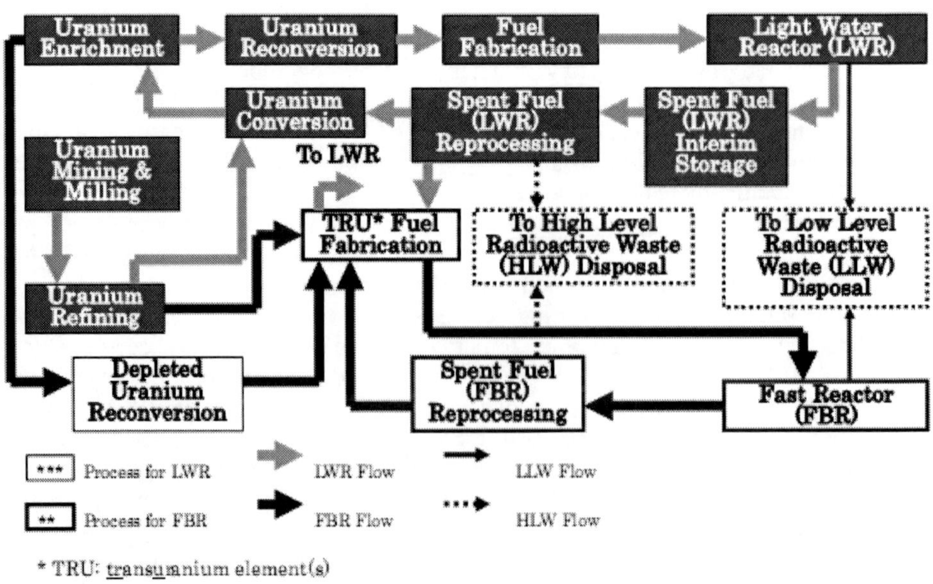

Figure 1-1. Nuclear Fuel Cycles of LWR and FBR.

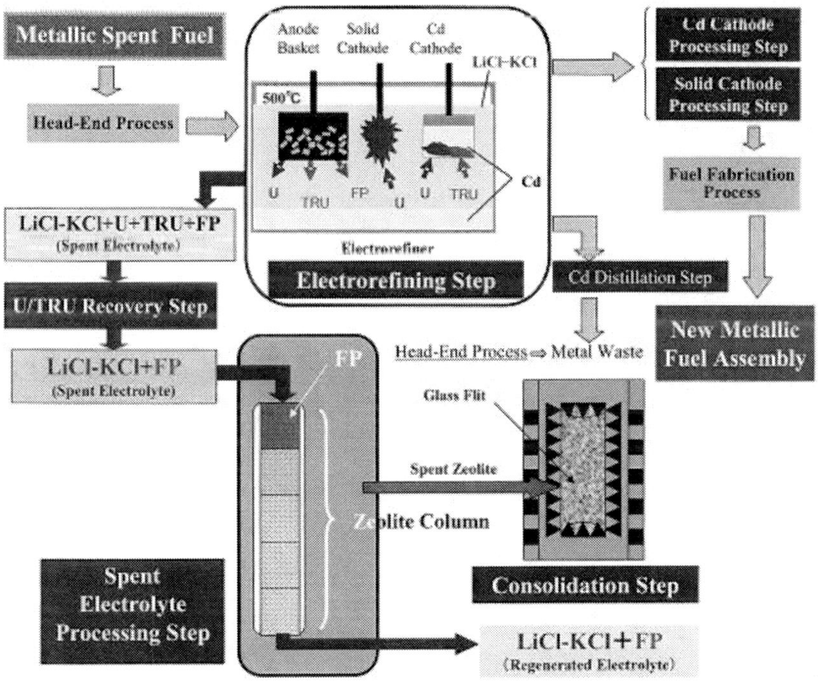

Figure 1-2. Metallic FuelElectrorefiningProcess (ANL Process).

In the technical development of pyroprocessing by the electrorefining method or electrowinning method, it is important to understand the behaviour of each constituent of the spent fuel in the spent electrolyte. The soluble constituents in the medium are mainly fission products (FP) of chlorides which possess metallic cations of alkali metals (AL), alkaline earth metals (ALE), rare earth elements (RE) and actinoid elements (Ac) . Meanwhile, the platinum group elements as FP precipitates remain in the electrolytic cell as insoluble residues [4]. In the case of soluble substances, they are mostly accumulated in the medium due to prolonged electrolytic operation. This can affect the chemical composition and physical properties of the electrolyte resulting in the rising of melting point of the electrolytic bath [5] and /or lowering the current efficiency [6]. Some measures e.g., its replacement by a virgin medium, etc., should be taken to maintain its stable condition. The constant replacement of the electrolyte as a waste salt, however, could lead to the generation of enormous volume of the high-level radioactive waste (HLW) because the spent electrolyte as a waste salt is classified as HLW. In terms of load reduction and economical improvement, it is desirable to have the spent electrolyte purified for recycling by removing its FP.

The conceptual methods which are still in its development in several countries to reduce the volume of HLW are described as below.

2. FP Removal by Zeolite Sorption Method

Among FP removal methods to refine the spent electrolyte arising from the pyroprocessing, the zeolite sorption method can be given as one of the most progressive. It applies the unique properties of zeolite as a sorbent. Zeolite is known as a crystalline porous alminosilicate and metallosilicate which is a combination of Si-O tetrahedra and Al-O tetrahedral and is widely used in the industrial circles. Its unique ability to function as an ion-exchanger, molecular sieve and sorbent is due mainly its chemical composition and/or framework. It is also used for refining of the spent electrolyte. Based on former study, it was found that among the various kinds of zeolites, Linde Type A (Framework Type Code (FTC): LTA) and Faujasite (FTC: FAU) possess higher FP sorptivities [7]. Na- LTA, which possesses sodium as exchange cation, and is the crystal by the noncovalent bond of the four-membered circles of the sodalite cages formed from the four and six-membered rings, is used for the treatment of the spent electrolyte. The composition formula can be expressed by [Na12(AlO2)12(SiO2)12·27H2O][8].

The sorption effect of zeolite is due to the cavity existed in its inside that can be obtained by the removal of crystalline water called zeolitic water. In the case of Na-LTA, the cavity, which has a sorption function of chlorides, can be formed upon being heated to dehydrate the zeolitic water.

By this heating, Na-LTA changes its formula to [Na12(AlO2)12(SiO2)12]8 without changing its framework. The sorption reaction of the unit cell of Na-LTA and NaCl which is one of chlorides can be expressed as below [9]:

$$[Na_{12}(AlO_2)_{12}(SiO_2)_{12}]+12NaCl \Rightarrow Na_{24}[(AlO_2)_{12}(SiO_2)_{12}]Cl_{12} \quad \text{Eq.(2-1)}$$

As whole three-dimensional framework presents the polyanions, $(AlO_2)^{2-}$, which are composed of the framework of zeolite, Na-LTA can reversibly occlude the variety of cations into its cavities.

The removal methods of FP from the spent electrolyte are applied the functions of Na-LTA such as sorption and ion-exchange action which made it possible to treat 24/x mol chlorides, i.e., 12/x mol by sorption and 12/x mol by ion-exchange as shown in Equation [2-2].

$$[Na_{12}(AlO_2)_{12}(SiO_2)_{12}]+12FP(x)Cl \Rightarrow FP(x)_{12/x}[(AlO_2)_{12}(SiO_2)_{12}] + 12NaCl \quad \text{Eq.[2-2]}$$

where FP(x) refers to the specific FP element which possesses x valence (x=1-3) and hereafter, the Roman numerals in the brackets express the valence of the specific FP element.

In the case of the spent electrolyte treatment by ANL method[10], the reaction mentioned above is generated by mixing the spent electrolyte and Na-LTA at the temperature of 500C.Simultaneously the borosilicate glass flit is added into this mixture. Na-LTA occluded FP is made to treat for the immobilization of FP. As shown in Figure2-1, Na-LTA is converted to the glass-bonded sodalite by pressurizing Na-LTA occluded FP with borosilicate glass flit at the temperature of 850-900C.

$$FP(x)_{24/x}[(AlO_2)_{12}(SiO_2)_{12}]Cl_{12}$$
$$\Rightarrow 2[FP(x)_{6/x}[(AlO_2)_6(SiO_2)_6]\cdot 2FP(x)_{1/x}Cl]+8FP(x)_{1/x}Cl \quad \text{Eq.[2-3]}$$

Based on experimental result [6], only one-fourth of $FP(x)_{1/x}Cl$ by 1 unit-cell of Na-LTA can be immobilized in this treatment. Therefore an additional 2unit-cell of Na-LTA is necessary to treat 8/x mol of $FP(x)_{1/x}Cl$. The final waste volume is expected to increase 3 times its original volume. The chemical formula of LTA occluded FP can be expressed as $Na_{0.1}K_{5.5}Li_{9.538}FP(I)_{11.862}[(AlO_2)_{12}(SiO_2)_{12}]Cl_{15}$. When this Na-LTA is converted as shown in Eq.2-2, the equilibrium molar amount of chlorides would be generated by the ion-exchange reaction between FP and cations in Na-LTA. Therefore final loading of the waste form becomes 3-10 wt% whose value is lower than 15wt% FP loading of borosilicate glass [11] which is used as the current waste form.

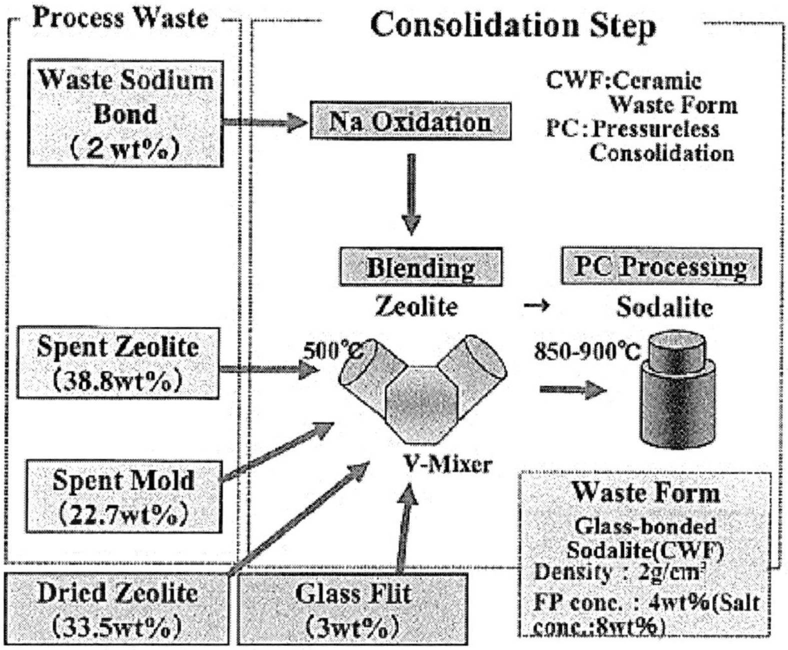

Figure 2-1. Conceptual block flow diagram of consolidation step.

3.FP Removal Methods Applying Phosphates Properties

3.1. Phosphate Precipitation Method

The FP separation methods can be divided broadly into two categories: the separation method of chlorides without changing chlorides to other compounds and the conversion method of FP chlorides to phosphates or oxides, etc.[12]. Generally, as the chemical formula of FP in the spent electrolyte are chlorides, the most rational separation method is to remove them as they are. Though zeolite sorption method comes under the first category, it has a serious problem in that the FP loading becomes lower after conversion of Na-LTA to sodalite for the immobilization of FP due to FP loading capacity of sodalite being lesser than that of LTA.

There is a conversion method which converts FP chlorides to other insoluble compounds in the molten salt, or the spent electrolyte. This method, which is based on a technique developed by RIAR is shown in Fig 3.1-1 [13]. It uses Na_3PO_4 as an additive to precipitate FP at the temperature of 630-650C in the spent electrolyte without changing the constituent of mediumas shown in Eq[3.1-1].

$$FP(III)Cl_3 + Na_3PO_4 \Rightarrow FP(III)PO_4 + 3NaCl \quad \text{Eq.[3.1-1]}$$

where FP(III) refers to the rare earth elements(RE) and/or theminor actinoids (MA).

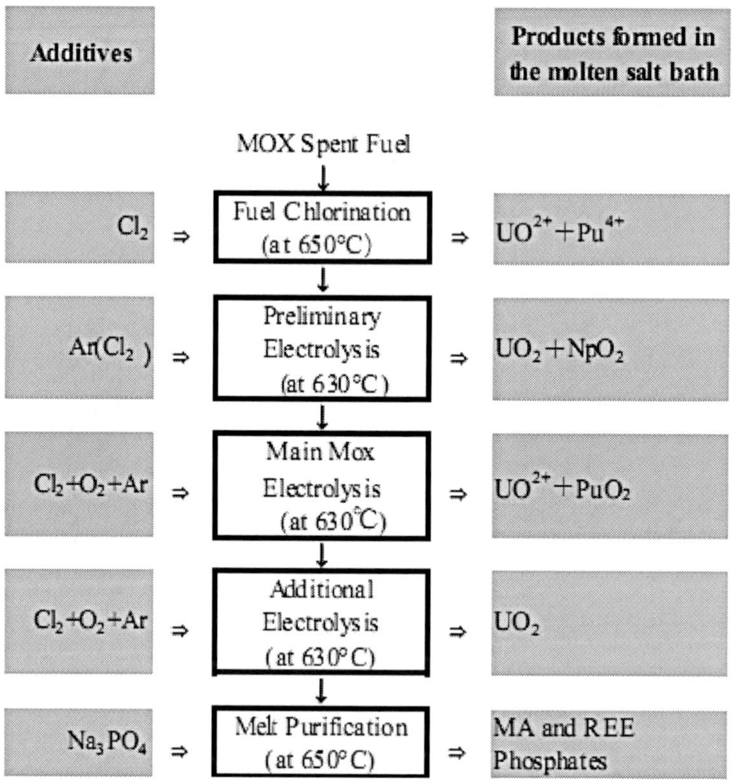

Figure 3.1-1. Conceptual block flow diagram of RIAR method.

Although FP(I) as AL and FP(II) as ALE exist in the spent electrolyte, most of these elements seldom form the phosphate precipitates which is not more than the detectable limit. As they form soluble residual in the medium, it is necessary to immobilize them as mentioned earlier. In the case of borosilicate glass, which is commonly used as a glass medium, its loading capacity of chlorides is very low, about 1wt% [14]. Subsequently, RIAR undertakes to convert waste chlorides to oxides by the reaction with boric acid, etc. before they are loaded into the borosilicate glass. RIAR uses the lead-iron phosphate (loading:27.5wt%) for the immobilization of waste phosphates. Furthermore, for the treatment of the waste salt medium, it was converted to phosphate precipitate using meta phosphoric acid (HPO_3) as shown in Eq.[3.1-2]. The phosphate as a product by this reaction is loaded into the fluoric-phosphate glass (loading: 60wt%) for the reduction of waste volume.

$$FP(I)Cl + HPO_3 \Rightarrow FP(I)PO_3 + HCl \quad \text{Eq.[3.1-2]}$$

3.2. GRSS Method

The reprocessing of the oxide spent fuel discharged from LWR is studied at the Korea Atomic Energy Research Institute (KAERI) in the Republic of Korea as shown in Figure 3.2-1 called Advanced Spent Fuel Conditioning Process (ACP)[15].ACP was developed by KAERI, which is composed of 3 main steps, i.e., Head-End Feed Preparation Step (Boroxidation Step), Electrolytic Reduction Step, and Refining Step[16].

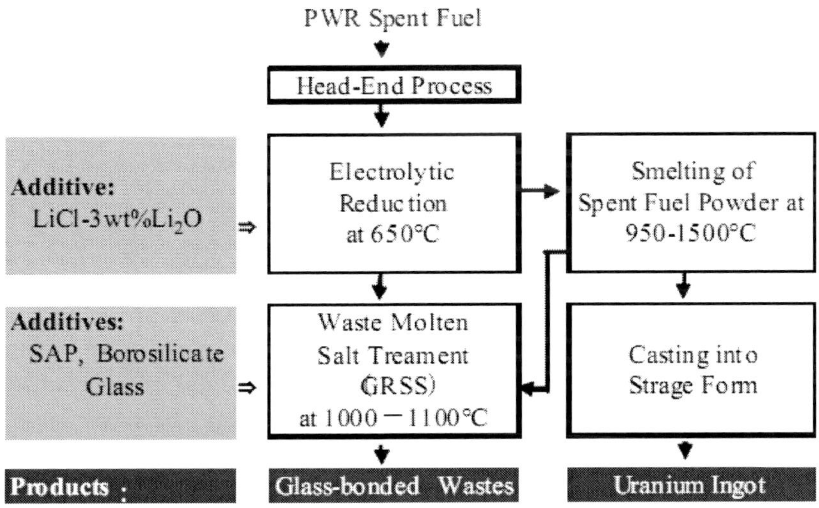

Figure 3.2-1. Conceptual block flow diagram of ACP for reprocessing the spent fuel generated from the pressurized water reactor (PWR) which is of LWR type.

The spent oxide fuel is pulverized to powder at the Pre-treatment Step and then it is reduced to metal at the Electrolytic Reduction. This reduction is carried out by immersing the spent fuel into the molten salt (LiCl-3wt%Li_2O as reductant) at the temperature of 650℃[17]. In this reduction, cesium and strontium are not reduced but remain as CsCl and $SrCl_2$ in the melt, therefore, HLW, mainly LiCl, is generated as the waste salt. To treat this waste salt, GRSS (Gel-Route Stabilization/Solidification) method [18]is developed at KAERI. The GRSS method is to convert metallic elements in the waste salt to aluminosilicates, aluminophosphates, or phosphates using powder called SAP (SiO_2-Al_2O_3-P_2O_5)[19] which is produced by the sol-gel method. The following reaction will occur when the mixture of the waste salt and SAP are heated at the temperature of 650-850℃.

$$LiCl + SAP \Rightarrow Li_3PO_4 + Li_yAl_ySi_{1-y}O_{2-y} + Cl_2 \quad \text{Eq.[3.2-1]}$$

$$LiCl + CsCl + SAP \Rightarrow Li_3PO_4 + Cs_2P_3O_{10} + (Li,Cs)\text{-aluminosilicate} + Cl_2 \quad \text{Eq.[3.2-2]}$$

$$LiCl + SrCl_2 + SAP \Rightarrow Li_3PO_4 + Li_yAl_ySi_{1-y}O_{2-y} + Sr_5(PO_4)_3Cl + Cl_2 \quad \text{Eq.[3.2-3]}$$

$$LiCl + CeCl_3 + SAP \Rightarrow Li_3PO_4 + Li_yAl_ySi_{1-y}O_{2-y} + CePO_4 + Cl_2 \quad \text{Eq.[3.2-4]}$$

All of the above-mentioned products, which presented a more stable chemical condition than that of metallic chlorides, are separated and are made to the glass waste form by mixing with borosilicate glass flit at the temperature of 1100℃.

KAERI also studies the removal of RE from the 3LiCl-2KCl medium of the spent electrolyte. In their study, the RE phosphates are precipitated by addition of $NH_4H_2PO_4$ into the spent electrolyte at the temperature of 650℃. The recovered precipitates, RE phosphates ($REPO_4$) is heated in the air at the temperature of 1000℃ to make Monasite (rare earth phosphate mineral) for the immobilization of rare earth element.

3.3. Phosphate Conversion Method

The phosphate conversion method is being studied by Japan Atomic Energy Agency (JAEA) together with some organizations and universities in Japan. It consists of three steps: 'Spent Electrolyte Regeneration Step', 'Phosphate Conversion Step', and 'Phosphate Immobilization Step' and is a unique method to treat the spent electrolyte. In the 'Spent Electrolyte Regeneration Step', soluble FP chlorides such as CsCl, $SrCl_2$, etc. in the spent electrolyte are removed by sorbent made of the iron phosphate glass system. At the 'Phosphate Conversion Step', redundant electrolytes are converted into phosphate with the use of orthophosphoricacid,etc.. Those phosphates arising from the former two steps are immobilized by the iron phosphate glass system at the 'Phosphate Conversion Step'. As the phosphates are the only substances to treat throughout these steps, we envisage the conceptual process to be relatively simpler compared to other methods.

The basic idea of the 'Spent Electrolyte Regeneration Step' to remove lanthanoid FP as phosphates was developed by RIAR in Russia, which has beenbriefly described in section 3.1.

3.3.1. Computational Evaluation

(1) Composition of Spent Electrolyte

Before we analyse the phosphate conversion behaviours of FP chlorides, the computational composition in the spent metal fuel was calculated by ORIGEN-2 code [20]which was developed at Oak Ridge National Laboratory for use in the calculation of the nuclide composition, etc.. The condition of the spent metal fuel employed for this computation is assumed that it is irradiated to 98GWd/t of burnup (average 48GWd/t, the breeding ratio of 1.2) and is stored for 4 years of cooling time. The results are summarized in Table 3.3-1[21][22][23] which shows the main composition of the spent electrolyte after removal of uranium, TRU, e.g., plutonium and MA, elements of platinum group, some heavy metals, etc. before treatment at the electrolyte recycle process. From this table, it found that the principal groups of FP which should be separated from the medium are the AL metals, ALE metals, and RE.

Based on Table 3.3-1, the component of surrogate spent electrolyte for this experiment is defined in Table 3.3-2.

Table 3.3-1. Calculated composition of the spent electrolyte

Group	symbol	atomic number	atomic/ molecular weight	m.p.	b.p. (°C)	density(g/cm^3)	elememts in S.S. (kg/d)	elememts in S.S. (mol/d)	mole fraction	Ref.
Alkali Metals	Rb	37	85.468	39.3	688	1.53	0.009866	0.115435	7.92244E-05	[21]
	Cs	55	132.905	28.5	671	1.93	0.170134	1.280117	0.000878558	[21]
Alkaline-Earth Metals	Sr	38	87.62	777	1382	2.64	0.221383	2.526631	0.001734053	[21]
	Ba	4	137.327	727	1867	3.62	0.808617	5.888256	0.004041171	[21]
Rare-Earth 1	Y	39	88.906	1522	3345	4.47	0.150853	1.696766	0.001164508	[21]
	Eu	63	151.964	822	1529	5.24	0.070543	0.464206	0.000318589	[21]
	Sm	62	150.36	1074	1794	7.52	0.618605	4.114157	0.002823588	[21]
Rare-Earth 2	Ce	58	140.116	798	3443	6.77	1.3891	9.91393	0.006804032	[21]
	Pr	59	140.908	931	3520	6.77	0.69455	4.929104	0.003382894	[21]
	Nd	60	144.24	1021	3074	7.01	2.313205	16.0372	0.011006492	[21]
	Pm	61	(145)	-	-	-	0.043145	-0.29755	-0.00020421	[22]
Rare-Earth 3	La	57	138.906	918	3464	6.15	0.683626	4.921502	0.003377677	[21]
	Gd	64	157.25	1313	3273	7.9	0.076374	0.485684	0.00033333	[21]
Transition Elements	Mo	42	95.94	2622	4639	10.2	0	0	0	[21]
	Tc	43	98	2157	4265	11	0	0	0	[21]
	Ru	44	85.468	39.3	688	1.53	0	0	0	[21]
	Rh	45	102.906	1963	3695	12.4	0	0	0	[21]
	Pd	46	106.42	1554.8	2963	12	0	0	0	[21]
	Ag	47	107.868	961.78	2162	10.5	0	0	0	[21]
Typical Elements	Cd	48	112.411	321.069	767	8.69	0	0	0	[21]
	Sn(white)	50	118.71	231.93	2602	7.265	0	0	0	[21]
Halogen	I	53	126.905	113.7	184.4	4.933	0.162388	1.279603	0.000878205	[21]
	Br	35	79.904	-7.2	58.8	3.1028	0.007612	0.095264	6.53804E-05	[21]
Chalcogen	Se(α)	34	78.96	221	685	4.39	0.023702	0.300172	0.000206011	[21]
	Sb	51	121.76	630.628	1587	6.68	0.013739	0.11284	7.74431E-05	[21]
	Te	52	127.6	449.51	988	6.24	0.312559	2.449522	0.001681132	[21]
Noble Gases	Kr	36	83.8	-157.375tp	-153.34	3.425g/?	0	0	0	[21]
	Xe	54	131.29	-111.745tp	-108.09	5.366g/?	0	0	0	[21]
Bond Na	Na	11	22.99	97.794	882.94	0.97	3.23	140.4959	0.096423752	[21]
							sub total	196.8087	0.13507183	[21]
Medium	LiCl-KCl	-	55.59(ave.)	352		1.648	70.01	-	-	[23]

Table 3.3-1.(Continued)

Group	symbol	atomic number	atomic/ molecular weight	m.p.	b.p. (°C)	density (g/cm³)	elememts in S .S . (kg/d)	elememts in S .S . (m ol/d)	m ole fraction	R ef.
59mol%	L iCl	-	42.394	610	1383	2.07	31.50726	-	-	[21]
41mol%	KCl	-	74.551	771	-	1.988	38.50274	-	-	[21]
composition	L i	3	6.941	180.5	1342	0.534	5.158558	743.201	0.510066453	[21]
	K	19	39.098	63.5	759	0.89	20.19262	516.4617	0.354452959	[21]
							total	1457.067	0.999999663	
Fuel	Z r	40	91.224	1854	4409	6.52	0	0	0	[21]
Halogen	C l	17	35.453	-101.5	-34.04	2.898g/?	-	-	-	[21]

Table 3.3-2. Composition of surrogate spent electrolyte

Group	Symbol	Atomic Number	Atomic Weight	Mole Fraction	Fomula of Clhorides	Molecular Weight	Boiling Point (°C)	Density(g/cm3)	Weight in 1m ol of Spent Salt (g)
Alkali Metal	Cs	55	132.905	0.000882	C sC l	168.37	645	3.988	0.126781473
Alkaline Earth	Sr	38	87.62	0.00174	S rCl_2	158.53	874	3.052	0.275849993
Metals	Ba	56	137.327	0.004055	B aC l_2	208.232	962	3.9	0.844410863
	Y	39	88.906	0.001169	Y Cl_3	195.264	721	2.61	0.228172803
	La	57	138.906	0.003389	LaC l_3	245.264	859	3.84	0.831287221
Rare-Earth Elements	Ce	58	140.116	0.006828	C eC l_3	246.474	817	3.97	1.682815938
	Pr	59	140.908	0.003395	P rCl_3	247.266	786	4	0.839367194
	Nd	60	144.24	0.011045	N dC l_3	250.6	758	4.13	2.767764498
	Sm	62	150.36	0.002833	Sm C l_3	256.72	682	4.46	0.72737808
	Eu	63	151.964	0.00032	EuC l_3	258.322	623	4389	0.082583256
Bond Na	Na	11	22.99	0.096757	N aC l	58.443	800.7	2.17	5.654779738
Medium	Li	3	6.941	0.51183	LiC l	42.394	610	2.07	21.69853036
	K	19	39.098	0.355679	K Cl	74.551	771	1.988	26.51619737
			Total	0.99992					62.27591879

(2)Thermodynamical Approach

The purpose of the phosphate conversion is to regenerate the spent electrolyte to a 'purified' medium by removing its FP chlorides as insoluble phosphate precipitates.Li_3PO_4 and K_3PO_4 which possess same metal ions as the eutectic medium are thus appropriate additive (phosphate converting reagent) candidates because they will have lesser effects on the chemical composition of the medium as shown in the following reaction formulas:

$$xMCl_y + zLi_3PO_4 = M_x(PO_4)_z + 3zLiCl \qquad \text{Eq.[3.3-1]}$$

or

$$xMCl_y + zK_3PO_4 = M_x(PO_4)_z + 3zKCl \qquad \text{Eq.[3.3-2]}$$

As the products from above reaction are some metal phosphates and LiCl or KCl, the composition of the medium changes little. Here, M refers to specific metals such as alkali metals (AL),alkaline earth metals (ALE) and rare earth elements (RE), y for their valences, and the relationship among x,y and z can be expressed as "y = 3z/x".

First, we try to discuss the tendency of the reactions shown in Eq.[3.3-1] and Eq.[3.3-2]by applying the thermodynamic equilibrium constant K on this evaluation.

When the formula shown as Eq.[3.3-1] or Eq.[3.3-2] has maintained the thermodynamic equilibrium condition, the equation shown as Eq.[3.3-3] can be established between K and $\Delta_r G^0$ (kJmol^{-1}) for the standard reaction Gibbs function.

$$RT\ln K = -\Delta_r G^0 \qquad \text{Eq.[3.3-3]}$$

where, R: gas constant [0.00831451 kJK^{-1}mol^{-1}]
T : temperature [K]

For this calculation, $\Delta_f G^0$ (kJmol^{-1}) for the standard Gibbs function of formation is employed from the known data for chlorides and the computational results by CALPHAD method for phosphates[24]. The relationship between $\Delta_f G^0$ and $\Delta_r G^0$ is shown in Eq.[3.3-4] as below.

$$\Delta_r G^0 = \sum_j \nu_j \Delta_f G^0 \qquad \text{where, } \nu\text{: stoichiometric coefficient} \qquad \text{Eq.[3.3-4]}$$

Let $\nu[M_x(PO_4)_z]$, $\nu[LiCl]$, $\nu[MCl_y]$ and $\nu[Li_3PO_4]$ denote the stoichiometric coefficient of each substance, then Eq.[3.3-4] can be expanded as Eq.[3.3-5].

$$\begin{aligned}\Delta_r G^0 &= \nu[M_x(PO_4)_z]\times\Delta_f G^0[M_x(PO_4)_z] + \nu[LiCl]\times\Delta_f G^0[LiCl] - \nu[MCl_y]\times\Delta_f G^0[MCl_y] \\ &\quad - \nu[Li_3PO_4]\times\Delta_f G^0[Li_3PO_4] \\ &= \Delta_f G^0[M_x(PO_4)_z] + 3z\times\Delta_f G^0[LiCl] - x\times\Delta_f G^0[MCl_y] - z\times\Delta_f G^0[Li_3PO_4]\end{aligned} \qquad \text{Eq.[3.3-5]}$$

Substituting the known $\Delta_f G^0$ for Eq.[3.3-5]and the acquired $\Delta_r G^0$ for Eq.[3.3-3], and the thermodynamic equilibrium K can be obtained. The relationship between K, a for the activity of each substance, γ for the activity coefficient and m for the molarity (or molar fraction) is generally expressed by Eq.[3.3-6].

$$K = (\prod_j (a_j)v^j)_{equilibrium} = (\prod_j (\gamma_j m_j)^j)_{equilibrium}$$ Eq.[3.3-6]

Therefore, equilibrium constant K can be expressed by

$$\begin{aligned} K &= a[MCl_y] \times a[Li_3PO_4]/a[M_x(PO_4)_z] \times a[LiCl] \\ &= (\gamma[MCl_y] \times \gamma[Li_3PO_4]/\gamma[M_x(PO_4)_z] \times \gamma[LiCl]) \\ &\quad \times (m[MCl_y] \times m[Li_3PO_4]/m[M_x(PO_4)_z] \times m[LiCl]) \end{aligned}$$ Eq.[3.3-7]

Let K_γ be $(\gamma[MCl_y] \times \gamma[Li_3PO_4]/\gamma[M_x(PO_4)_z] \times \gamma[LiCl])$ and K_m be $(m[MCl_y] \times m[Li_3PO_4]/m[M_x(PO_4)_z] \times m[LiCl])$,

then K can be expressed by Eq.[3.3-8].

$$K = K_\gamma \times K_m$$ Eq.[3.3-8]

At the present condition, as the activity coefficient γ has not been known yet, such presumption was taken that all γ would be nearly 1, which meant K_γ could be assumedas1[25]. Based on this condition, i.e., $K = K_m$, the progress of the chemical reaction can be discussed from the angle of the change of molality in this section.

The thermodynamic equilibrium constants for the phosphate converting reaction of the target chlorides under the condition at 0.101325MPa of pressure and at 27C(≈300K) to 727C(≈1000K) of temperature range is summarized in Figure3.3-1 and 3.3-2.

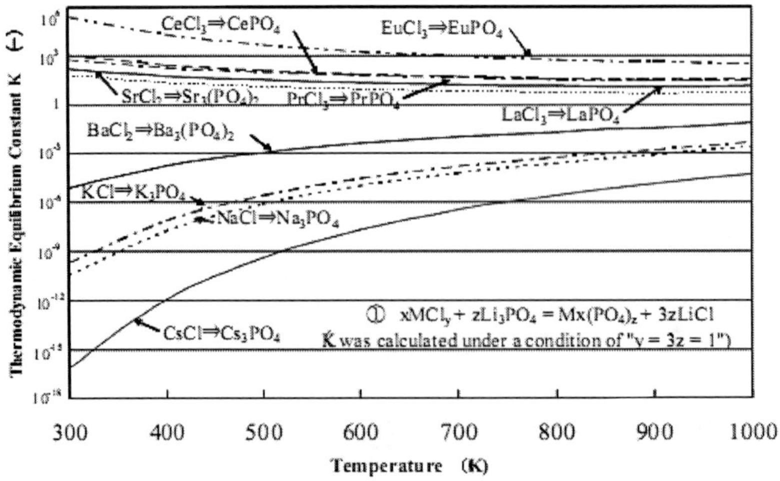

Figure 3.3-1. Thermodynamic equilibrium constants for phosphate converting reaction of chlorides by Li_3PO_4 as additive at 0.101325MPa of pressure and 300 to 1000K of temperature range.

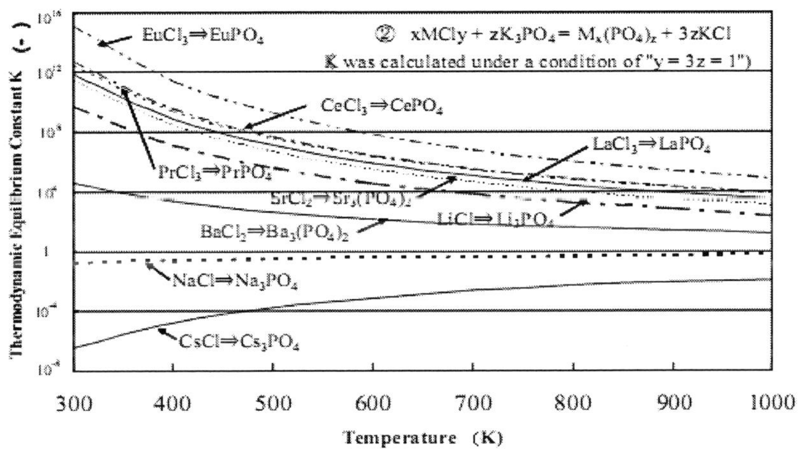

Figure 3.3-2. Thermodynamic equilibrium constants for phosphate converting reaction of chlorides by K_3PO_4 as additive at 0.101325MPa of pressure and 300 to 1000K of temperature range.

Figure 3.3-1 is a result of calculation on each K of phosphate converting reaction shown in Eq.(1) that is "$xMCl_y + zLi_3PO_4 = M_x(PO_4)_z + 3zLiCl$". The molarity of chlorine was set for constant (y=3z=1) to let comparison of each K be easier in this calculation.

The acquired each K was evaluated by the general concept [26] defined by the reaction zones that $1 \leq K$ for the forward reaction zone, and $K \leq 1$ for inverse reaction zone. From this definition, $SrCl_2$ and $LnCl_3$ should be able to convert by Li_3PO_4 under whole temperature range whereas other alkali or alkaline earth chlorides are unlikely to be converted.

Based on the above calculation, Figure3.3-2 shows the result from evaluation of K obtained from Eq.[3.3-2] which uses K_3PO_4 as the additive. Under this condition, all chlorides except NaCl and CsCl can be converted to phosphates such as Li_3PO_4, $Ba_3(PO_4)_2$, $Sr_3(PO_4)_2$, $LaPO_4$, $CePO_4$ and $EuPO_4$.

3.4. Phosphate Conversion Behaviours of Related FP Elements

3.4.1. Experimental

As shown in Table 3.4-1, the surrogate spent electrolyte for the simple salt experiment was made up by mixing 20g of 3LiCl-2KCl eutectic (Aldorich-APL Co.,U.S., 99.99% purity) and 0.05mol of simple chloride such as LiCl (Sigma-AldorichCo.,U.S., 99.995% purity), CsCl (Kojundo Chemical Laboratory Co. Ltd., Japan 99.999% purity), $SrCl_2$ (Aldorich-APL Co., 99.99% purity) or $LaCl_3$(Kojundo Chemical Laboratory Co.Ltd. 99.9% purity) in an alumina crucible (Nikkato Corp. Japan, 52mmOD×67mmH, 99.6% purity). Then 3 or 4.5 times of the required equivalent of Li_3PO_4 (Kojundo Chemical Laboratory Co. Ltd., 99% purity) or K_3PO_4 (Kojundo Chemical Laboratory Co. Ltd., 99% purity) as additive for conversion was added on the mixture.

In the case of the complex salts experiment, all salts mentioned above together with NaCl (Aldorich-APL Co., 99.99% purity), $BaCl_2$(Aldorich-APL Co., 99.99% purity), $CeCl_3$ (Kojundo Chemical Laboratory Co. Ltd., 99.9% purity) and $EuCl_3$(Kojundo Chemical Laboratory Co. Ltd., 99.9% purity) were mixed to prepare the surrogate spent electrolyte with the rough composition ratio calculated from Table 1 and the additive, i.e., Li_3PO_4 or

$K_3PO_4$was added. The component of the surrogate spent electrolyte and the added amount of additive are shown in Table 3.4-1.This preparation was carried out in a glove-box under an argon gas atmosphere at ambient temperature.

Table 3.4-1.Amount of chlorides and phosphates used for experiment

formula of chloride	molecular weight	weight of chloride (g)	molar amount (mol)	additive w eight (g)	
				Li_3PO_4	K_3PO_4
C sCl	132.905	0.0585693	0.000440685	0.03401386	0.06236169
SrCl 2	158.53	0.13788978	0.000869802	0.13426966	0.24617264
B aCl 2	208.232	0.42209766	0.002027055	0.31291238	0.57369973
Y Cl 3	195.264	0.11405725	0.000584118	0.13525373	0.24797686
LaCl 3	245.264	0.41553745	0.001694246	0.39230595	0.71926147
C eCl 3	246.474	0.84119311	0.003412908	0.79026569	1.4488887
PrCl 3	247.266	0.41957639	0.001696862	0.39291189	0.7203724
N dCl 3	250.6	1.3835289	0.005520866	1.27836745	2.34378408
S mC l 3	256.72	0.36359615	0.001416314	0.32795036	0.60127065
EuCl 3	258.322	0.04128102	0.000159805	0.03700305	0.06784213
N aCl	58.443	2.82666781	0.048366234	3.7330994	6.84433801
LiCl	42.394	10.8464945	0.255849756	0	0
K C l	74.551	13.2547129	0.177793898	0	0
total (g)	-	31.1252022	0.499832549	7.60706861	13.8759684

After that, the crucible was placed in an electric furnace of the glove-box and was heated at 520̊C or 600̊C for 2h following preheating at 300̊C for 1h. During experiment, the mixture was occasionally stirred with the stainless steel wire to promote reaction. The temperature in the crucible was measured by the type K (chromel-alumel) thermocouple and was appropriately controlled manually. Subsequently, the crucible was cooled spontaneously, and then it was taken out of the glove-box to recover the precipitates.

The solidified mixture was rinsed with 500mL of distilled water repeatedly for 5 times to separate the insoluble phosphate and soluble substances such as LiCl, KCl, etc.. The supernatant solution with precipitate was filtrated with the suction filter using 0.2μm pore-size membrane filter (ADVANTEC, Japan).

The particles, which were obtained by drying the precipitate in the electric oven at 100̊C for 24h, were measured their weight with the analytical balance (AB204–S, METTLER-TOLEDO Inc., Switzerland) and observed with the scanning electron microscope (SEM, JSM-5700, JEOL, Japan) which is also equipped with anenergy dispersion X-ray spectrometer (EDS) for elemental mapping. Their sizes and distributions were measured with the automatic centrifugal particle analyser (CAPA300, Horiba, Japan) and their crystal structures were determined with the X-ray powder diffraction (XRD) analyzer (RINT-2000, RIGAKU, Japan, X-ray source: Cu Kα).

3.4.2.Results

(1) Precipitation Behavioursof Phosphates in Simple Salt Experiment

(I)Formation of Alkali Metal Phosphates

From the conversion result of AL chloride i.e., LiCl or CsCl which interacted with 4.5 times of required quantity of K_3PO_4, it was found that only simple Li_3PO_4, was recovered as precipitate after filtering the dissolved sample in water. Figure 3.4-1 shows the analytical results of crystal structure of particle with XRD. Even using Li_3PO_4 as a converting reactant, Cs_3PO_4 did not form the precipitate as mentioned earlier in our paper[27]. These formation behaviours agree with results of theoretical analysis. Their recovery ratio were approximately 34.9% at 520°C and 59.6% at 600°C, respectively, corresponding to the whole LiCl quantity in the sample which means higher recovery could be expected at higher temperature. At the same time, the particle size would grow bigger at higher temperature as shown in Figure3.4-2, approximately 2μm on the average at 520°C and approximately 2.3μm on the average at 600°C.

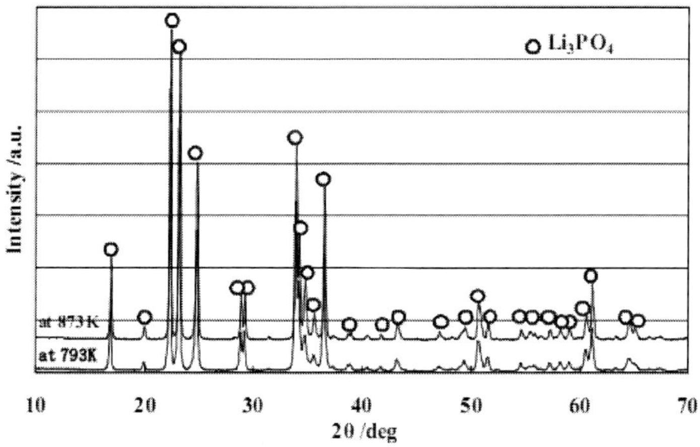

Figure 3.4-1. Crystal structure of precipitate with XRD formed at 520C and at 600 C.

(1) particle formed at 520°C (2) particle formed at 600°C

Figure 3.4-2. Li_3PO_4 particle by the reaction with K_3PO_4at 520°C and at 600C .

To measure the actual particle size, a polycarboxylic high molecular weight dispersant (Geosper K, FlowricCo.Ltd.Japan), diluted with water to 0.1wt.% was added into the sample solution containing the particle. The mean particle size after dispersing became approximately 1μm at 600°C. Figure 3.4-3, which shows its integral distribution of Li_3PO_4 with/without adding the dispersant at 600°C, indicates that 40-50% of particle in whole after dispersing is more than 1μm.

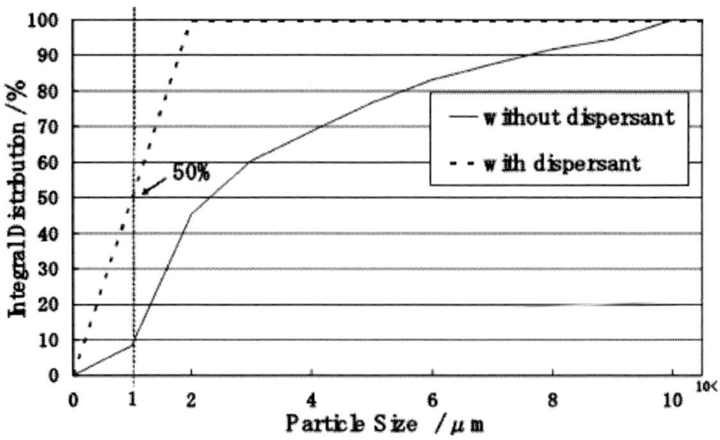

Figure 3.4-3. Integral distribution of Li_3PO_4 formed at 600C.

(II) Formation of Alkaline Earth Metal Phosphate

In this experiment, $SrCl_2$, which was assumed theoretically, to be able to form the phosphate, $Sr_3(PO_4)_2$ usingLi_3PO_4 as the additive, was used as a representative sample of ALE.However, judging from the XRD pattern, it could not be detected in the precipitate as reported in past literature[28] nor with K_3PO_4. This phenomenon should be further investigated from the viewpoint of accuracy of thermodynamic properties of $Sr_3(PO_4)_2$, e.g., the activity of strontium compounds in the melt, the validity of experimental condition, etc..

(III) Formation of Rare-Earth Phosphate

The experimental result shows that $LaPO_4$ could be observed in the precipitate with Li_3PO_4 as shown in Figure3.4-4using Li_3PO_4 orK_3PO_4 as the additive. The two samples of washing water obtained by filtration of dissolving electrolytes following conversion by the addition of Li_3PO_4 or K_3PO_4 were analysed with the ICP-AES but La couldn't be detected in both of them which means the phosphate conversion ratio of La would be 100% with any additives.

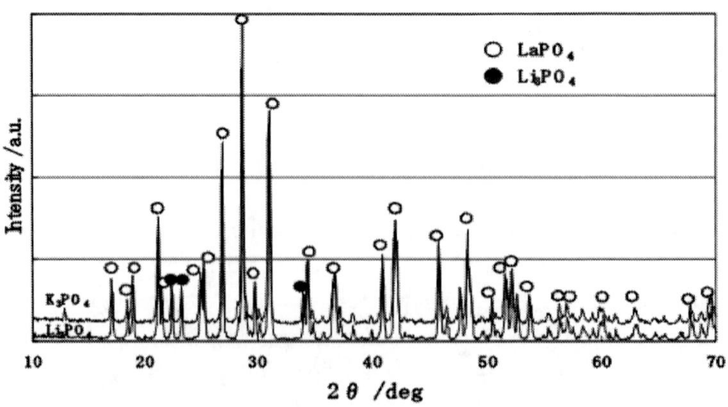

Figure 3.4-4.Crystal structure of precipitate with XRD formed by conversion of $LaCl_3$ with $Li_3PO_4$and with K_3PO_4.

(1) formed by the reaction with Li_3PO_4 (2) formed by the reaction with K_3PO_4

Figure 3.4-5. Particle of $LaPO_4$ formed by the reaction with Li_3PO_4 and withK_3PO_4 at 600℃.

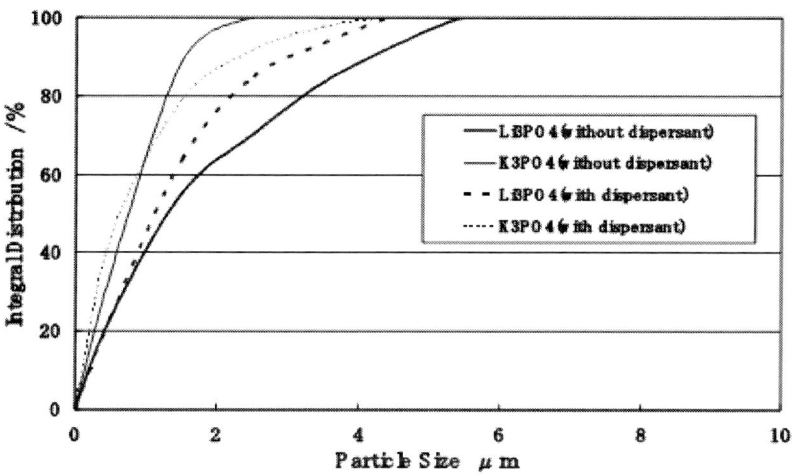

Figure 3.4-6. Integral distribution of $LaPO_4$ formed at 600℃.

Figure 3.4-5 shows the particles of precipitate which were observed with SEM. From this figure the particle size of $LaPO_4$ formed by the reaction with K_3PO_4 was relatively smaller than that of Li_3PO_4. The automatic centrifugal particle analyser was used to measure the actual particle sizes and the distribution of the $LaPO_4$. The mean particle sizes using Li_3PO_4 or K_3PO_4 as the additive are 1.84μm and 1.28μm, respectively. With/without using dispersant, their particle size distributions are shown in Figure 3.4-6. In view of these results, the particle size of $LaPO_4$ shows a tendency to be smaller than that of Li_3PO_4 and some of them are limited to submicron size. The formation of different particle sizes depends to certain extent from the kind of additives used.

(2) Precipitation Behavioursof Phosphates in Complex Salts Experiment

The formation of each precipitate as a result of reaction of the surrogate spent electrolyte with Li_3PO_4 or K_3PO_4 was observed with SEM as shown in Figure3.4-7 and found that most particle size of both samples were in the range of several to some ten micron, some particles were even presented in submicron size. The particle formed by Li_3PO_4 reaction showed the idiomorphic face whilst particle obtained from reaction with K_3PO_4 presented something that looked agglomerate. The constituents of these particle mixtures were proven by elements mapping with EDS and found to be basically composed of Na, Ln and P whilst the large particles were Li_3PO_4 as shown in Figure3.4-8. Judging from the analytical result with XRD as shown in Figure3.4-9, the molecular formula of those minute particles could be determined as $NaCe_2(PO_4)_3$, and $LnPO_4$. It can be said there is the existence of even very small quantity of Na in the 3LiCl-2KCl medium, some kind of Ln tend to form the double phosphates such as $NaCe_2(PO_4)_3$ as observed in the NaCl-KCl medium[29].

(1) formed by the reaction with Li_3PO_4 (2) formed by the reaction with K_3PO_4

Figure 3.4-7. Mixture particles of phosphates formed by the reaction with Li_3PO_4 and with K_3PO_4 at 600°C.

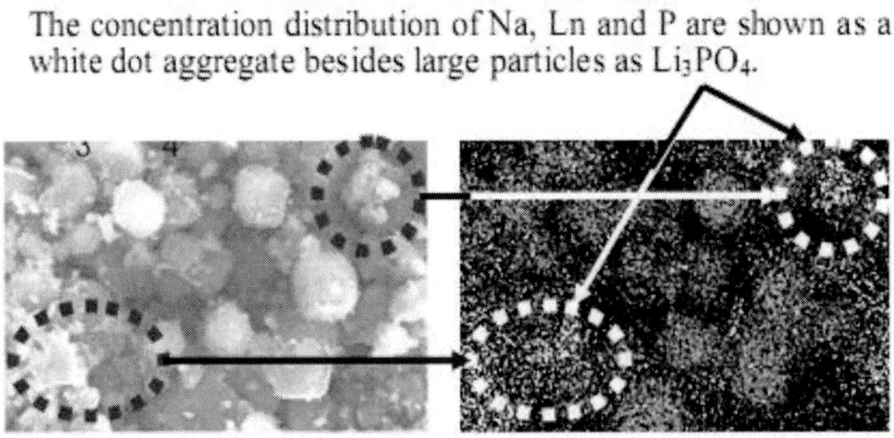

Figure 3.4-8. Analytical result with EDS of precipitate formed by conversion of surrogate spent electrolyte with Li_3PO_4.

These results indicated the possibility that Li_3PO_4 could be a better additive than K_3PO_4 from the viewpoint of formation of larger particle, and sodium, which does not form the precipitate from the theoretical or experimental result, could be precipitated as double salts. If so, it will be possible to remove a variety of FP as precipitate from the actual spent electrolyte.

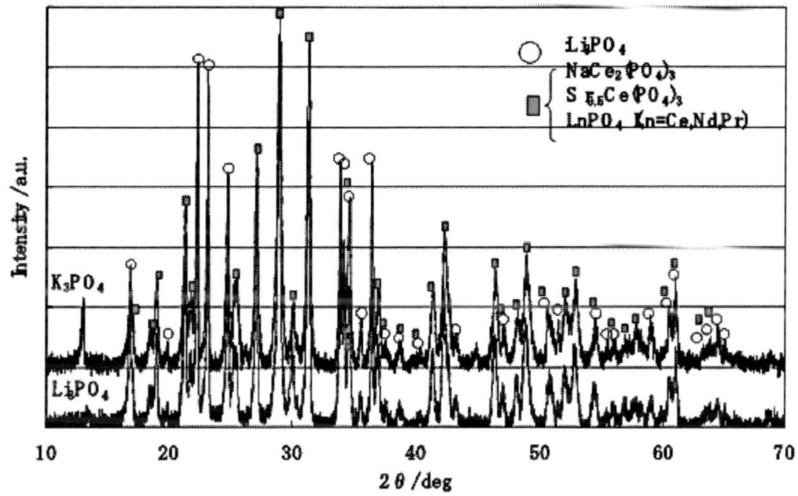

Figure 3.4-9. Crystal structure of precipitate with XRD formed by conversion of surrogate spent electrolyte with Li_3PO_4 and with K_3PO_4.

3.4.3. Conclusion

The phosphate conversion method was applied to remove FP in the spent electrolyte of pyroprocessing. The feasibility of the formation of FP precipitates was estimated by thermodynamical analysis and preliminary experiment. It was found that Li and RE formed orthophosphate particles and sodium formed insoluble double salts which could be recovered. Based on these results, Li_3PO_4 is considered a more suitable converting reagent than K_3PO_4 which despite being used as an additive, its first reaction would occur with LiCl which will lead to the formation of Li_3PO_4.

3.5. Separation Methods of FP in the Spent Electrolyte

In the case of applying the phosphate conversion method to remove FP in the spent electrolyte, the most reasonable idea is to use the phosphate glass system for immobilization of FP. The phosphate glass system, despite possessing a higher FP loading than borosilicate glass (BG) system, thus making it possible to reduce the final HLW disposal volume, is considered relatively inferior in terms of water-resistant quality to current waste form[30].Prof. Day (University of Missouri-Rolla) has recently developed the water-resisting phosphate glass system called the iron phosphate glass (IPG) system[31] as shown in Figure3.5-1 [32], for the purpose of reducing the disposal volume arising from the reprocessing of defence waste. The physical properties of this IPG is almost the same as that of BG, see Table 3.5-1 and it has also overcome some problems such as chemical reliability, material corrosiveness, etc..

In Japan, the fundamental study of IPG system for processing and disposal of HLW has also been undertaken by the Japan Atomic Energy Agency (JAEA) together with organizations and universities in Japan [33]. We are trying to use the IPG system not only as a glass medium but also as a sorbent for FP in the spent electrolyte. The developmental results are presented in the following section.

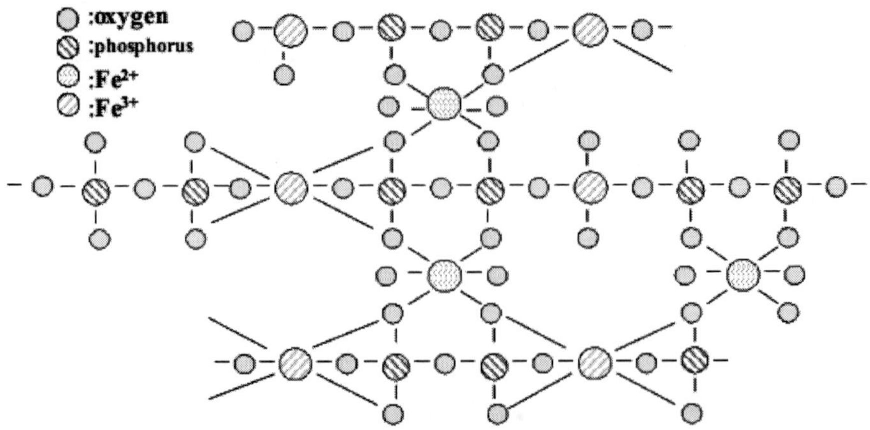

Figure 3.5-1. Illustration of the glass network structure of theIPG.

Table 3.5-1. Comparison of properties in borosilicate glass and IPG

Glass	Iron phosphate glass	Borosilicate glass
Composition	24.3～32.5	14
[w t l%]	of $Fe_2O_3/(Fe_2O_3+P_2O_5)$	of $B_2O_3/(B_2O_3+SiO_2$+others)
Density	2.90～3.45	2.7～2.8
[g/cm3]	(at ambient temp.)	(at20°C)
Specific heat	0.96～1.28	0.78～0.95
[J/g/K]	(at 300°C)	(at□00°C)
Viscosity	0.3～0.6	4～5
[Pa・s]	(at 1100°C)	(at1150°C)
Glass trasition temp.[°C]	550	501
Slightly soluble elements	Zr(phosphate), platinum group	P,F,Cl,Cr,Mo, platinum group
Furnace material	high-purity Al_2O_3, ZrB_2	high-purity Cr brick

3.5.1. Use of IPG as Filtration Medium and Sorbent

(1) Experimental

(I) Precondition

The fundamental experiments were carried out to observe the possibility of separation of soluble FP in the spent electrolyte by the IPG sorbent following the removal of insoluble phosphate particles by the IPG filtration medium.

Prior to carrying out the experiments, it is apparent from the results presented in section 3.3 that:

(a)Lithium phosphate (Li_3PO_4) is a suitable phosphate converting reagent, because K_3PO_4 would react with LiCl to form Li_3PO_4 at the earlier stage of phosphate conversion. Consequently, the direct reaction between K_3PO_4 and other chlorides is unlikely to occur in the spent electrolyte;

(b)As alkali chlorides except LiCl and alkaline earth chlorides would not form the precipitate by the reaction with Li_3PO_4 but $RECl_3$ could, target elements for removal are AL except Li and ALE by sorption and Li and RE by filtration, respectively;
(c)To precipitate RE in the spent electrolyte, excess amount of additive is needed.

(II) Preparation of THE IPGs

The IPG consisting of P_2O_5 as network former and Fe_2O_3 as the network modifier possesses good chemical durability, low viscosity, high-loading concentration of other elements, etc. In addition, the ionic valence of iron is changeable depending on the melting temperature of the raw material of the IPG[34]. Reflecting such characteristics of IPG, some samples of IPG were prepared as a filtration medium as well as FP sorbents under three different melting temperatures of raw materials, i.e., 1100C, 1200C and 1300C, respectively. According to previous study[35], IPG manufactured under higher temperature forms more ferrous ion (Fe^{2+}) than that at lower temperature. Also, such IPG would be in a rather unstable state [36] compared with the ferric (Fe^{3+})-rich IPG. It is expected that this unstable state of IPG would precipitate other ions that are incorporated into this glass network. Here, IPG sample made at the melting temperature of 1200C was used solely as a filtration medium.

For this preparation, appropriate amounts of 85% purity of orthophosphoric acid (Kishida Chemical Co.,Ltd.,Japan) and ferric oxide (Kanto Chemical Co.,Inc.,Japan, above 99%) were mixed in the Pt crucible to obtain $2Fe_2O_3$-$3P_2O_5$ (molar ratio) known as a composition of durable IPG [37]. Then it was melted at the temperature of 1100C-1300C for 1h in the air after preheating for 3h at 150C. The melt was poured onto the graphite board to make it vitrified.

The finished product was crushed and classified their sizes to 1.2-1.18 mm, 850-425 μm, 425-250 μm, and 250 μm or less. These glass powders were dried over in a vacuum electric furnace at 60C for 24h. In the next stage, these glasses were measured their softening points and crystallization temperatures with the differential thermalgravimetricanalyser (TG8120, Rigaku Corp., Japan), and coefficient of thermal expansion and glass transition temperatures with the thermomechanical analyzer (TMA8310, Rigaku Corp.)

(III) Preparation of the Surrogate Spent Electrolytes

The surrogate spent electrolyte for the filtration experiment was made up by mixing 20g of 3LiCl-2KCl eutectic (Aldorich-APL Co.,U.S., 99.99% purity) and 0.5mol $LaCl_3$ (Sigma-AldorichCo.,U.S., 99.995% purity) in an alumina crucible (Nikkato Corp. Japan, 52mmOD×67mmH, 99.6% purity). Then 4 times of the required equivalent of Li_3PO_4 (Kojundo Chemical Laboratory Co. Ltd., Japan, 99% purity) as additive was added on the mixture. Meanwhile, the surrogate spent electrolyte for the sorbent experiment was prepared by mixing 5 g of 3LiCl-2KCl with phosphates (Hakushin Chemical Laboratory Co,Ltd., Japan, 99.9% purity) each containing 1 mmol (0.095g) of phosphate anion (PO_4^{3-}), i.e., 1 mmol (0.2278 g) for Cs_3PO_4, 0.5 mmol (0.1388 g) for $Sr_3(PO_4)_2$ and 1 mmol (0.2338 g) for $LaPO_4$ in an alumina crucible (Nikkato Corp. Japan, 38 mmOD×45 mmH, 99.6% purity) in the argon gas atmosphere.

All of the above-mentioned reagents were measured their weight with the analytical balance (AB204-S, Mettler-Toledo Inc., Switzerland) in a glove-box under an argon gas atmosphere at ambient temperature. After that, the crucibles were placed in an electric furnace (FO-410, Yamato Scientific Co. Ltd., Japan) and were heated at 600°C for 4h to prepare samples for the filtration experiment and 24h for the sorption experiment following preheating at 300°C for 1h under the argon gas atmosphere. Then surrogate spent electrolyte samples for the filtration experiment were taken out from the crucibles.

(IV) Experimental

The filtration experiment of the insoluble particle from the spent electrolyte was carried out with the high-temperature suction filtration equipment as shown in Figure3.5-2. The surrogate spent electrolyte was set on a filtration medium to separate the medium and particle by heating for 1h at a temperature range of 450 to 550°C under 0.15 MPa of argon gas atmosphere. The IPG powder (grain size:2.0-1.18 mm) was used as a filtration medium that was dried in advance at 350°C for 24h in an electric furnace. Following filtration, the recovered samples as the treated medium were analysed with a *wave dispersive X-ray fluorescence*spectrometer (WDXRF, System 3270 Rigaku Industrial Corp., Japan).

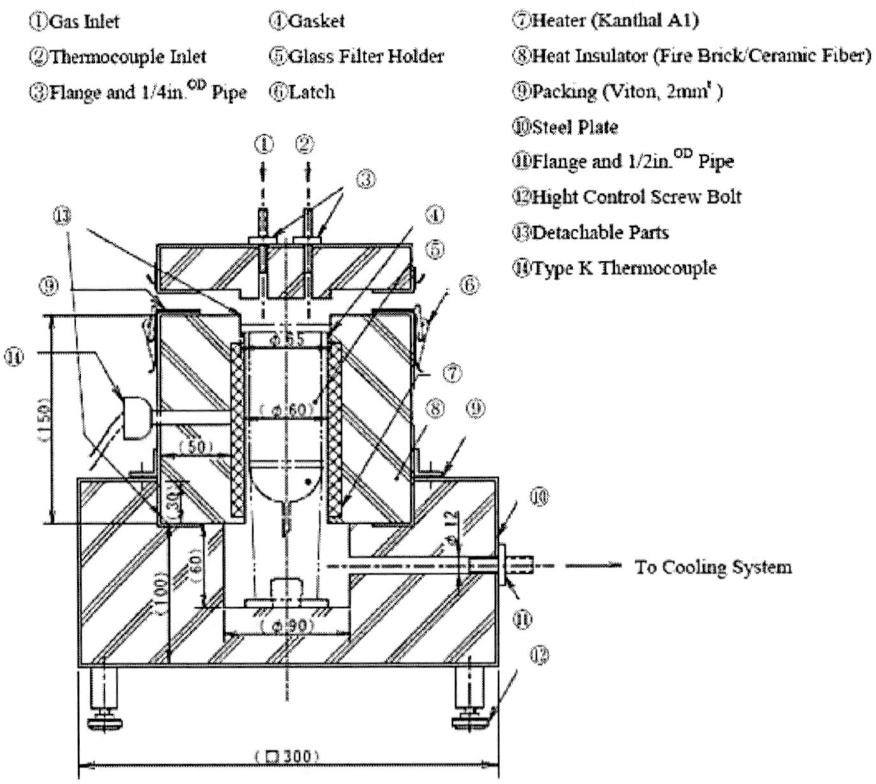

Figure 3.5-2. Schematic design of the high-temperature suction filtration equipment.

In the case of the FP sorption experiment, it was carried out by adding 0.5 g of IPG powder (425-250 μm) to each crucible which contained the surrogate spent electrolyte. After that, the crucibles were placed in a muffle furnace and were heated at the appropriate laying temperatures from 480 to 540°C for 18h under an argon gas atmosphere. Subsequently, the crucibles were cooled spontaneously and then they were taken out of the furnace for analysis of the FP sorption efficiency of IPG.

The solidified mixture was rinsed with 0.5 L of distilled water repeatedly for five times to separate the insoluble precipitates and soluble substances such as 3LiCl-2KCl, etc. The supernatant solution accompanying precipitates was filtrated with the suction filter using 0.2μm pore-size membrane filter (ADVANTEC, Japan). After removing the particle, the supernatant solution was made up to 2.5 L exactly with the distilled water to determine the FP amount quantitatively with the inductively coupled plasma optical emission spectrophotometer (ICP-OES, SPS5510CCD, SII Nano Technology Inc.,Japan).

The analytical values were used to calculate the sorption efficiency (Se) of the IPG sorbent by the equation as shown below:

$$\mathrm{Se}(\%) = 100 - (\text{FP amount in the filtrate/dosage of FP}) \times 100 \quad \text{Eq.[3.5-1]}$$

(V) Results

Prior to the experiments, the IPG samples were analyzed their crystal structures with the X-ray powder diffraction (XRD) analyzer (RINT-2200V, RigakuCorp.,X-ray source: Cu Kα) and confirmed they were in the glass state because all samples manufactured at 1100, 1200, and 1300C of melting temperatures presented broad amorphous peaks in the range from 20 to 30 degrees called *halo patterns* as shown in Figure3.5-3.

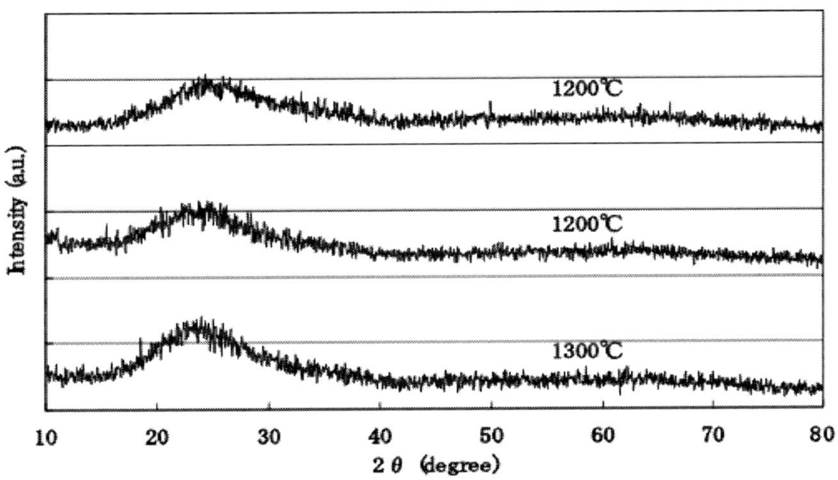

Figure 3.5-3. XRD pattern of the IPG samples under different manufacturingtemperature.

Table 3.5-2. Thermal properties of IPG samples manufactured under different temperature

Manufacturing Temperature(K)	$\alpha_{303\text{-}573}$(×10^{-7}/K)	Glass Transition Temperature (K)	Softening Tepmerature (K)	Crystallization Temperature (K)
1373	58	761	858	(i) 1106
1473	59	758	848	(i)961 (ii) 1124
1573	65	758	undetectable	(i) 934 (ii) 1163

The thermal properties of each IPG samples were also taken as shown in Table 3.5-2. From this table, coefficients of thermal expansion ($\alpha_{303-573}$) and crystallization temperatures (Tc) are apt to increase with higher manufacturing temperature, whilst the glass transition temperatures(Tg), softening points(Ts) decrease. Therefore, some changes of the IPG structure would occur due to differences in their manufacturing temperatures.

Table 3.5-3. Filtration result of precipitates in the spent electrolyte

Bed Height(mm)		16	8	4
Recovery rate at 823K(%)		38	38.6	55
Composition of the recovery medium except Li with WDXRF (mass%)	Cl	55.6	46.2	66.8
	K	25.4	20.0	25.5
	P	0.09	0.34	1
	Fe	18.9	33.3	6.77
	S i	0.05	0.09	0
	A l	0	0.07	0
	La	N.D.	N.D.	N.D.
Recovery rate at 723K(%)		-	51.4	56.1
Composition of the recovery medium except Li with WDXRF (mass%)	Cl	-	60.1	55.8
	K	-	33.4	37.8
	P	-	1.81	0.626
	Fe	-	4.72	5.61
	S I	-	0	0.0984
	A l	-	0	0
	La	-	N.D.	N.D.
Recovery rate at 773K(%)		-	52.6	61.5
Composition of the recovery medium exceptLi with WDXRF (mass%)	Cl	-	55.1	64.3
	K	-	32.8	30.5
	P	-	1.81	1.03
	Fe	-	10.8	4.14
	S i	-	0.333	0.00479
	A l	-	0.221	0.037
	N i*	-	0.334	0
	La	-	N.D.	N.D.

* Ni was excuded from the structural material of experimental equipment.

* A wave dispersive X-ray fluorence spectrometer (WDXRF System3270 Rigaku Industrial Corp. Japan) was used for this measurement.

* N.D.: Not-detected.

Using one of the IPG samples manufactured at the temperature of 1200C, the filtration experiment was tried out to separate $LaPO_4$, which is the representative of RE precipitate, from the medium. In this experiment, the temperature range and the bed height of the filtration medium are employed as parameters.

The result is shown inTable 3.5-3. From this Table, it can be deduced that:

- The surrogate spent electrolyte could be successfully purified to regenerate the medium, i.e. no $LaPO_4$ was detected in the recovered medium with the WDXRF;
- The medium could be purified more efficiently at the lower temperature range due to less reaction between the filtration medium and the spent electrolyte;

- The amount of IPG used should be of an appropriate amount and the bed height of the filtration medium, preferably to be shallow in depth, taking into consideration that some medium is apt to remain as residue in the filtration medium.

To remove soluble FP in the spent electrolyte, IPG was selected as the sorbent material due to its easily changeable iron valence which is a part of IPG network modifier. It is known in general that the iron valence is likely to shift from ferric(Fe^{3+}) to ferrous (Fe^{2+}) ion under higher temperature. The IPG manufactured under higher temperature would become rather unstable due to the loss of some cations. Under this precondition, we prepared 3 kinds of IPG manufactured at different temperatures, i.e. 1100C,1200C and 1300C respectively and measured the ferrous ion with theMoessbauer spectrometer (Laboratory Equipment Corp., Japan) which can determine the relative abundance of ferrous ion in the IPG.

The ^{57}Fe Moessbauer spectrometry method is to obtain information on the properties of target element through the state of atomic nucleus and to observe the recoil-free resonant absorption of ^{57}Fe atomic nucleus by the gamma ray emitted from the standard radiation source of ^{57}Co which is the parent nucleus of ^{57}Fe. In this case, the ^{57}Fe Moessbauer spectrum was measured to acquire the iron valences and their relative abundance. The curve-fitting of absorption line to the acquired spectrum was made by the Lorentzian function. Then the relative abundance of Fe^{2+} as ($Fe^{2+}/(Fe^{2+}+Fe^{3+})$) was calculated by the absorption area intensity ratio as one of the Moessbauer parameters as well as their valences by the isomeric shift and the quadrupole splitting between ferrous ion and ferric ion as other parameters.

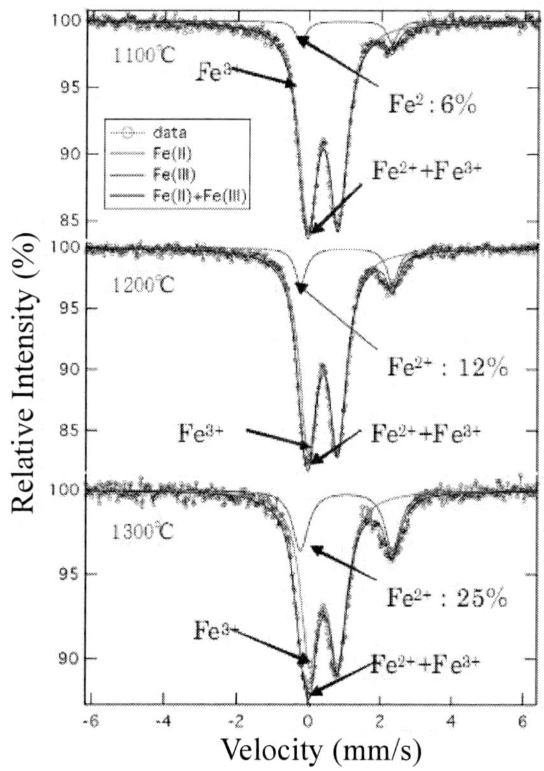

Figure 3.5-4. Ratio of $Fe^{2+}$to total iron in the IPG samples measured with the Moessbauer spectrometer.

The Moessbauer measurement was carried out under the condition of ±6 mm/s of the Doppler velocity at ambient temperature and 15 to 20 mg-Fe/cm^2 of thickness for each of the test samples.

The result is shown in Figure 3.5-4. From this figure, it can be said that the ferrous ion ratio in the IPG increases with its rising manufacturing temperature, e.g., 6% of Fe^{2+} in the IPG manufactured at 1100°C, 12% of Fe^{2+} at 1200°C and 25% of Fe^{2+} at 1300°C. Accordingly, the ferrous-rich IPG could be produced at higher melting temperature of the raw materials.

3.5.2. Sorption Effect of Metal Cations by IPGs

It can be postulated that the metal cations of Cs_3PO_4 and $Sr_3(PO_4)_2$ are the target elements to be sorbed in the IPG whilst $LaPO_4$ and Li_3PO_4 can be separated by filtration. For this experiment, three kinds of IPG sample manufactured at the different temperatures were prepared in anticipation of their potential function as sorbent. The IPG manufactured at higher temperature might function better as a sorbent from the viewpoint of its ferrous ion content.

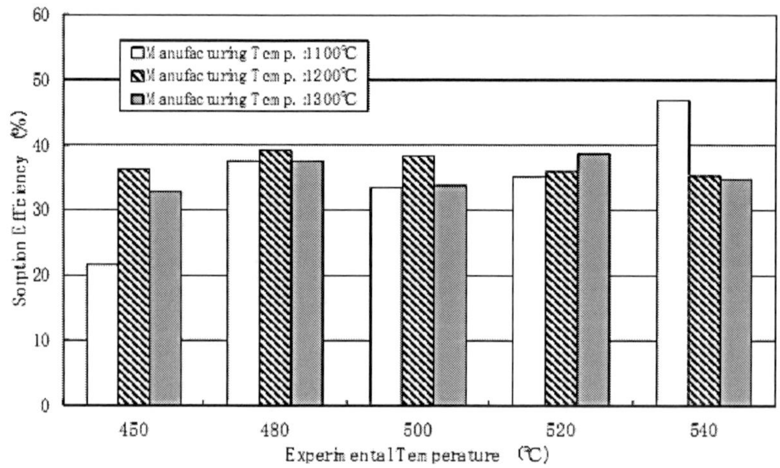

Figure 3.5-5. Caesium sorption efficiency by IPG as a sorbent.

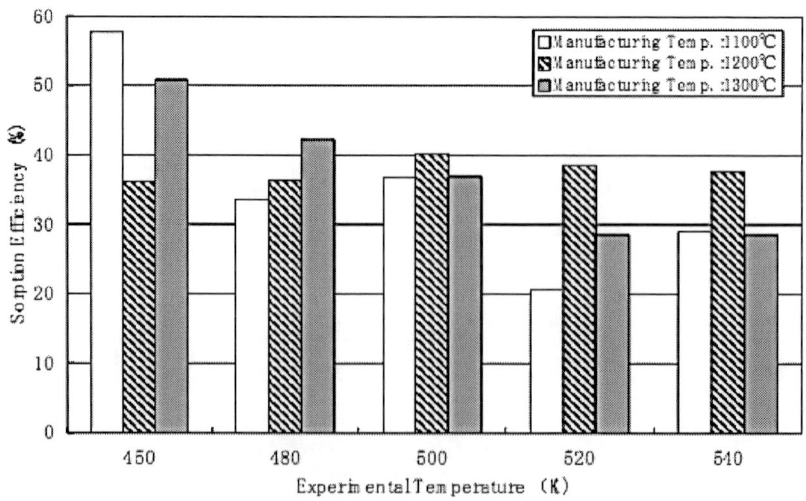

Figure 3.5-6. Strontium sorption efficiency by IPG as a sorbent.

The experimental results are shown in Figure 3.5-5 for the sorption efficiency of cesiumand Figure 3.5-6 for that of strontium. These experiments were carried out at temperatures of 450 to 540̊C at intervals of 20̊C.

From Figure 3.5-5 and 3.5-6, it can be said that:

- All IPGs used for this experiment worked as a sorbent for metal cations and their sorption efficiencies were between 20 and 40 % regardless of the type of metal cation;
- The sorption efficiency of Cs showed little differences despite changes in experimental temperatures whilst that of strontium seemed to decrease following increase of temperature;
- We had initially estimated that the IPG manufactured at 1300̊C, with more ferrous ion produced, would be able to sorb more metal cations. However, the IPG containing too much ferrous ions became very unstable to the extent that they became soluble in the medium together with the sorbed metal cations. Based on the experimental results, the IPG manufactured at 1200̊C displayed the best sorption efficiency among the three kinds of IPG samples.

3.6. IPG Forms

The glass structure of the IPG consists of the glass network governed by $P_2O_7^{4-}$ with the chain of the Fe^{2+}-Fe^{3+}-O_n polyhedral as shown inFigure3.5-1. About 20% of the oxygen in the IPG is the bridging oxygen which makes P-O-P binding while the rest forms mainly Fe-O-P binding.

The IPG possesses high chemical durability compared with the ordinary glass such as the borosilicate glass, the ratio of the nonbridging oxygen in the IPG is also greater as shown inFigure3.6-1.

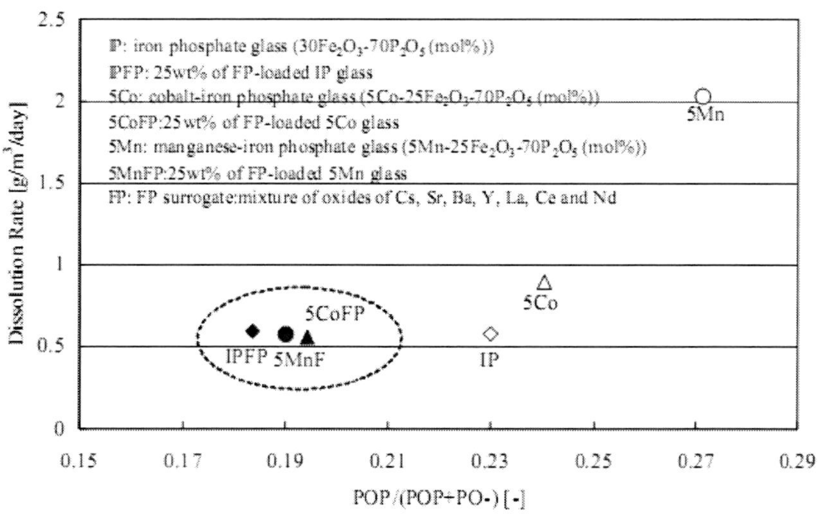

Figure 3.6-1. Dissolution rates of various composition of IPG after immersing in distilled water at 90°C for 3 days.

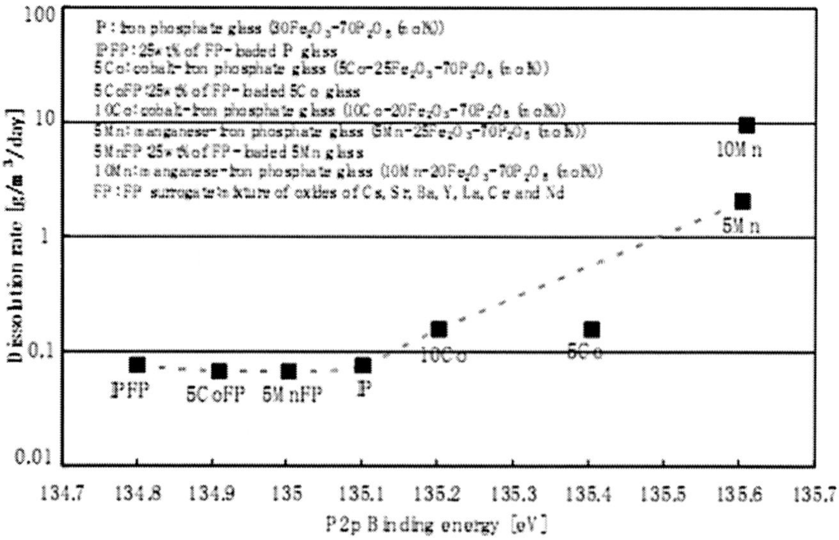

Figure 3.6-2. Relationship between dissolution rates and P2p binding energy of various composition of IPG after immersing in distilled water at 90°C for 3 days.

TheX-ray Photoelectron Spectroscopy (XPS) Analyzer (ESCA5500MT, UlLVAC-PHI Inc., Japan) is used to elucidate the factor which contributes to the chemical durability of the IPG. The binding energy between phosphorous (P2p) and oxygen (O1s) which are the constituent of the network former of the IPG was measured by XPS and it was found that the chemical shift of P2p of the IPG is smaller hence, presented better chemical durability as shown in Figure3.6-2[38]. The chemical durability of the IPG which was loaded with approximately 35 wt % of FP was confirmed [39]. As there is also the positive correlation between the FP loading amount and improvement of the chemical durability, the IPG presented good characteristics as a glass form.

REFERENCES

[1] L.Burris,R.K.Steunenberg and W.E.Miller,:*Proc. Annual AICHE Meeting*, (Miami, FL, 1986),pp.135-142.

[2] O.V.Skiba, Yu.P.Savochkin, A.V.Bychkov, et al.:*Proc. Int.Conf.on Future Nuclear Systems, Global'93*, (Seattle WA, 1993), pp.1344- 1350.

[3] A,V.Bychikov, S.K.Vavilov, P.T.Porodnov,et al.:*Proc. Int.Conf.on Future Nuclear Systems, Global*'93, (Seattle WA, 1993), pp.1351- 1356.

[4] Y.Ito (Ed):*Yoyuen no Oyo EnerugiKankyogijutu eno Tenkai* (IPC□□Tokyo, 2003),pp.243-244.[in Japanese].

[5] O.Amano and K.Yasui: Proc.2001 Fall Meeting of the Atomic Energy Soc.of Japan (Sapporo, Japan, 2001),pp.810. [in Japanese].

[6] T.Tsukada,:*CERIEPI Rep.*,L04009,2005. [in Japanese].

[7] I.Amamoto, M.Myochin,S.Hayashi.et al.: *Proc.22nd Fall Meeting of The Ceram,Soc.of Japan*(Ehime,Japan,2009), pp.405. [in Japanese].

[8] Y.Ono and T.Yashima(Ed.):*Zeolite no Kagaku to Oyo*,(Kodansha Scientfic,2000),pp.6. [in Japanese].

[9] N.Hara and *H.Takahashi(Ed.):Zeolite-Kiso to Oyo*, (Kodansha Scientfic, Tokyo, Japan, 1975), pp.3. [in Japanese].

[10] Committee on Electrometallurgical Techniques for DOR Spent Fuel Teratment,et al.:*Electrometallurgical Techniques for DOE Spent Fuel Treatment*, (National Academy Press, Washington, D.C.,US,2000),pp38-42.

[11] M.Yamane,et al(Eds.):*Garasu Kogaku Handbook*,(Asakura Publishing Co., Ltd, Tokyo, Japan, 1999), pp.661. [in Japanese].

[12] V.A,Volkovich,TR.Griffiths,R.C.Thied:*J.Nucl.Mater.*,323(2003), 49-56.

[13] G.R.Choppin and A.K.Khankhasayev (Eds.):Chemical Separation Technologies and Related Methodsof Nuclear Waste Management: *Applications, Problems, and Research Needs* (NATO Science Partnership Sub-Series:2),(Kluwer Academic Publishers, Dordrecht, Netherlands,1999),pp90-92.

[14] L.L. Kahl, M.C. Ruiz-Lopez, et al., Rep. *Kernforschungszentrum Karlsruhe KfK-*3251,1982.

[15] J.M.Hur,C.S.Seo,et al.:*Proc.WM'03 Conf.*(Tuson, az,USA,2003).

[16] Y.H.Kim,H.J.Lee,et. Al:*J.Nucl.Sci.Technol.*45[7](2008),617-624.

[17] B.H.Park, S.B.Park, et al.:*J.Radioanal.Nucl.Chem.*,270[3](2006), 575-583.

[18] H.S.Park,I.T.Kim, et al.:*Environ.Sci.Tech.*,41(2007),1345.

[19] H.S.Park,I.T.Kim, et al.: *Proc.16th Pacific Basin Nucl.Conf.*, (Aomori,Japan.2008), P16P1385.

[20] A.G.Croff, ORIGEN2: A versatile computer code for calculating the nuclide compositions and characteristics of nuclear materials, *Nucl.Technol.*, 62, 335 (1983).

[21] D.R.Lide(Editor-in-chief), "*CRC Handbook of Chemistry and Physics 85th Edition*", CRC Press LLC, Boca Laton, FL, U.S.A. 4-37-4-96 (2004).

[22] N.N.Greenwood and A. Earnshow, "*Chemistry of the Elements 2nd Edition*", Oxford, UK, 2006.

[23] The Electrochemical Society of Japan (Ed.),"*Denkikagakubinran 4th Edition"*, pp.105, Maruzen,Japan (1993). [in Japanese].

[24] I.Amamoto,H.Kofuji and et al.,*Fundamental Study on Electrolyte Recycle Process by Phosphate Conversion Technique, GLOBAL* 2007,pp.748-756(2007).

[25] P.W.Atkins and J.De Paula, *Physical Chemistry 8th Edition*, pp.317,Oxford UP, Oxford,UK (2006).

[26] P.W.Atkins, *The Elements of PHYSICAL CHEMISTRY*, pp.128,Oxford UP, Oxford, UK (1994).

[27] H.Kofuji, I.Amamoto and et al., Phosphate conversion behaviours of FP chlorides for spent electrolyte recycle, *J.Nucl.Mater.*[389]pp173-178 (2009).

[28] V.A.Volkovich, T.R.Griffiths et al.,, "Treatment of molten salt wastes by phosphate precipitation: removal of fission product elements after pyrochemical reprocessing of spent nuclear fuels in chloride melts" *J.Nucl.Mater.*, 323[1], pp46-56 (2003).

[29] V. Volkovich, T. R.Griffiths,et al., Formation of lanthanide phosphates in molten salts and evaluation for nuclear waste treatment, *Phys. Chem. Chem. Phys*, 5, 3053- 3060 (2003).

[30] S.Sakka(Ed.):*Garasu no Jiten,* (Asakura Publishing Co.,Ltd, Tokyo,Japan,1985), pp.13.

[31] D.E.Day and C.S.Ray,et al.:*DOE DE-FG07*-96 ER 45618,2000.

[32] M.Sazarashi,K.Maruyama,et al.: *Proc.Global 2005 Nuclear Energy Systems for Future Generation and Global Sustainability*, (Tsukuba, Japan, 2005),Paper500.

[33] I.Amamoto,H.Kofuji, at al.: *Proc.2010 Annual Meeting of the Atomic Energy Soc.ofJapan* (Mito, Japan, 2010),pp.441.

[34] S.T.Reis, A.Mogus-Mlankovic, V.Licina, et al.: J.Non-Cry Solids,353,151-158(2007).

[35] X.Yu, D.E.Day, J.Long , et al.: *J.Non-Cry Solids*,215,21-31(1997).

[36] D.E.Day, C.S.Ray, et al.: *DE-FG07-96 ER* 45618 (2000).

[37] M.Sazarashi, K.Maruyama, K.Suzuki et al.: "Vitrified HLW Waste Form in an Iron-Phosphate Glass Matrix",*Proc. Int,Conf. on Nuclear Energy systems for Future Gemneration and Global Sustainability*, Global 2005, Tsukuba, Japan, Oct. 9-13 (2005).

[38] N.Akagi, T.Yano, I.Amamoto,et al.:*Proc.3rd Int.Conf. on Sci.and Technol. for Advanced Ceram.*(Yokohama,Japan,2009),pp.95.

[39] M.Sazarashi,K.Maruyama,et al.: *Proc.Global 2003*, (New Orleans, LA,USA,2003).

In: Radioactive Waste: Sources, Types and Management
Editors: Satoshi Yuan and Wenxu Hidaka
ISBN: 978-1-62100-188-1

Chapter 5

SOURCES OF RADIOACTIVE CONTAMINATION OF FROZEN TAIGA LANDSCAPES IN SOUTHERN YAKUTIA

A. P. Chevychelov and P. I. Sobakin*
Institute for Biological Problems of Cryolithozone SB RAS,
Yakutsk, Russia

ORIGIN OF RADIOACTIVE CONTAMINATION

In the second half of the 20th century the large-scale exploration of radioactive raw material was carried out in southern Yakutia. The extensive exploration work led to the discovery of promising Elkon uranium ore district in the area of Elkon Horst of Aldan Plateau.

Elkon Horst is very well studied from a geological prospective [5,8]. The scale and the volume of mining work performed in this region is not simply impressive (Figure 1,2), it is astonishing [3,4]. The overall length of all holes bored there was about 1 million meters. Individual holes reached 2 km in depth. 60 km of underground mining and 1.3 millions m^3 of trenches had been driven. The number of geological and geophysical studies has been carried out as well. During which hundreds of thousands of thin sections of ore and rocks have been studied [4].

According to the data of Far Eastern division of Federal Service for Nuclear Supervision of Russia (FSNS) [3] after 25 years of mining in this area over 1 million tons of Uranium ore have been mined and stored on the surface as mine dumps. The total content of uranium in this amount of ore is approximately 2000 tons. This suggests high levels of radioactive contamination in this territory. Moreover, because these radioactive mine dumps have been around for nearly 40 years, one can suppose that anthropogenically contaminated territories, characterized by high levels of natural radioactive nuclides in soil, vegetation cover and surface waters, occupy large areas. It was revealed that the territory of Yuzhnoye deposit was the most contaminated one (Figure 3,4).

* Correspondence: chev.soil@list.ru.

Figure 1. Construction of the mining village on Elkon Horst in southern Yakutia circa 1960.

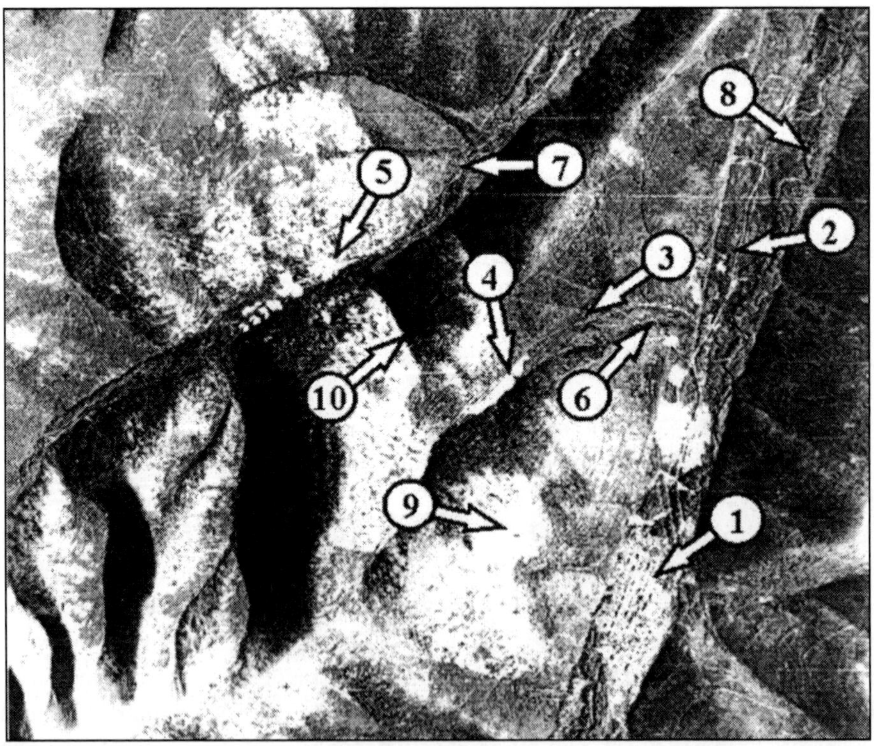

Figure 2. Aerial photograph of the Yuzhnoye deposit area 1 – Zarechniy; 2 – the road from Zarechniy to Tommot; 3 – the road to Kurung-1 site; 4 – Kurung-1 site; 5 – Akin site; 6 – Propadayushiy Creek; 7 – Akin Creek; 8 – Kurung Creek; 9 – tundra-goletz and stone barrens ; 10 – watershed slopes with taiga vegetation.

When exploration works and conservation of mines and adits were completed the huge mass of rocks was left on the surface and stored as radioactive mine dumps. For the first time FSNS experts of the Sakha Republic (Yakutia) investigated uranium mine dumps for the inventory purposes of at Elkon Horst in 1993.

Figure 3. Mine dumps at Kurung-1 site.

Figure 4. Mine dumps at Holodnaya site.

Among the expected 14 sites with mine dumps resulting of exploration works to be studied, only 10 had been studied, since 3 sites were not easy to reach and 1 site was not found. In total, the studied area was about 300 000 m^2. And the recorded values of background gamma radiation on the surface of radioactive mine dumps varied from 13-20 to 86-650 mR/h under specific effective activity from 94-116 up to 472-12712 Bq/kg [6]. The later demonstrates significant radiation intensity and high levels of radioactive contamination in anthropogenic zones which requires comprehensive analysis and detail study on the level of all major components in these frozen taiga landscapes.

PHYSICAL AND CHEMICAL PROPERTIES OF THE ROCK FROM THE MINE DUMPS

Based upon the content of basic oxides, the rock from the mine dumps belongs to the group of the basic rocks of sub-alkaline row of potassium-sodium or potassium lines according to the classification of magmatic rocks [2]. The average values of the main characteristics of studied rocks are as follows: silica content – 65.6%, total content of alkaline oxides – 8.1%, sodium-potassium ratio – 0.4. The average iron content is 5.8% with the variability within 3.7-8.1%, and the coefficient of alumina content is 1.9 and 1.5-2.7 accordingly, that allows us to classify them as highly- and very highly aluminous. Of all studied mine dumps (Table 1) rocks from the dump R1 of Russkaya site are particularly distinguished, as they contain the minimum amount of calcium oxide, sodium oxide and especially magnesium oxide, but high content of silica. We think that the reason is obvious. This mine dump is the oldest among all studied mine dumps, and the rock here, which is laying on the surface in the hypergenesis zone is the most weathered. Its composition is primarily the result of acid oxidizing humid weathering mainly involved in the process of ferciallitization, and in initial stage of formation of eluvial crust of weathering accompanied by the intensive debases i.e. by the loss of bases, exactly CaO, MgO and Na_2O and by the relative enrichment of rock with SiO_2 at the expense of accumulation of residual quartz.

Concerning the assessment of physical properties of mine dumps of the examined rocks one should point out that they are characterized by the high density of composition (the mean value of volume weight is about 2.7 g/cm^3), and, as a rule, by the high degree of stoniness and detritusness as well as by minor relative content (11-23%) of fine earth i.e. by the particles of less than 1 mm in size. However exactly the particles of fine earth are mostly exposed to water and windy dispersion (eolomotion), and consequently they finally define the levels and scales of eolian and hydrogenic migration of radionuclides in the studied anthropogenic landscapes. Therein the grain-size composition of fine earth of mine dumps (Table 2) is generally considered as sandy (mine dumps K13, K14, K16, K17, K18, M1, N11 and N13, R1 and H1), rarely as sandy-loam (dumps D2, R2, H21 and H22). And the heaviest fine earth with specific gravity (SG) of 2.84 g/cm^3 was found in the dump N13 of Neprokhodimy site in Severnoye deposit. As a whole the values of SG of fine earth of the dumps vary within the limits of magnitudes 2.58-2.84 g/cm^3, that makes 2.70 g/cm^3 on the average, i.e. at the level of average density of primary aluminum silicates.

In the five of the most radioactive rock dumps in Kurung-1 site of Yuzhnoye deposit 10 samples of the ground were taken from the surface layer (0-20 cm) for the next screen-mechanical analysis (Table 3) of the ratio of fractions of fine earth (< 1mm) and coarse earth (>1 mm). The following fractions of the mechanical elements of rock dumps have been determined: >10 mm, 10-7, 7-5, 5-3, 3-2, 2-1, 1-0.5, 0.5-0.25 and <0.25 mm. The results of analysis showed that there was the predominance of the particles of coarse earth in the composition of the surface layer of dumps that was totally 77.0-89.3%, fine earth however amounted only 10.7-23.0%, i.e. less a quarter by weight. Course earth mostly consists of stones i.e. the elements > 3 mm in size, their content changed from 60.7 to 75.7%. Among stones the great bulk were large stones of > 10 mm in size, they amounted 34.9-55.5% of the total mass of the analyzable rocks. Gravel fraction (3-1 mm) was 10.5-16.6%, and the fraction of large (1-0.5mm) and middle (0.5-0.25 mm) sand was totally 5.2-13.0%.

Table 1. Bulk composition of fine earth and the content of ^{238}U in samples taken from dumps of uranium mines in southern Yakutia

Deposit, site	Mine dump	The content of oxides (%) per calcinated sample (air-dry basis)										^{238}U, mg/kg
		SiO_2	Al_2O_3	Fe_2O_3	CaO	MgO	Na_2O_3	K_2O	TiO_2	P_2O_5	MnO	
Yuzhnoye, Kurung-1	K13	59.04	14.64	5.32	5.16	1.69	1.82	8.69	1.10	0.47	0.06	1680
Yuzhnoye, Kurung-1	K17	64.46	15.33	3.72	4.28	1.50	3.71	4.16	0.47	0.16	0.05	21
Yuzhnoye, Kurung-2	K23	62.40	14.62	7.35	4.04	1.22	3.04	4.29	0.59	0.28	0.08	9
Yuzhnoye, Russkaya	R1	68.94	12.59	4.47	1.06	0.11	0.44	9.70	0.94	0.54	0.35	110
Yuzhnoye, Russkaya	R2	65.92	11.59	5.10	4.52	0.97	0.70	7.71	0.67	0.39	0.10	294
Yuzhnoye, Russkaya	R3	72.12	12.99	5.74	8.32	2.31	3.59	2.34	0.72	0.29	0.05	6
Severnoye, Neprokhodimy-1	N12	60.82	16.64	7.74	2.49	0.88	1.33	8.11	0.72	0.38	0.14	153
Nevskoye, Drozhevoy	D4	64.05	14.87	4.43	4.50	1.61	3.07	4.71	0.66	0.22	0.05	125
Tayozhnoye, Holodnaya-1	H1	65.20	14.07	8.10	1.00	1.53	1.45	6.62	0.77	0.38	0.04	31
Tayozhnoye, Holodnaya-2	H21	75.96	11.64	3.76	1.18	0.05	0.53	5.41	0.46	0.13	0.06	83
Tayozhnoye, Holodnaya-2	H22	62.90	13.45	6.58	4.43	2.28	1.81	5.88	0.74	0.38	0.10	103
Tayozhnoye, Mineevskiy	M1	65.42	12.91	7.57	2.58	0.03	1.61	6.91	1.45	0.64	0.02	57

Along with the total ^{238}U its migratory- capable forms were defined in the composition of fine earth of 10 radioactive rock dumps in the three uranium deposits in central Aldan. The total ^{238}U was determined by the method of X-ray spectrum analysis, and the forms- by the method of laser-luminescent analysis. According to method [1] (Table 4) the following uranium forms have been revealed: 1) water-soluble – extragent (solvent) - destilled water; 2) exchangeable – ammonium acetate buffer solution pH 4.8; 3) acid-soluble – 1N HCl; 4) residual (nonextractive) obtained by the difference of total ^{238}U and its amount that was totally included in the composition of forms 1-3. When extracting forms 1-3 the ratio of solid and liquid phase 1:5 was used, and the exposure time was one day. In the examined rock dumps the total uranium content varied from 9 up to 940 mg/kg, herewith the content of water soluble form of this radionuclide was considerably changed from 0,04 up to 0,46 %, exchangeable - 7,2 -38,9 %, acid-soluble – 7,0-50,0 %, and the percentage of -+ uranium was from 10.7% to 84.8%. The results of analysis shows that according to the ratio of migratory-capable forms of ^{238}U the presented rocks of three deposits are very variable even within the limits of different mine dumps of the same deposit, and therewith the contrast increases, i.e. it increases with the age of dumps, making up 52.6-84.8% for Yuzhnoye deposit, 31.6-85.1% – for Severnoye, and 10.7-85.8% – for Tayozhnoye.

Table 2. Grain-size composition of fine earth and content of ^{238}U in the rock from radioactive mine dumps of uranium deposits in Southern Yakutia

Specific weight, g/cm^3	Fractions, %; size, mm 1-0.25	0.25-0.05	0.05-0.01	0.01-0.005	0.005-0.001	<0.001	<0.01	^{238}U, mg/kg
Yuzhnoye deposit/Kurung-1 site/dump K13								
2.73	48.4	36.4	10.6	4.1	0.4	0.1	4.6	1680
Kurung-1/dump K14								
2.69	45.9	39.7	9.0	3.7	1.2	0.5	5.4	308
Kurung-1/dump K16								
2.72	44.4	39.6	11.4	1.7	2.4	0.5	4.6	314
Kurung-1/dump K17								
2.75	45.4	30.9	17.6	0.4	3.3	2.4	6.1	21
Kurung-1/dump K18								
2.60	45.2	31.9	14.3	2.5	2.4	3.7	8.6	63
Russkaya/dump R1								
2.58	41.8	39.8	13.0	3.3	1.6	0.5	5.4	110
Russkaya/dump R2								
2.67	43.8	33.8	10.6	6.5	2.5	2.8	11.8	294
Severnoye deposit/Neprokhodimy site/dump N11								
2.71	54.0	31.2	10.2	2.5	1.6	0.5	4.6	230
Neprokhodimy/dump N13								
2.84	45.1	41.3	5.4	5.7	0.8	1.7	8.2	19
Nevskoye deposit/Drozhevoy site/dump D2								
2.77	39.6	34.7	15.1	3.3	4.0	3.3	10.6	363
Drozhevoy/dump D3								
2.73	38.2	37.3	15.9	3.7	1.2	3.7	8.6	49
Tayozhnoye deposit/holodnaya-1 site/dump H1								
2.67	37.8	38.9	15.5	4.0	2.1	1.7	7.8	31
Holodnaya-2/dump H21								
2.63	33.2	36.9	15.1	8.2	4.1	2.5	14.8	83
Holodnaya-2/dump H22								
2.70	38.9	32.9	17.6	3.7	4.1	2.8	10.6	103
Mineevskiy/dump M1								
2.66	37.6	41.9	13.5	4.1	2.8	0.1	7.0	57

Table 3. Mechanical composition of rocks of mine dump, % to air-dry basis

Particle size, mm											
>10	10-7	7-5	5-3	3-2	2-1	1-0.5	0.5-0.25	<0.25	>3	>1	<1
44.1	10.3	6.8	8.0	5.7	7.0	3.4	4.3	10.3	69.2	81.8	18.2
50.4	8.8	7.9	8.6	5.3	6.5	2.8	3.0	6.7	75.7	87.5	12.5
40.5	7.2	6.0	8.2	6.3	8.8	4.8	5.9	12.3	61.9	77.0	23.0
48.6	9.5	6.4	7.9	4.5	7.4	3.2	4.2	8.4	72.4	84.3	15.7
44.9	8.2	6.7	7.7	5.3	8.7	3.7	4.4	10.5	67.5	81.5	18.5
55.6	8.6	6.1	8.5	4.7	5.8	2.3	2.9	5.5	78.8	89.3	10.7
42.7	9.4	8.3	9.9	6.5	7.2	3.3	3.8	8.9	70.3	84.0	16.0
37.7	9.6	7.7	9.0	6.3	9.4	4.0	5.7	10.7	64.0	79.7	20.3
34.9	9.5	7.5	8.8	6.8	9.5	4.3	5.9	12.8	60.7	77.0	23.0
41.1	6.1	6.1	7.5	6.1	10.5	5.2	7.8	9.5	60.8	77.4	22.6

Table 4. The total content of ^{238}U and the ratio of its migratory forms in the rock from mine dumps of uranium deposits in southern Yakutia

#	Deposit	Site	Mine dump	Total content of ^{238}U, mg/kg
1	Yuzhnoye	Kurung-1	K16	38
2	Yuzhnoye	Kurung-1	K12	420
3	Yuzhnoye	Kurung-1	K11	940
4	Severnoye	Neprokhodimy	H21	60
5	Severnoye	Neprokhodimy	N12	280
6	Severnoye	Neprokhodimy	N23	400
7	Yuzhnoye	Russkaya	R1	338
8	Tayozhnoye	Holodnaya	H1	53
9	Tayozhnoye	Holodnaya	H2	149
10	Tayozhnoye	Mineevskiy	M1	9

#	Migratory forms*, % from total content				Residual content, %
	1	2	3	Total	
1	0.04	21.1	17.2	38.3	61.7
2	0.05	6.3	8.8	15.2	84.8
3	0.15	9.6	11.2	21.0	79.0
4	0.02	27.1	41.3	68.4	31.6
5	0.17	34.2	40.2	74.6	25.4
6	0.01	7.3	7.6	14.9	85.1
7	0.46	23.2	23.7	47.4	52.6
8	0.01	7.2	7.0	14.2	85.8
9	0.23	21.8	13.1	35.1	64.9
10	0.35	38.9	50.0	89.3	10.7

* 1 – water-soluble, 2 – exchangeable, 3 – acid-soluble.

Besides with the age of the examined dumps the total content of migratory- capable forms of ^{238}U increases, totally making 89.3% for the dump M1 of the Tayozhnoye deposit which began excavating in the early 60s of the last century. Thus one may state that the landscape-climatic conditions of hypergenesis zone, the location of the examined deposits, promote the active migration of ^{238}U in the process of weathering of rocks of mine dumps, however the total regularity of mosaic structure of radioactive contamination of the different components of landscapes in the anthropogenic zones applies either to the content or to the ratio of migratory-capable forms of this radionuclide.

Radiation Characteristics of Rocks from Mine Dumps

To characterize the sources of radioactive contamination we measured power of exposure dose of gamma radiation (PED), and the content and the effective specific activity (A_{eff}) of the natural radionuclides in fine earth of the mine dumps [9]. The power of exposure dose of gamma radiation was measured with radiation meter SRP-68-01 in the search mode near the surface of mine dumps with detection of anomalous points or by the definite observation network. The results of gamma radiation background measurements, which were carried out 1m above the surface of mine dumps were ambiguous. Sometimes at the distance of 1m above the surface there were higher values of PED of gamma radiation in contrast to the results recorded near the surface (0.1 m), and vice versa in other cases. Therefore we used the values recorded near the surface. Due to the geological peculiarities of the studied territory the natural radiation background of the anthropogenic-undisturbed sites of landscapes varied wide-ranging. The results of the route gamma survey revealed that on the bedrocks yield of Archaean intrusions (Alaskan granites, biotite -amphibolos, tourmaline, dykes of pyroxenites and some others) and on their debris the gamma background ranged from 15 up to 100 mR/h and more. The lowest radiation background was common to metamorphized (schists, gneisses and quartzites) and sedimentary (dolomites and dolomitic limestones) rocks. The level of gamma radiation background for these rocks usually ranges from 4 up to 16 mR/h. Since in the course of radioecological works the observable area was not completely covered with gamma-survey the data obtained on natural radiation background of rocks was approximate to a considerable degree.

For more detailed characterization of gamma radiation background of studied area the data [12] given in table 5 were used. In Aldan shield the level of background of gamma radiation varied in the wide range from 5 up to 70 mR/h, i.e. there were more than 15 fold increase from minimum to maximum level (Table 5). Herewith the mean values of background of gamma radiation in the studied area fluctuate within ranges of 5-50 mR/h, i.e. within the limits of one order of values.

The lowest mean values of PED (5-6 mR/h) were registered on the surface of the basic intrusive rocks (gabbro, diabases), and the maximum ones – 50 mR/h, as it was mentioned earlier, were typical to pegmatites particularly to alaskan granites. It should be also noted that the rocks of fedorovskaya suite exactly hypersthene - bipyroxene- diopside schists and gneisses as well as biotitic gneisses and schists are also characterized by maximum background of gamma radiation where the average limits of PED variation are 26-33 mR/h.

The data on background of gamma radiation of the studied anthropogenic landscapes are shown in Table 10, it was gathered from 38 mine dumps of the dormant mine workings (adits and mines). The data of the table show that background of gamma radiation on the mine dumps changes from 11 to 2150 mR/h. The most active mine dumps are located in the basin of the Kurung River in the sites with the conditional names Akin and Kurung. In the territory of the sites in some definite points the background of gamma radiation may reach 3000 mR/h. Taking into account that mean values background of gamma radiation on the surface of radioactive mine dumps in Elkon horst vary within the range of 11-896 mR/h (Table 6), and within the territory of Aldan shield – 5-50 mR/h accordingly, one should state that the values of PED in the mine dumps on average 2-18 times more than those in the condition of the regional background. It should be also noted that if the natural radiation background, measured in the open territory of Russia, generally varies within the range of 5-20 mR/h [7], whereas the radiation background registered on the surface of mine dumps and outside the limits was higher, so it may be considered to be the maximum.

Table 5. Background gamma radiation of rocks of Aldan shield [12]

Rock	Power of exposure dose, mR/h		
	Minimal	Average	Maximal
Upper Aldan suite			
Quartzites	7	13	17
Garnet-cordierite-sillimanite-biotite gneisses and schist	14	20	25
Biotite-garnet gneisses, garnet gneisses and schist	14	22	28
Amphibolous, biotite-amphibolous schist and Gneisses	8	19	28
Fedorovskaya suite			
Hypersthene- bipyroxene- diopside schist and gneisses	7	33	35
Amphibolous, biotite –amphibolous schist and gneisses	14	15	22
Biotite gneisses and schist	17	26	52
Diopside, diopside- scanolithic metasomatites	6	8	12
Intrusive rocks, archaean, proterozoic			
Biotite, leukocratoic granites	12	28	48
Alaskan granites, pegmatites	17	50	70
Diabases	4	5	6
Gabbro	5	6	8

As for the assessment of the regional radio-geochemical background within the limits of Aldan shield it should be noted that the minor contents of radioactive elements as a whole are common to metamorphic formations in Central Aldan. Herewith the highest concentrations of uranium and thorium are characteristic for different types of granitoids and mylonites, and the lowest concentrations were revealed in weakly granitized differences of gneisses and schist. Thus for schist of fedorovskaya series the average content of uranium is 3.5×10^{-4}% at the dispersion of $0.6\text{-}11\times10^{-4}$, the average content of thorium – 5×10^{-4} at the dispersion of 2.0-

8.3×10^{-4}. For gneisses of the same series the average content of uranium is 2.7×10^{-4} at the dispersion of $0.6\text{-}3.8\times10^{-4}$%, and thorium – 5×10^{-4} at dispersion of $2.0\text{-}22.6\times10^{-4}$% (Table 7).

Table 6. The levels of gamma radiation on the surface of studied mine dumps

#	Dump	Power of exposure dose, mR/h		
		Minimal	Average	Maximal
Akin site (A), Yuzhnoye deposit				
1	A1	210	896	2150
2	A2	180	580	1000
3	A3	104	322	540
4	A4	54	110	250
5	A5	62	100	210
6	A6	39	52	71
7	A7	34	43	180
Kurung-1 site (K1), Yuzhnoye deposit				
8	K11	660	882	1600
9	K12	200	508	740
10	K13	46	140	250
11	K14	46	74	100
12	K15	56	120	1000
13	K16	26	44	46
14	K17	36	72	110
15	K18	40	115	220
Kurung-2 site (K2), Yuzhnoye deposit				
16	K21	11	16	25
17	K22	27	35	40
18	K23	10	11	11
Drozhevoy site (D), Nevskoye deposit				
19	D1	50	96	120
20	D2	46	123	180
21	D3	52	59	66
22	D4	42	126	170
Neprokhodimy-1 site (N1), Severnoye deposit				
23	N11	21	152	320
24	N12	30	70	175
25	N13	34	167	420
Neprokhodimy-2 site (N2), Severnoye deposit				
26	N21	40	52	68
27	N22	52	109	280
28	N23	52	86	115
Russkaya site (R), Yuzhnoye deposit				
29	R1	72	119	157
30	R2	54	157	322
31	R3	22	25	28
Holodnaya site (H), Tayozhnoye deposit				
32	H1	41	57	76
33	H2	90	116	157
34	H3	142	228	330
Mineevskiy site (M), Tayozhnoye deposit				
35	M	28	97	157
Elkon site (E), Yuzhnoye Deposit				
36	E1	25	46	80
37	E2	23	29	40
38	E3	21	26	38

The high contents of these radioactive elements reflecting the effect of scanning process are characteristic for zones of iron- magnesium-calcium metasomatism of fedorovskaya series (phlogopite- amphibole-diopside rocks and magnesian scarns). Their concentration of uranium is 2.5-4.5×10^{-4}%, and thorium – 5-20×10^{-4}% [4].

In the process of ultrametamorphism the following stable tendency of increase of uranium content in the row of granitoid rocks was determined: charnockites – granulites, and gneiss-granites – Alaskan granites – quartz-feldspar metasomatites and blastomylonites (Table 7).

Table 7. The content of uranium and thorium in the rock from Elkon uranium-ore district, n×10^{-4}% [4]

Type of rock	U	Th	Th/U
Schist and amphibolites of Fedorovskaya series	3.5 0.6-11.0	5.0 2.0-8.3	1.43
Weaklygranitizied gneisses of different composition of fedorovskaya series	2.7 0.6-3.8	5.0 2.0-8.6	1.83
Amphibolites, amphibolitic gneisses, schists of fedorovskaya series	0.9	6.0	6.7
Biotite- garnet gneisses of fedorovskaya series	1.3	4.0	3.1
Gneisses of different composition of iengra series	2.2 0.8-7.2	11.0 2.0-22.5	5.4
Charnockite granites, granodiorites, plagiogranites	3.2 1.0-10.0	2.5 2.0-3.6	0.77
Microclinic granites, granodiorites, diorites	3.2 0.5-14.0	1.5 0.5-6.6	2.62
Alaskan granites, pegmatoids	5.2 0.6-14.0	61.7	11.85
Pegmatoid granites	4.8 0.7-13.0	134.0 1.9-475.0	27.90
Gray plagiogranites	3.2 1.0-10.0	2.5 2.0-3.6	0.80
Quartz diorites	4.9 1.0-16.0	12.7	2.60
Blastomylonites and mylonites in granitoids	5.8 0.6-18.5	14.8 2.0-40.0	2.52
Mylonites in gneisses, diorites and schists	3.1 0.6-6.7	10.0 1.2-25.9	3.27

Note: In numerator – average content of U and Th, in denominator – limits of U and Th variations.

In "grey" plagiogranites and pink granites the average content of uranium is stable – 3.2×10^{-4}% on average. Among granitoids the alaskan granites and pegmatites are rich in uranium (5.2×10^{-4}% on average). The concentration of thorium in granitoids (excluding grey plagiogranites) increases by 2-3 times as much in comparison with schist and gneisses (up to 62×10^{-4}% in alaskan granites).

Besides bright red alaskan granites of the second stage of granitization the pink mylonites, blastomylonites and metasomatic formations are also characterized by the anomalous content of uranium (up to 5.8×10^{-4}%) in the zones of the ancient fractures, that twice exceeds the Clarke value. The limits of fluctuations of uranium concentrations in blastomylonites on granitoids are rather high ($0.6\text{-}18.5\times10^{-4}$% and more), while the thorium content varies within the limits of $4.4\text{-}10\times10^{-4}$%. In the examined types of rocks however the correlation between uranium and thorium is not constant. For all types of rocks the thorium-uranium ratio varies from 1 to 10 on average. The most fluctuations of ratio of value are inherent in the different types of granitoids (Table 7).

The results of laboratory gamma spectrum analysis showed that the content of radionuclides in fine earth of rocks of mine dumps varied within wide limits: uranium (on radium) from 2 up to 0.1888%, thorium – $2\text{-}80\times10^{-4}$%, and potassium from 0.8 up to 4.5% (Table 8). At that the change of uranium concentration is especially variable, almost within the limits of three orders of magnitudes (a 1000 fold). It should be reminded that the limits of variations of the average uranium content in rocks of Elkon uranium-ore region within the range of the regional background (Table 7) is $0.9\text{-}5.8\times10^{-4}$% (there is a 6 fold difference, within the limits of one order of magnitude). Therefore in rocks from mine dumps the average uranium contents increase by 2-326 times as much. The limits of variation of thorium content in mine dumps of rocks ($2\text{-}80\text{x}10^{-4}$ or by 40 times) do not exceed the limits of fluctuations of its concentrations ($1.5\text{-}134.0\times10^{-4}$% or nearly by 90 times) in the rocks of the studied area within the limits of regional background (Table 7). It should be also pointed out that if in rocks of background landscapes the content ratio of Th:U, as a rule, is more than 1 (Table 7), then in rocks of mine dumps of anthropogenic landscapes in Elkon uranium-ore region (the content ratio) Th:U is practically always less than1. The exclusion in this respect are the rocks of only four of thirty eight mine dumps (mine dumps K21, N11, R3 and E3), where the uranium content is less 10×10^{-4}%.

Thus, thorium and potassium concentration in fine earth of mine dumps varies insignificantly, and at the average it corresponds to their level of Clarke value in rocks. At the same time the uranium content (radium) in rocks of mine dumps may exceed its Clarke values in rocks by several orders of magnitudes. The data obtained showed that gamma-radiation of uranium (radium) and its radioactive-decay products is the main source of gamma background on the surface of the mine dumps. Therefore in the mine dumps the higher uranium concentrations are common to the points with the highest magnitudes of gamma background (Table 8).

Besides measuring the background of gamma radiation and determining the radioactive element concentration in mine dumps, the effective specific activity of radionuclides was also estimated by the following formula accepted according to safety standards (SS-99):

$$A_{eff} = A_{Ra} + 1.3A_{Th} + 0.09A_{K},$$

where A_{Ra} and A_{Th} – specific activities of ^{236}Ra and ^{232}Th, which are balanced with the other components of uranium and thorium rows, A_k – specific activity of ^{40}K (Bq/kg).

The effective specific activity in fine earth of the studied mine dumps varies from 68 to 23640 Bq/kg. In the majority of radioactive ore mine dumps the percentage ratio of order 1-5% of the effective specific radionuclides activity is attributed to ^{40}K and ^{232}Th. As expected the highest values of A_{eff} are typical to the mine dumps located in the basin of Kurung river of Yuzhnoye deposit also characterized by the highest values of gamma radiation background.

Table 8. Content and effective specific activity of radionuclides in fine earth of rock of mine dumps in studied deposits

#	Dump	K %	U (Ra) 10^{-4}%	Th	A_{eff} Bq/kg
Akin site (A), Yuzhnoye deposit					
1	A1	Not detected	1637	59	20441
2	A1	Not det-ed	724	24	9029
3	A2	2.3	120	5	1537
4	A2	Not det-ed	906	25	11273
5	A3	2.3	285	13	3633
6	A3	3.0	322	15	4117
7	A3	2.2	140	8	1871
8	A4	1.4	191	12	2448
9	A5	1.1	81	8	1066
Kurung-1 site (K1), Yuzhnoye deposit					
10	K11	Not det-ed	205	2	2531
11	K11	1.4	695	8	8626
12	K11	Not det-ed	1888	80	23640
13	K12	2.4	290	19	3728
14	K12	3.9	596	16	7516
15	K12	0.8	776	18	9658
16	K13	2.8	45	8	668
17	K13	2.8	94	10	1298
18	K13	3.6	554	27	7049
19	K13	1.8	586	25	7384
20	K14	1.8	193	11	2477
21	K15	1.2	351	25	4478
22	K16	2.3	38	16	611
23	K17	1.3	16	3	245
24	K18	2.1	62	2	827
Kurung-2 site (K2), Yuzhnoye deposit					
25	K21	1.1	2	3	68
26	K22	1.1	3	3	77
27	K23	2.8	13	3	248
Drozhevoy site (D), Nevskoye deposit					
28	D2	1.6	513	17	6439
29	D3	2.9	21	2	345
30	D4	3.6	149	10	1979
Neprokhodimy-1 site (N1), Severnoye deposit					
31	N11	1.1	8	16	210
32	N12	1.1	267	17	3372
33	N13	3.4	75	5	1038
Neprokhodimy-2 site (N2), Severnoye deposit					
34	N21	4.5	63	9	939
35	N22	2.0	15	4	257
36	N23	Not det-ed	388	21	4881
Russkaya site (R), Yuzhnoye deposit					
37	R1	3.7	58	27	951
38	R2	3.7	338	22	4369
39	R3	2.0	3	10	697

Table 8. (Continued)

#	Dump	K %	U (Ra) 10^{-4}%	Th	A_{eff} Bq/kg
Holodnaya site (H), Tayozhnoye deposit					
40	H1	4.8	53	2	788
41	H2	3.8	47	12	741
42	H3	4.4	149	9	1995
Mineevskiy site (M), Tayozhnoye deposit					
43	M	4.3	93	11	1314
Elkon site (E), Yuzhnoye Deposit					
44	E1	1.7	208	8	2405
45	E2	3.9	34	7	569
46	E3	3.1	5	9	194

Ranking of the Anthropogenic Radionuclide Contamination on Assessment Criteria of Radioecological State of the Territories

At present there are no official documents approved by State Sanitary and Epidemiological Supervision of Russia on classification and radiation risk level for solid minerals of mined and dormant deposits. To estimate the radiation condition and anthropogenic radionuclide contamination of the studied area we used two methodological approaches that are the most accepted and used in solving such kind of issues today. In the first case the ranking of studied anthropogenic territories was based on mean values of power of exposure dose (PED) estimated for certain mine dumps (Table 9). Herewith the three radioecological criteria of assessment of the territory condition were presented: satisfactory - relatively safe (RS) with the value of the power of exposure dose (PED) up to 20 mR/h, emergency ecological state (EES) with the value of PED within the limits of 200-400 mR/h, and ecological disaster (ED) with the value of PED over 400 mR/h. According to this ranking of 38 studied mine dumps only 2 of them (or 5.3%) belonged to the category of RS; 31 (the major part or 81.6%) – to the category of EES, and 5 (or 13.2%) – to the category of ED.

For the recent years under the recommendations of BSRRS-99 [10] Khaikovich I.M. and co-authors [11] made the classification on magnitude Aeff for all industrial types of deposits of solid minerals on the degree of radiation hazard (Table 10).

According to this classification all the deposits are subdivided into four categories (classes):

(1) – specially dangerous, (2) – dangerous, (3) – potentially dangerous, and (4) – nonhazardous. On this classification 28.6% of all studied mine dumps belong to the 1st category, 25.7% and 40.0% – to the 2nd and 3rd categories respectively, and the rest 5.7% belong to the 4th category. Herewith 19 of 35 mine dumps (more than a half) (or 54.3%) are referred to the category of dangerous or specially dangerous based on the mean values of the effective specific activity of natural radionuclides (Table 11).

Since the improvement of the ecological situation in the studied territory is of great importance, we split the mine dumps into several groups in accordance with the necessity of conducting some special rehabilitation works (Table 12). Herewith the mean values of PED obtained earlier were also used as the basic criterion.

Table 9. Ranking of mine dumps of studied deposits on criteria of assessment of radioecological state of the territory

#	Site, Deposit	Radioecological state		
		Relative safety (RS)	Emergency ecological situation (EES)	Ecological disaster (ED)
1	Akin,Yuzhnoye	-	A3, A4, A5, A6, A7	A1, A2
2	Drozhevoy, Nevskoye	-	D1, D2, D3, D4	-
3	Kurung-1, Yuzhnoye	-	K13, K14, K15, K16, K17, K18	K11, K12
4	Kurung-2, Yuzhnoye	K21, K23	K22	-
5	Mineevskiy, Tayozhnoye	-	M1	-
6	Neprokhodimy-1, Severnoye	-	N11, N12, N13	-
7	Neprokhodimy-2, Severnoye	-	N21, N22, N23	-
8	Russkaya, Yuzhnoye	-	R1, R2, R3	-
9	Holodnaya, Tayozhnoye	-	H1, H2	H3
10	Elkon, Yuzhnoye	-	E1, E2, E3	-

Table 10. Limiting values of A_{eff} for classification of deposits of solid minerals on the degree of radiation hazard [11]

Category of deposit (type of danger)	A_{eff}, Bq/kg From	to
1. Specially dangerous	More than 3500	
2. Dangerous	1000	3500
3. Potentially dangerous	100	1000
4. Nonhazardous	Less than 100	

Table 11. Ranking of dumps of studied deposits on the degree of radiation hazard

#	Site, Deposit	Type of radiation hazard			
		Non-hazardous (NH)	Potentially dangerous (PD)	Danger ous(D)	Specially dangerous (SD)
1	Akin,Yuzhnoye	-	-	A3, A4, A5	A1, A2
2	Drozhevoy, Nevskoye	-	D3	D4	D2
3	Kurung-1, Yuzhnoye	-	K16, K17, K18	K14	K11, K12, K13, K15
4	Kurung-2, Yuzhnoye	K21, K22	K23	-	-
5	Mineevskiy, Tayozhnoye	-	-	M1	-
6	Neprokhodimy-1, Severnoye	-	N11	N13	N12
7	Neprokhodimy-2, Severnoye	-	N21, N22	-	N23
8	Russkaya, Yuzhnoye	-	R1, R3	-	R2
9	Holodnaya, Tayozhnoye	-	H1, H2	H3	-
10	Elkon, Yuzhnoye	-	E2, E3	E1	-

In accordance with recommended criteria the sites with PED values of less than 30 mR/h are considered to be relatively safe. The sites with PED values of 30-300 mR/h should be reclamated, and in the sites with PED of more than 300 mR/h radioactive wastes must be removed to the special burial place. According to these criteria 27 of 38 studied mine dumps (or 71.1%) should be reclamated, and rock mass of 6 mine dumps (15.8%) must be buried in the proper way.

Table 12. Ranking of mine dumps of studied deposits on criteria of need of rehabilitation works to be conducted

#	Site, Deposit	Requirements according to the criteria of radio-ecological state		
		Relatively Safe (RS)	Mine Reclamation (MR)	Dump burial (DB)
1	Akin,Yuzhnoye	-	A4, A5, A6, A7	A1, A2, A3
2	Drozhevoy, Nevskoye	-	D1, D2, D3, D4	-
3	Kurung-1, Yuzhnoye	-	K13, K14, K16, K17, K18	K11, K12, K15
4	Kurung-2, Yuzhnoye	K21, K23	K22	-
5	Mineevskiy, Tayozhnoye	-	M1	-
6	Neprokhodimy-1, Severnoye	-	N11, N12, N13	-
7	Neprokhodimy-2, Severnoye	-	N21, N22, N23	-
8	Russkaya, Yuzhnoye	R3	R1, R2	-
9	Holodnaya, Tayozhnoye	-	H1, H2, H3	-
10	Elkon, Yuzhnoye	E2, E3	E1	-

In conclusion of this section we should also note that 38 mine dumps in 17 anthropogenic sites, unequally distributed in the area of 500 km^2 of taiga forest in the region of central Aldan uranium deposits, are the major source of radioactive contamination of soil, vegetation cover and surface water. In the territory of Yuzhnoye deposit there are 21 mine dumps, 6 - at Severnoye, 4 -at Nevskoye, 3 - at Snezhnoye, 2- at Tayozhnoye, 1 - at Interesnoye and 1- at Nadezhninskoye. Examination of 21 mine dumps of the most developed Yuzhnoye deposit on the level of their radiation hazard showed that only 2 dumps (9.5%), based upon their radiation characteristics (PED – Power of Exposure Dose and A_{eff} – Effective Specific Activity) belong to the category of nonhazardous (NH), while 6 dumps (28.6%) – belong to the category of potential danger (PD), another 6 dumps (28.6%) were classified as dangerous (D), and the rest 7 mine dumps (33.3%) were placed to the category of specially dangerous (SD).

Thus, more than 90% of all studied mine dumps of Yuzhnoye deposit are dangerous in terms of radiation, herewith more than 60% of dumps belong to the category of dangerous and specially dangerous. Therefore it is no coincidence that in accordance with criteria (Table 12) only in the territory of Yuzhnoye deposit, where the most active mine dumps are located, six of them are recommended for burying.

It should be also mentioned that mine dumps of Akin site (A1 and A2) as well as Kurung-1 (K11, K12, K13) are ore or nearly ore mine dumps per se (Table 12). Due to the beginning of commercial development of these uranium deposits it is Yuzhnoye deposit that is to be launched first. We think that reprocessing of rock from these highly radioactive mine dumps by autoclave technology with simultaneous extraction of gold will be the most rational method of their utilization and at the same time it will favor the rehabilitation of the contaminated territory.

REFERENCES

[1] Arkhipov N.P., Fedorova T.A., Fevralyova L.T. Correlation of forms of compounds of heavy natural radionuclides in soil. *Pochvovedeniye 1,* 69-73 (1986).

[2] Bogatikov O.A., Gonyshakova V.I., Efremova S.V. et al. Classification and nomenclature of magmatic rocks. *Reference textbook. Nedra*, 160 (1980).

[3] Burtsev I.S., Stepanova S.K., Kolodeznikova E.N., Arkhipov N.D. The operational experience on survey of underground nuclear explosions and mine dumps of uranium-containing ore in the territory of Yakutia. *Radiation safety in the Republic of Sakha (Yakutia).* Publishing House SB RAS, 56-67 (2004).

[4] Goroshko M.V., Malyshev Yu.F., Kirillov V.E. Metallogeny of uranium in the Far East of Russia. *Nauka*, 372 (2006).

[5] Krupennikov V.A. Uranium (branneritic) deposits in the basement faults of epiarchaean Cratons. *Hydrothermal uranium deposits*, 128-138 (1978).

[6] Lozhnikov B.N. Survey of uranium-containing mine dumps of rocks in the territory of Yakutia. *Radiation safety in the Republic of Sakha (Yakutia)*. Publishing House SB RAS, 284-292 (2004).

[7] Margulis U.Ya., Bregadze Yu.I. Radiation safety. *Principles and safety provision.* Editorial URSS, 120 (2000).

[8] Naumov S.S., Shumilin M.V. Uranium deposits of Aldan region. *Otechestvennaya geologia* 11/12, 20-23 (1994).

[9] Radiation standards (RS-99): *Hygienic regulations.* The center of sanitary and epidemiological standardization, hygienic certification and examination of Health Ministry of the Russian Federation, 116 (1999).

[10] Basic sanitary regulations for radiation safety (BSRRS-99). *Ionizing radiation, radiation safety.* Health Ministry of the Russian Federation, 98 (2000).

[11] Khaikovich I.M., Mats N.A., Kharlamov M.G. Classification of mineral deposits according to the radiation hazard. *Regional Geology and Metallogeny* 8, 131-140 (1999).

[12] Chevychelov A.P., Sobakin P.I. *Migration of natural radionuclides in anthopogenic taiga-frozen landscapes in Southern Yakutia.* Publishing House SB RAS, 138 (2008).

In: Radioactive Waste: Sources, Types and Management ISBN: 978-1-62100-188-1
Editors: Satoshi Yuan and Wenxu Hidaka

Chapter 6

CEMENTITIOUS WASTEFORMS FOR IMMOBILIZATION OF LOW-ACTIVITY RADIOACTIVE WASTES

Dawn M. Wellman,[1] Chase C. Bovaird,[1] Shas V. Mattigod,[1] Kent E. Parker,[1] Libby N. Clayton,[2] Laura Powers,[3] and Marcus I. Wood[4]

[1]Pacific Northwest National Laboratory, Richland, WA, US
[2]BP Exploration, Anchorage, AK, US
[3]Wiss, Janney, Elstner Associates, Inc, Northbrook, IL, US
[4]CH2M HILL Plateau Remediation Company, Richland, WA, US

ABSTRACT

Solidification of low-activity wastes with cementitious materials is a widely accepted technique that contains and isolates waste from the hydrologic environment. The radionuclides ^{129}I, ^{75}Se, ^{99}Tc, and ^{238}U are identified as long-term dose contributors. The anionic nature of these radionuclides in aqueous solutions allows them to readily leach into the subsurface environment. Any failure of concrete encasement may result in water intrusion and consequent mobilization of radionuclides from the waste packages via mass flow and/or diffusion into the surrounding subsurface environment. In order to assess the long-term performance of waste grouts for encasement of radionuclides, the diffusion of radionuclide species must be measured when contacted with vadose zone porewater or groundwater. A series of sediment-concrete half-cell diffusion experiments were conducted under unsaturated conditions (4%, 7%, and 15% by weight moisture content) to quantify the diffusion of radionuclides from surrounding subsurface sediment into concrete wasteforms. The data from this investigation provides valuable information for understanding the speciation, behavior, and fate of radionuclides immobilized within concrete wasteforms under vadose zone conditions. Compilation of this data with previous data quantifying the diffusion of radionuclides from concrete wasteforms into the surrounding subsurface environment will allow quantification of the complete diffusion profile and predictive simulation of the long-term performance of concrete wasteforms and the fate of contaminants immobilized in cementitious waste forms.

INTRODUCTION

Solidification of low activity wastes with cementitious materials is a widely accepted technique, and has the advantages of readily accessible materials, low cost, high physical strength, and easy tailoring for various waste streams. Concrete encasement contains and isolates the waste from the hydrologic environment. However, any failure of concrete encasement may result in water intrusion and consequent mobilization of radionuclides from the waste packages.

The geochemistry of pore fluids in contact with cementitious materials is characterized by highly alkaline pH values. As such, anionic radionuclides identified as long-term dose contributors in low activity waste include ^{129}I, ^{75}Se, ^{99}Tc, and ^{238}U (Mann et al. 2001; Wood et al. 1995). Because of their anionic nature in aqueous solutions, ^{129}I, ^{75}Se, ^{99}Tc, and carbonate-complexed ^{238}U may readily leach into the subsurface environment (Serne et al. 1993; Serne et al. 1992; Serne et al. 1995; Serne et al. 1989) by mass flow and/or diffusion. Therefore, the leachability and/or diffusion of radionuclide species must be quantified to assess the long-term performance of concrete waste forms when contacted with vadose zone porewater or groundwater, and the ability of the surrounding sediment to retard radionuclide migration.

Although significant research has been conducted on the design and performance of cementitious wasteforms, one limitation in the current protocol (i.e. Toxicity Characteristic Leaching Procedure, Method 1311 Federal Registry, (*Toxicity Characteristic Leaching Procedure* 1992) and ANSI/ANS-16.1 leach test (ANSI 1986) is that most tests are conducted under water-saturated conditions. This testing protocol does not evaluate the performance of cementitious wasteforms within the context of waste depositories which are located in hydraulically unsaturated environments. This limitation has been recently recognized by a number of investigators who have conducted a limited number of studies under arid and semi-arid conditions (Sanchez et al. 2003; Sanchez et al. 2002; Al-Khayat et al. 2002; Garrabrants and Kosson 2003; Garrabrants et al. 2002; Garrabrants et al. 2004; Gervais et al. 2004).

Structurally, it has been noted that microcracks may form within the cement wasteform during carbonation (Campbell et al. 1991; Lange et al. 1996b; Lange et al. 1996a). Microcracks could serve as preferential flow paths for mass flow and/or diffusion of contaminants; however, there have been no known investigations conducted to this end. Exposure of cementitious waste forms to wet-dry cycles has also been show to afford significant changes to the carbonation of the cement wasteforms. Chemically, carbonation of cement wasteforms has been shown to inhibit the diffusion of cationic radionuclides through interaction with calcite formed during carbonation (Sanchez et al. 2003; Sanchez et al. 2002; Hartmenn et al. 1999; Garrabrants and Kosson 2003; Garrabrants et al. 2002; Garrabrants et al. 2004; Gervais et al. 2004; Hills et al. 1999; Lange et al. 1996b; Lange et al. 1996a; Smith and Walton 1991; Walton et al. 1997). However, the same is not true for anionic contaminants, which exhibit no inhibition and in some instances have displayed an increase in diffusion or leachability (Lange et al. 1996b; Lange et al. 1996a; Garrabrants and Kosson

2003; Garrabrants et al. 2002; Garrabrants et al. 2004; Gervais et al. 2004; Sanchez et al. 2003; Sanchez et al. 2002; Venhuis and Reardon 2001).

The results of recent investigations conducted under arid and semi-arid conditions provide valuable information suggesting structural and chemical changes to concrete waste forms which may affect contaminant containment and waste form performance. However, despite the significant advances in understanding the performance and behavior of cementitious wasteforms provided by wet-dry leaching studies, quantification of contaminant leaching is still conducted under water-saturated conditions. Results of such investigations may still provide a conservative estimate of *in situ* performance. To more accurately quantify the mass flow and/or diffusion of contaminants from cementitious wasteforms, it is critical to conduct diffusion investigations under water-unsaturated conditions.

Wellman et al. (2008) recently completed a set of sediment-concrete half-cell diffusion experiments under hydraulically unsaturated conditions. Results of this investigation suggested the diffusivity was a function of sediment moisture content, in that, a 3% increase in moisture content afforded one to two orders of magnitude increase in diffusivity. Moreover, the diffusion of iodine and rhenium from concrete waste forms was dependent on the treatment of the concrete monolith (i.e. with or without iron, and carbonated or non-carbonated). Release of I and Re increased 1-3 orders of magnitude from concrete monoliths that had been carbonated versus those which were not carbonated. Additionally, the inclusion of iron in non-carbonated monoliths reduced the diffusion of I and Re suggesting the presence of iron will reduce I and Re to reduced species, which are less soluble and more tightly retained within the concrete monolith. The most substantial release of I and Re was quantified from carbonated, iron-bearing monoliths. Petrographic analyses and scanning electron microscopy revealed enhanced formation of microcracks in the presence of iron, which provides preferential paths for contaminant migration. These results support previous hypotheses presented by Campbell and Lange et al., (1991; 1996b; 1996a) regarding the formation of microcracks serving as a means for preferential diffusion and migration of contaminants, and has provided critical clarifying information regarding the effects of carbonation on the diffusion of anionic contaminants. Furthermore, these results underscore the significance of quantifying the diffusion of radionuclides under relevant environmental conditions.

The objective of this investigation was to quantify the diffusion of 1) I and Re from sediment into concrete waste forms and 2) Tc from concrete waste forms into sediment. Both sets of half-cells were conducted as a function of concrete properties likely to influence radionuclide migration including carbonation and iron content. Concrete-sediment half-cell experiments were conducted using carbonated and non-carbonated concrete-sediment half cells under unsaturated conditions (4, 7, and 15% moisture content by weight). The results correlate the diffusion of I, Re, and Tc with sediment moisture content and concrete properties (i.e. carbonation and iron content). Compilation of this data with previous data, quantifying the diffusion of I and Re from concrete wasteforms into the surrounding subsurface environment (Wellman et al. 2008), will allow quantification of the complete diffusion profile and predictive simulation of the long-term performance of concrete wasteforms and the fate contaminants immobilized in cementitious waste forms.

Materials and Methods

Specified Concrete Composition for Encasement

The concrete composition for the burial encasement was specified in *Specification for Concrete Encasement for Contact-Handled Category 3 Waste.* ("Specification for Concrete Encasement for Contact-Handled Category 3 Waste 1998) [(a)] This specification was used as the basis to prepare a concrete for fabrication of test specimens. The composition includes sulfate-resistant Portland Type I or Type II cement, a pozzolanic material (Class F fly ash), fine and coarse aggregates, and steel fiber. Additional specifications include a water-to-cement ratio of 0.4 and an air content 6.0 ± 1.5%. The nominal proportions and material specifications based on this initial design are listed in Table 2.1.

Table 2.1. Material Specifications and Composition

Material	Specifications	Specified Field Mix	Normalized Specification Design
Cement	Portland Type I or Type I/II sulfate-resistant cement	381 kg/m3	0.27
Fly Ash	Class F fly ash; nominal 15% of cement by volume	54 kg/m3	0.04
Coarse Aggregate	No. 676 or equivalent (3/4" nominal size)	55% by volume	0.04
Fine Aggregate	Sand	45% by volume	0.51
Water	Nominal water:cement ratio: 0.4	399 kg/m3	0.10
Steel Fiber	Deformed Type I, nominal length 2.5 – 3.8 cm (1 – 1.5")	59 kg/m3	0.04
Air Content		6.0±1.5%	

Materials and Laboratory-Scale Mixture Design

A laboratory-scale concrete mixture (Table 2.2) was prepared based on specifications shown in Table 2.1. Because of the required small dimensions of laboratory test specimens, the size of the coarse aggregate and the dimensions of the steel fiber specified in Table 2.1 were proportionately reduced. This was accomplished by decreasing the 2-cm (~0.75 in.) coarse aggregate size in the original specification to a particle size ranging from 2.83 mm to 2 mm in the laboratory mix. Aggregate passing a 7-mesh sieve and retained on a 10-mesh sieve met this particle size specification. Iron particles were used in the laboratory mix in place of the steel fibers. Based on these modifications, a concrete mix was prepared that consisted of Portland Cement (Type I and II, American Society for Testing and Materials C-150 compliant), Class F fly ash, scaled-down coarse aggregate, fine aggregate, iron particles, and a water-entraining agent (Polyheed 997). The water-entraining agent was included in the mix to facilitate the workability of the concrete. The volumes of the Polyheed 997 and the air-entraining agent, MB AE 90, were not included in the normalization calculations because of their negligible contribution to the overall mix volume. The material specification and composition for the laboratory-scale concrete mixture is given in Table 2.2.

Table 2.2. Laboratory-scale Material Specification and Composition

Material	Material Specifications for Field Mix	Normalized Laboratory Design	Material Specifications Used in Revised Laboratory Mix Comparison
Cement	Portland Type I or Type I/II sulfate-resistant cement	0.27	Portland Type I and II
Fly Ash	Class F fly ash; nominal 15% of cement by volume	0.04	Class F fly ash; nominal 20% of cement by volume
Coarse Aggregate	No. 676 or equivalent (3/4–in. nominal size)	0.04	Sieve size +7 to -10 (2.83 – 2 mm size)
Fine Aggregate	Sand	0.51	Sand -10 sieve size (< 2 mm)
Water	Nominal water:cement ratio: 0.4	0.10	Water-to-cement ratio: 0.5
Iron Powder	Iron particles	0.04	-200 mesh
Polyheed 997		0.00375	Water-entraining agent
Air Content	6.0±1.5%	6.0±1.5%	--

CONCRETE MIX AND SPECIMEN PREPARATION

In general, concrete monoliths were prepared with mix components added in the order: water, steel (if applicable), coarse aggregate, fine aggregate, fly ash, cement, Polyheed 997, and MB AE 90. The concrete was mixed on medium speed using a Hobart three-speed, bench-top mixer in a 4 L steel bowl. The concrete molds for casting specimens were fabricated from Schedule 40 polyvinyl chloride (PVC) piping material. Gaskets were glued to the bottom of the molds and leak tested before use. The PVC molds were filled in the vertical position. After filling, the molds were lightly tapped on the laboratory bench until a significant decrease in the release of air bubbles was observed. The forms were stored in a humidity chamber for 28 days to provide moisture while the concrete set. The concrete monoliths were subsequently removed from the molds and the respective monoliths were carbonated by soaking for 7 days in a saturated sodium-bicarbonate solution. Specific treatments for concrete monoliths (e.g. carbonation, iron content, contaminant content) used in concrete-sediment half-cell experiments are discussed below within the context of the respective half-cell preparation.

Solid and Fractured Concrete-Sediment Half-Cells Spiked with Iodine and Rhenium

Concrete monoliths were prepared in two separate batches based on the laboratory-scale specifications for the concrete (Table 2.2): one batch contained iron particles, the other batch did not contain any added iron. Within the two batches of concrete, with iron and without iron, half the monoliths were carbonated by submersion in a saturated sodium bicarbonate solution for seven days. The resulting compositions consisted of four concrete specimens: 1) no iron, carbonated, 2) no iron, not carbonated, 3) iron, carbonated, and 4) iron, not carbonated. The characteristics of the solid concrete monoliths and concrete-Sediment diffusion half cells are listed in Table 2.3.

In order to assess the effect of fractures within the concrete monolith on the diffusion of iodine and rhenium concrete monoliths were encased in shrink-wrap and struck with a hammer to prevent the formation of rubble. Each fractured core possessed a single fracture extending the length of the core, perpendicular to the concrete-sediment interface. The characteristics of the specimens used are listed in Table 2.4.

SOLID CONCRETE-SEDIMENT HALF-CELLS SPIKED WITH TECHNETIUM

The concrete half cells were prepared using the preparation detailed above. Two separate batches, one with and one without iron particles, were prepared based on the laboratory-scale specifications for the concrete (Table 2.2). None of the concrete monoliths were subjected to carbonation because the process of carbonating the cores via submersion in a saturated sodium bicarbonate solution would result in leaching of Tc prior to half-cell testing. The resulting compositions consisted of two concrete specimens: 1) no iron, not carbonated and 2) iron, not carbonated. The characteristics of the solid concrete monoliths and concrete-sediment diffusion half cells are listed in Table 2.5.

Half-Cell Preparation, Testing, and Analysis

Diffusion tests were conducted to assess the effects of carbonation and the inclusion of colloidal iron on the rate of diffusion of key, long-lived, mobile contaminants (I, Re, and Tc), in unsaturated Hanford sediments (~4, 7, and 15% moisture content by weight). The experiments were conducted using a sediment half cell ~4 cm in diameter and 20-cm long in contact with a concrete monolith ~4 cm in diameter and ~4-cm long. One concrete core of each composition was encased in a Schedule 40, 14 cm long, PVC pipe. The remaining volume of each pipe was filled with Hanford Site sediment, spiked with stable I and Re or Tc added to the water component prior to mixing with the sediment, at the respective moisture content. Medium coarse sand obtained from the sidewall of the W-5 burial ground located on the Hanford Site in Southeastern Washington State was used for the sediment half cell. The physical, chemical and mineralogical properties of this sediment have been previously characterized by Serne *et al.* (1993).

Table 2.3. Characteristics of the Solid Concrete-Sediment Half Cells used to Assess the Effects of Concrete Iron Content and Carbonation on the Diffusion of I and Re from Sediment to Concrete

Cell No.	Concrete Treatment	Length Concrete Half Cell (cm)	Diameter (cm)	Surface Area (cm^2)	Volume of Concrete Half-Cell (cm^3)	Density Concrete (g/cm^3)	Volume of Sediment Half-Cell (cm^3)	Volume of Half-Cell (cm^3)	Density Sediment (g/cm^3)	Sediment Moisture Content (%)	Test Duration (days)
C-4-4-3	Iron, Carbonated	4.21	4.32	86.26	61.5	2.21	239.26	300.76	1.58	4	348
NC-4-4-3	Iron, Not Carbonated	4.33	4.32	87.99	63.37	2.18	237.39	300.76	1.56	4	348
C-4-0-3	No Iron, Carbonated	4.26	4.32	87.23	62.55	2.08	238.209	300.76	1.58	4	348
NC-4-0-3	No Iron, Not Carbonate	4.39	4.33	89.09	64.55	2.04	236.206	300.76	1.56	4	348
C-7-4-3	Iron, Carbonated	4.34	4.32	88.16	63.55	2.23	237.212	300.76	1.61	7.1	354
NC-7-4-3	Iron, Not Carbonated	4.26	4.33	87.22	62.54	2.21	238.22	300.76	1.62	7.1	354
C-7-0-3	No Iron, Carbonated	4.15	4.32	85.69	60.89	2.1	239.876	300.76	1.53	7.1	354
NC-7-0-3	No Iron, Not Carbonate	4.38	4.32	88.72	64.16	2.06	236.599	300.76	1.56	7.1	354
C-15-4-3	Iron, Carbonated	4.27	4.32	87.15	62.46	2.23	238.297	300.76	1.72	15	355
NC-15-4-3	Iron, Not Carbonated	4.27	4.32	87.39	62.73	2.19	238.035	300.76	1.7	15	355
C-15-0-3	No Iron, Carbonated	4.15	4.33	85.76	60.95	2.08	239.806	300.76	1.72	15	355
NC-15-0-3	No Iron, Not Carbonate	4.43	4.32	89.58	65.09	2.06	235.675	300.76	1.7	15	355

Table 2.4. Characteristics of the Fractured Concrete-Sediment Half Cells used to Assess the Effects of Concrete Iron Content, Carbonation, and Fractures on the Diffusion of I and Re from Sediment to Concrete

Specimen No.	Concrete Treatment	Length (cm)	Diameter (cm)	Surface Area (cm^2)	Volume (cm^3)	Density Concrete (g/cm^3)	Volume of Sediment Half-Cell (cm^3)	Volume of Half-Cell (cm^3)	Density Sedimen t (g/cm^3)	Sediment Moisture Content (%)	Test Duration (days)
NC-4-0-4	No Iron, No Carbonation	4.15	4.33	85.89	61.10	2.09	236.7	297.79	0.91	4	698
NC-4-4-4	Iron, No Carbonation	3.89	4.34	82.44	57.35	2.2	243.2	300.50	0.92	4	698
C-4-0-4	No Iron, Carbonation	4.15	4.33	86.04	61.25	2.12	236.7	297.94	0.93	4	698
C-4-4-4	Iron, Carbonation	4.02	4.33	84.04	59.10	2.2	239.9	298.98	0.96	4	698
NC-7-0-4	No Iron, No Carbonation	4.24	4.33	87.02	62.32	2.09	234.7	296.99	0.92	7	698
NC-7-4-4	Iron, No Carbonation	3.98	4.34	83.75	58.77	2.2	240.8	299.61	0.92	7	698
C-7-0-4	No Iron, Carbonation	4.18	4.34	86.63	61.89	2.13	236.0	297.93	0.92	7	698
C-7-4-4	Iron, Carbonation	4.12	4.34	85.68	60.87	2.2	237.4	298.28	0.96	7	698
NC-15-0-4	No Iron, No Carbonation	4.37	4.34	89.05	64.51	2.10	231.6	296.10	0.82	15	698
NC-15-4-4	Iron, No Carbonation	3.95	4.34	83.31	58.29	2.2	241.7	299.98	0.78	15	698
C-15-0-4	No Iron, Carbonation	4.76	4.34	94.54	70.48	2.13	221.1	292.57	0.90	15	698
C-15-4-4	Iron, Carbonation	4.52	4.34	91.26	66.91	2.2	228.0	294.87	0.87	15	698

Table 2.5. Characteristics of the Solid Concrete-Sediment Half Cells used to Assess the Effects of Concrete Iron Content on the Diffusion of Tc from Concrete to Sediment

Specimen No.	Concrete Treatment	Length (cm)	Diameter (cm)	Surface Area (cm^2)	Volume (cm^3)	Density Concrete (g/cm^3)	Volume of Sediment Half-Cell (cm^3)	Volume of Half-Cell (cm^3)	Density Sediment (g/cm^3)	Sediment Moisture Content (%)	Test Duration (days)
NC-4-0-5	0% Fe	4.17	4.32	85.84	61.05	2.04	236.32	297.37	0.63	4	699
NC-4-4-5	4% Fe	4.26	4.31	86.90	62.20	2.23	234.14	296.33	0.63	4	699
NC-7-0-5	0% Fe	4.34	4.31	88.01	63.39	2.07	232.14	295.53	0.62	7	699
NC-7-4-5	4% Fe	4.23	4.31	86.51	61.77	2.23	234.95	296.72	0.62	7	699
NC-15-0-5	0% Fe	4.39	4.32	88.74	64.18	2.08	231.04	295.22	0.58	15	699
NC-15-4-5	4% Fe	4.37	4.31	88.41	63.82	2.23	231.56	295.38	0.58	15	699

The ends of each pipe were machined and fit with o-ring gaskets to ensure the test cells were sealed. The diffusion cells were allowed to set horizontally and undisturbed for the test duration, nominally 2 years, with periodic rotation of the cell by 90 degrees.

At the conclusion of the test period, the end caps of the cells were removed and the sediment was extruded at approximately 1 cm intervals along the length of the half cell Figure 2.1. The moisture content of each section was quantified found to be uniform throughout the length of the half-cell and consistent with the starting moisture content. The sediment samples were weighed and extracted with deionized water. One-to-one water extracts were done on sediment fractions. The concentrations of iodine rhenium and technetium were measured via inductively coupled plasma-mass spectrometry (ICP-MS).

Concrete half cells were sectioned parallel to the concrete-sediment interface using a Buehler slow-speed saw fitted with a diamond blade. Concrete cores used in half-cells to assess the effects of fractures in concrete on the diffusion of I and Re were sectioned such that one half of the concrete half-cell was sliced parallel to the concrete-sediment interface and the other half of the concrete half-cell was sliced parallel to the fracture (perpendicular to the concrete-sediment interface). Cutting was done without water cooling to prevent the leaching of soluble I, Re, and Tc during the cutting process. The concrete slices were then ground using an agate mortar and pestle. One-to-one (by mass) water extracts were done on both sediment and concrete fractions. The concentrations of I, Re, and Tc were measured via ICP-MS in sediment and concrete thin slices.

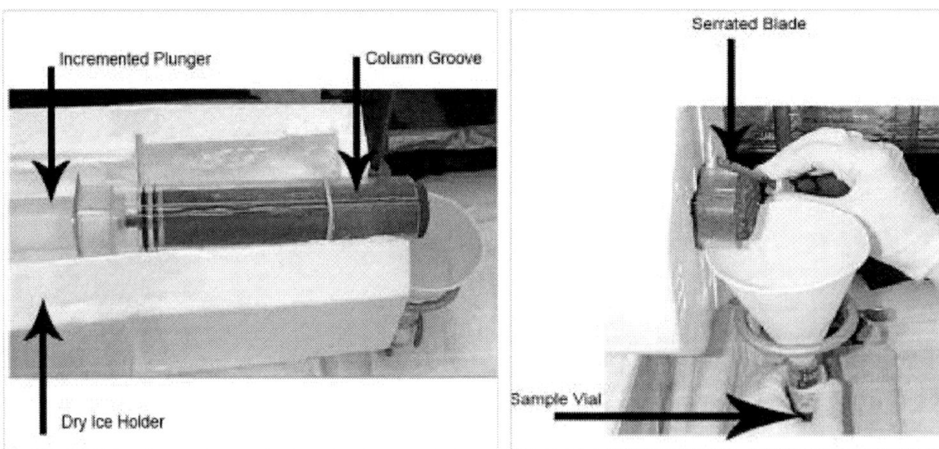

Figure 2.1. a) Mechanism for Sediment Sampling from a Concrete-Sediment Diffusion Half Cell. b) Sediment Sampling from a Diffusion Half Cell.

RESULTS AND DISCUSSION

Concrete-Sediment Half-Cell Experiments to Determine the Diffusion of Iodine and Rhenium into Solid Concrete

Figure 3.1 and Table 3.1 present the concentration profiles for I and Re within the sediment half of concrete-sediment half-cell diffusion experiments. The concentration of Re decreased as a function of increasing moisture content within the sediment half-cell regardless of iron or carbonate treatment of the concrete monolith. Comparable behavior was previously observed in concrete-sediment half-cell experiments wherein diffusion of I and Re from concrete to sediment was quantified as a function of moisture content (Mattigod et al. 2001; Wellman et al. 2006; 2008).

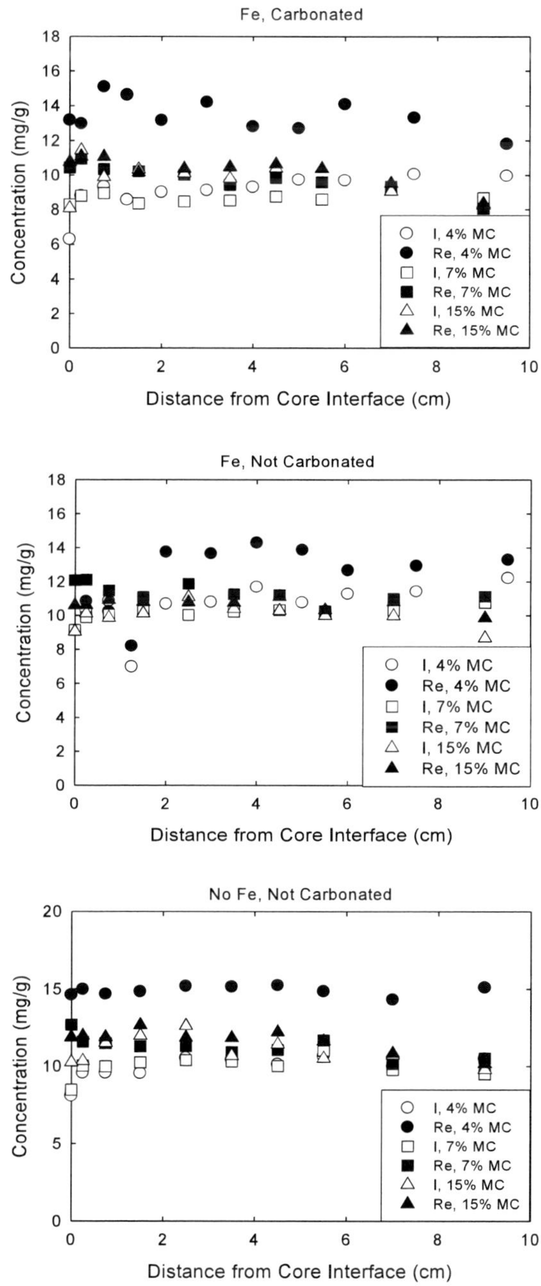

Figure 3.1. Concentration Profiles for Sediment Half-Cell I and Re Profiles at 4, 7, and 15% Moisture Content by Weight in a) Iron, Carbonated, b) Iron, Not Carbonated, c) No Iron, Carbonated, and d) No Iron, not Carbonated Concrete Monoliths.

A similar trend for I was not evident in results of half-cell diffusion tests presented here. Figure 3.2 presents the mass balance in sediment and concrete half-cells. The fraction of I within the sediment half-cell consistently ranged from 0.6 to 0.8 irrespective of moisture content. The relative fraction of I remaining within the sediment decreases as a function of moisture content from 7% to 15% moisture content. However, the fraction of I in sediment at 4% moisture content is highly variable and significantly less than that quantified for Re (Figure 3.2). Upon extraction of the half cells, it was noted that the inside walls of the PVC pipe were reddish-brown. It is postulated that vapor-phase diffusion of I onto the walls of the PVC pipe may have occurred at 4% moisture content. There was very minimal discoloration of the PVC walls noted at 7% and 15% moisture content. To determine whether vapor-phase diffusion of I had occurred, 1:1 isopropyl alcohol extractions were conducted on slices of the PVC pipe. 200 to 850 ppb I was extracted from various segments of the PVC pipes, whereas no measurable concentrations of Re were detected in the extracts. It appears that under higher moisture contents, I may remain in the aqueous phase; whereas, at very low moisture content (4%), it tended to partially transform into vapor phase.

Diffusion profiles suggest there was no quantifiable effect of iron or carbonation on the diffusion of I and Re from the sediment to concrete half-cell (Figure 3.1). However, the overall mass loss in the sediment half-cells shows that carbonated concrete half-cells, regardless of Fe content, have the most significant diffusion of Re into the concrete half-cell (Figure 3.2).

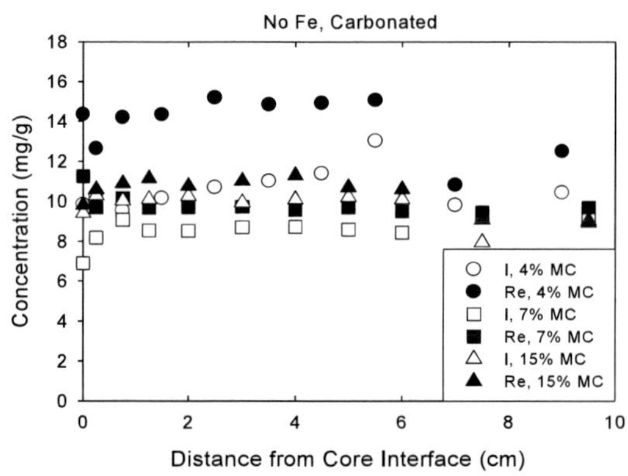

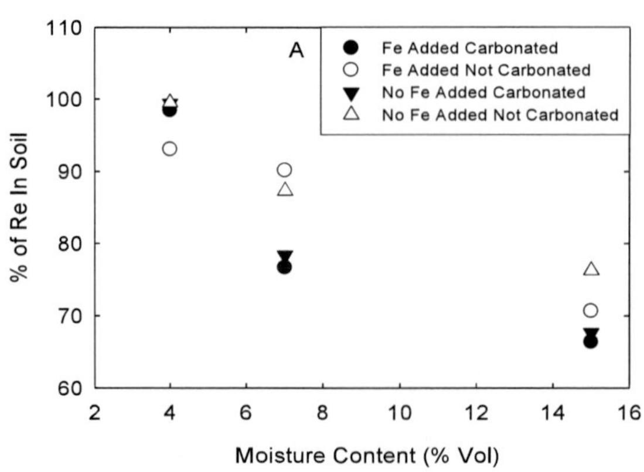

Figure 3.2. (continued).

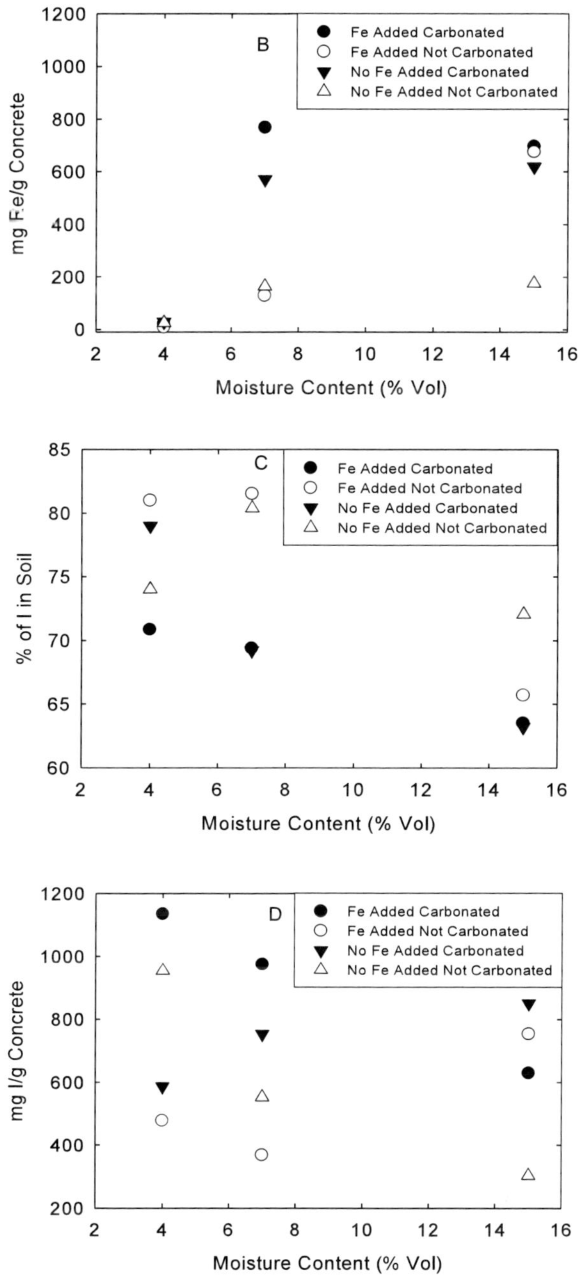

Figure 3.2. Mass Balance for Sediment Half-Cell I and Re Profiles at 4, 7, and 15% Moisture Content by Weight in a) Fraction of Re Remaining in Sediment, b) Measured Mass of Re Diffused into Concrete, c) Fraction of I Remaining in Sediment, and d) Measured Mass of I Diffused into Concrete.

Vapor-phase diffusion resulted in greater variability of mass balance results for I than exhibited for Re. Therefore, mass balance results for I at 4% sediment moisture content do not exhibit the same correlation of greater diffusion in carbonated monoliths. However, the diffusion of I in carbonated concrete monoliths is greater than or equal to, within analytical error, which exhibited in non-carbonated half-cells at 7% and 15% sediment moisture content (Figure 3.2).

Table 3.1. Concentration of Iodine and Rhenium (mg/g) for Sediment Half-Cell Profiles at 4%, 7%, and 15% Moisture Content by Weight

4% Fe-C			4% Fe-NC			4% No Fe-C			4% No Fe-NC		
Distance from Core, cm	I, mg/g	Re mg/g	Distance from Core, cm	I, mg/g	Re mg/g	Distance from Core, cm	I, mg/g	Re mg/g	Distance from Core, cm	I, mg/g	Re mg/g
9.5	9.9	11.8	9.5	12.2	13.3	9	10.4	12.5	9	10.5	15.1
7.5	10.0	13.3	7.5	11.4	12.9	7	9.8	10.8	7	10.0	14.3
6	9.7	14.1	6	11.3	12.7	5.5	13.0	15.0	5.5	10.6	14.8
5	9.7	12.7	5	10.8	13.9	4.5	11.4	14.9	4.5	10.1	15.2
4	9.3	12.8	4	11.7	14.3	3.5	11.0	14.8	3.5	10.4	15.1
3	9.1	14.2	3	10.8	13.6	2.5	10.7	15.2	2.5	10.5	15.2
2	9.0	13.2	2	10.7	13.7	1.5	10.1	14.3	1.5	9.5	14.8
1.25	8.6	14.6	1.25	7.0	8.2	0.75	9.5	14.2	0.75	9.6	14.7
0.75	9.3	15.1	0.75	10.8	10.2	0.25	10.3	12.6	0.25	9.6	15.0
0.25	8.8	13.0	0.25	10.3	10.8	0	9.8	14.3	0	8.1	14.6
7% Fe-C			7% Fe-NC			7% No Fe-C			7% No Fe-NC		
Distance from Core, cm	I, mg/g	Re mg/g	Distance from Core, cm	I, mg/g	Re mg/g	Distance from Core, cm	I, mg/g	Re mg/g	Distance from Core, cm	I, mg/g	Re mg/g
9	8.7	8.1	9	10.8	11.1	9.5	9.1	9.7	9	9.5	10.5
7	9.2	9.3	7	10.3	11.0	7.5	9.3	9.4	7	9.8	10.2
5.5	8.6	9.6	5.5	10.3	10.2	6	8.4	9.5	5.5	11.0	11.7
4.5	8.8	9.8	4.5	10.3	11.2	5	8.6	9.7	4.5	10.0	11.1
3.5	8.5	9.4	3.5	10.2	11.3	4	8.7	9.6	3.5	10.3	10.9
2.5	8.5	10.0	2.5	10.0	11.9	3	8.7	9.7	2.5	10.4	11.3
1.5	8.4	10.2	1.5	10.3	11.1	2	8.5	9.7	1.5	10.2	11.3
0.75	8.9	10.4	0.75	10.0	11.5	1.25	8.5	9.7	0.75	10.0	11.5
0.25	8.8	11.0	0.25	9.9	12.1	0.75	9.1	10.2	0.25	10.0	11.6
0	8.3	10.4	0	9.1	12.1	0.25	8.2	9.7	0	8.5	12.7
						0	6.9	11.3			
15% Fe-C			15% Fe-NC			15% No Fe-C			15% No Fe-NC		
Distance from Core,cm	I, mg/g	Re mg/g	Distance from Core, cm	I, mg/g	Re mg/g	Distance from Core, cm	I, mg/g	Re mg/g	Distance from Core, cm	I, mg/g	Re mg/g
9	8.4	8.3	9	8.7	9.9	9.5	9.0	8.9	9	9.8	10.2
7	9.1	9.5	7	10.0	10.8	7.5	8.0	9.1	7	10.7	10.8
5.5	10.4	10.4	5.5	10.0	10.3	6	10.1	10.6	5.5	10.5	11.6
4.5	10.4	10.6	4.5	10.3	11.1	5	10.2	10.7	4.5	11.4	12.2
3.5	9.8	10.5	3.5	10.4	10.8	4	10.1	11.3	3.5	10.7	11.8
2.5	10.1	10.4	2.5	11.1	10.8	3	9.9	11.0	2.5	12.6	11.9
1.5	10.3	10.2	1.5	10.1	10.8	2	10.3	10.8	1.5	12.0	12.6
0.75	9.9	11.1	0.75	9.9	10.9	1.25	10.1	11.2	0.75	11.5	11.9
0.25	11.4	11.1	0.25	10.1	10.6	0.75	10.0	10.9	0.25	10.3	12.0
0	8.1	10.8	0	9.1	10.6	0.25	10.3	10.6	0	10.3	11.9
						0	9.4	9.8			

Figure 3.3 and Table 3.2 displays plots showing the concentration profile for I and Re within the concrete half of the concrete-sediment half-cell diffusion experiments at 4% moisture content. The behavior of I and Re were markedly different within a given half-cell test. The diffusion of I into concrete was 40 to 60 mg I/g concrete within the first 5 mm of the concrete half-cell in contact with sediment. However, the diffusion of Re was limited to < 5 mg Re/g concrete within the same depth of concrete. As noted above, it is hypothesized that

vapor-phase diffusion of I may be occurring under conditions of 4% moisture content, which could afford greater diffusion of I into the concrete core. Additionally, the peak concentration of I at 4% moisture content in the concrete core has moved away from the interface. Concentration profiles suggest that diffusion of iodine and rhenium are influenced primarily by the effects of carbonation on the concrete monoliths and secondarily by iron content, diffusion rates are greatest in carbonate iron-bearing sample > carbonate, iron-free sample > non-carbonated, iron > non-carbonated, iron-free sample. Also, the slight elevation in the concentration of Re at the interface of concrete monoliths not containing iron suggests iron may exert a limited influence on the diffusion of Re.

Figure 3.4, Table 3.2 displays the concentration profiles for I and Re within the concrete half of the concrete-sediment half-cell diffusion experiments at 7% and 15% moisture content. For all monolith treatments (i.e. with/without iron and carbonated/non-carbonated), the greatest concentration of I and Re were measured within the first 10 mm of the concrete monolith (~10-20 mg/g) followed by a gradual decrease in concentrations of both elements over the remaining 20 to 30 mm of concrete monolith. The noted exception to this was the concrete monoliths that contained no iron and were not carbonated. This set of concrete monoliths displayed a more rapid decrease in concentration within the first 10 to 20 mm. In the concrete monoliths without iron and not carbonated, the diffusion of I appears to be greater at 7% than that observed at 15%. Conversely, in the concrete monolith containing iron, but not carbonated, the diffusion of iodine is noticeably greater at 15% than that at 7%. This behavior is also evident for Re. Diffusion of I and Re in carbonated monoliths do not exhibit such deviations; rather the diffusion of I and Re in carbonated monoliths are comparable, irrespective of iron content.

As demonstrated previously, the inclusion of iron and the effect of carbonation result in the formation of microcracks within the concrete monoliths (Wellman et al. 2008) that apparently facilitate diffusion. The absence of iron in the uncarbonated core likely limited the potential formation of any microcracks, thus limiting diffusion of I and Re into the concrete monolith from the contaminated sediment. Increased micro-cracking in the presence of Fe and for concrete monoliths that had undergone carbonation agrees with results of previous investigations (Wellman et al. 2006).

The concentrations of I and Re quantified within the monoliths at 7% and 15% sediment moisture content were much lower than the interfacial concentration measured at 4% moisture content (40-60 mg/g at 4% moisture content versus 10-20 mg/g at 7% and 15% moisture content). Furthermore, the depth of diffusion of I and Re diffusion into the concrete monoliths was greater at 7% and 15% Sediment moisture content than at 4% moisture content. At 4% moisture content, the concentration of I was greatest within the first 5 mm and displayed a rapid asymptotic decrease in concentration to ~1 mg/g within the first 10 mm. The behavior of I and Re at 7% and 15% moisture content was generally greatest within the first 10 mm, with a more gradual decrease over the remaining 20 to 30 mm. This accords with previous results, suggesting a direct relationship between diffusion within concrete monoliths and the degree of moisture content (Wellman et al. 2006; Wellman et al. 2007; Mattigod et al. 2001; Wellman et al. 2008).

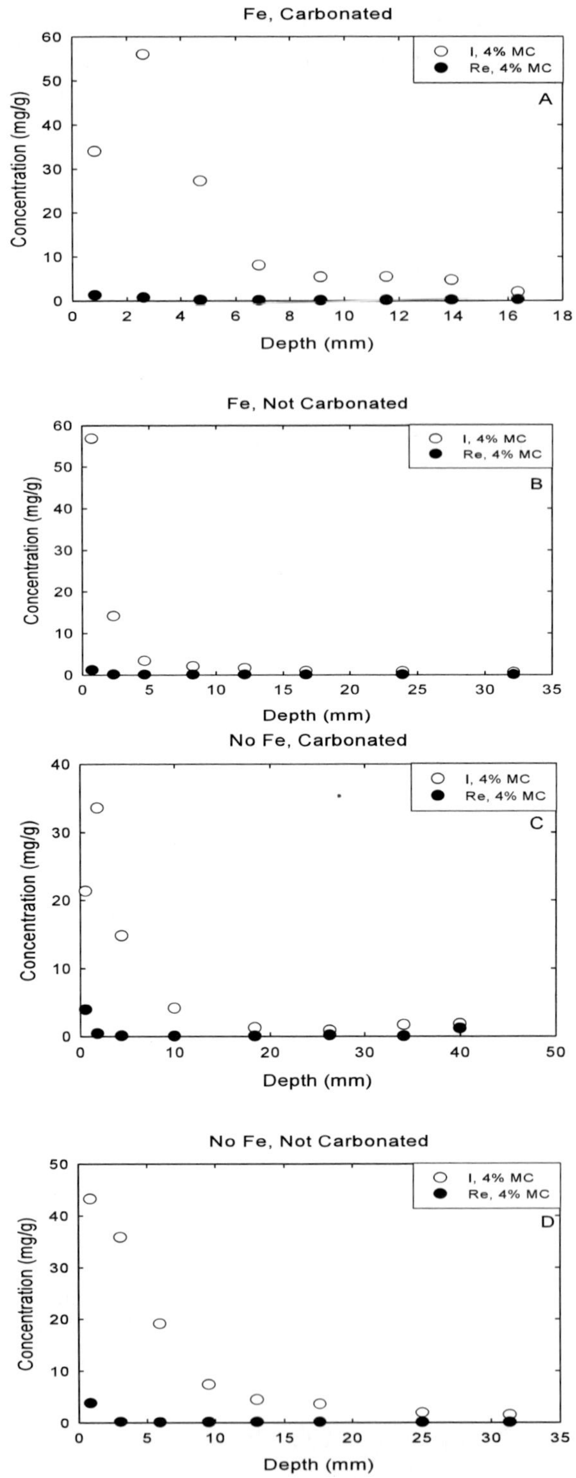

Figure 3.3. Concentration Profiles for Concrete Half-Cell I and Re Profiles at 4% Moisture Content by Weight in a) Iron, Carbonated, b) Iron, Not Carbonated, c) No Iron, Carbonated, and d) No Iron, Not Carbonated Concrete Monoliths.

Table 3.2. Concentration of Iodine and Rhenium for Concrete Half-Cell Profiles at 4%, 7%, and 15% Moisture Content by Weight

4% Fe-C			4% Fe-NC			4% No Fe-C			4% No Fe-NC		
Distance from Interface, mm	I, mg/g	Re mg/g	Distance from Interface, mm	I, mg/g	Re mg/g	Distance from Interface, mm	I, mg/g	Re mg/g	Distance from Interface, mm	I, mg/g	Re mg/g
1.3	33.93	1.23	1.5	56.79	1.13	1.2	21.28	3.90	1.8	43.20	3.78
2.9	55.87	0.69	3.21	14.07	0.08	2.51	33.53	0.38	4.41	35.78	0.12
6.1	27.11	0.10	6.12	3.32	0.03	6.32	14.73	0.02	7.52	19.06	0.03
9.2	8.01	0.06	10.53	2.01	0.02	13.73	4.08	0.00	11.63	7.36	0.04
13.8	5.34	0.06	13.74	1.59	0.02	23.14	1.20	0.00	14.44	4.46	0.07
18.5	5.39	0.08	19.75	0.89	0.02	29.65	0.82	0.13	20.85	3.59	0.07
27.1	4.66	0.09	28.06	0.80	0.02	38.56	1.65	0.01	29.26	1.87	0.03
35.2	1.93	0.19	36.27	0.56	0.01	41.47	1.76	1.12	33.47	1.45	0.01
7% Fe-C			7% Fe-NC			7% No Fe-C			7% No Fe-NC		
Distance from Interface, mm	I, mg/g	Re mg/g	Distance from Interface, mm	I, mg/g	Re mg/g	Distance from Interface, mm	I, mg/g	Re mg/g	Distance from Interface, mm	I, mg/g	Re mg/g
2.9	17.65	14.95	1.4	18.21	18.42	1.54	10.89	11.99	1.3	14.89	11.19
5.71	19.66	16.14	4.11	12.27	2.84	4.74	14.89	11.47	6.99	14.69	5.50
8.52	16.94	15.96	7.12	7.36	0.36	7.24	15.34	12.84	9.46	8.76	0.74
12.33	12.04	10.14	9.43	4.56	0.39	10.29	12.74	9.96	12.3	7.67	0.52
17.14	9.30	4.20	13.54	2.44	0.30	13.15	2.50	1.43	16.96	3.37	0.28
20.55	5.62	2.62	17.55	1.17	0.23	22.17	7.13	4.41	20.41	1.07	0.13
29.16	2.05	1.59	20.16	0.80	0.20	27.33	3.92	2.80	28.4	1.12	0.14
38.37	1.81	2.16	29.27	0.58	0.15	31.22	2.87	2.40	34.2	1.46	0.38
15% Fe-C			15% Fe-NC			15% No Fe-C			15% No Fe-NC		
Distance from Interface, mm	I, mg/g	Re mg/g	Distance from Interface, mm	I, mg/g	Re mg/g	Distance from Interface, mm	I, mg/g	Re mg/g	Distance from Interface, mm	I, mg/g	Re mg/g
1.97	8.41	10.22	1.49	18.95	27.39	1.4	7.25	8.89	1.69	12.91	16.22
4.14	13.35	13.15	3.87	19.08	21.22	4.11	12.28	11.76	3.58	10.90	5.83
7.96	12.42	11.01	5.95	17.18	18.77	7.12	13.60	11.45	5.88	7.54	1.77
10.69	10.23	9.62	7.91	17.13	17.02	9.43	10.73	11.34	7.87	6.16	1.15
13.85	9.32	5.68	9.78	16.54	14.21	13.54	11.57	7.98	10.44	4.07	0.86
18.14	6.35	3.59	11.77	13.99	7.19	17.55	10.62	5.58	12.67	3.09	1.28
22.68	3.38	2.91	14.57	8.51	1.53	20.16	8.39	4.12	15.23	2.55	1.62
-	-	-	16.56	4.22	0.42	29.27	6.33	3.51	17.54	1.84	1.77

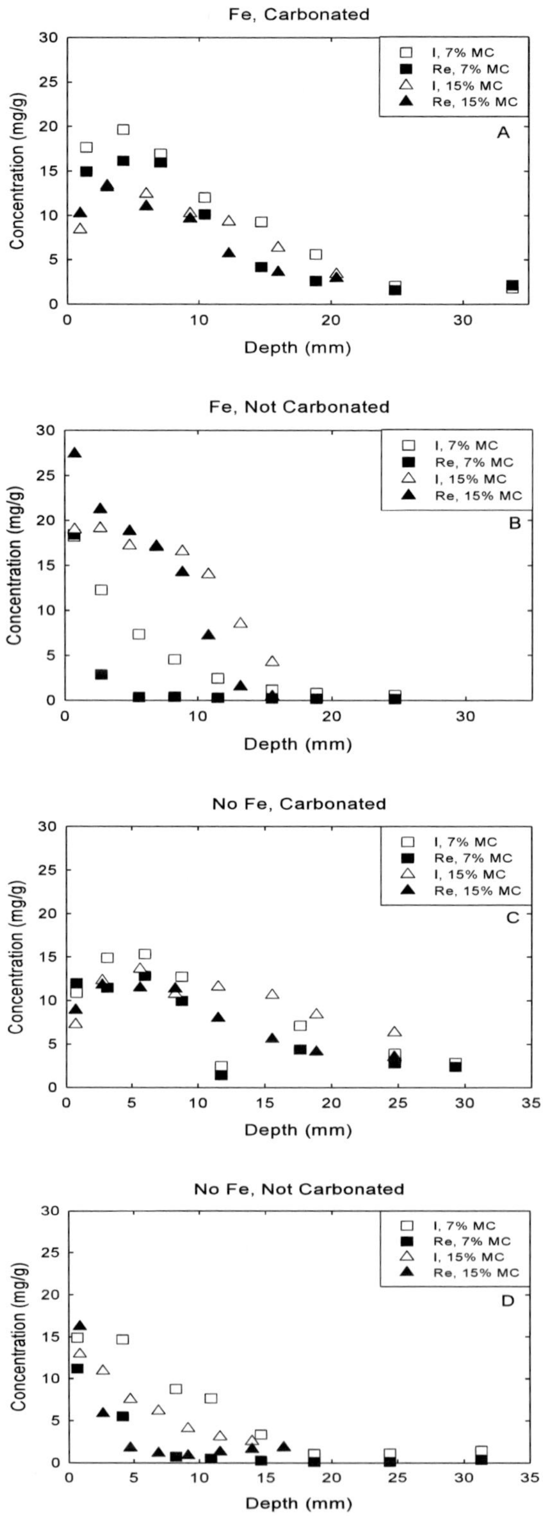

Figure 3.4. Concentration Profiles for Concrete Half-Cell I and Re Profiles at 7%, and 15% Moisture Content by Weight in a) Iron, Carbonated, b) Iron, Not Carbonated, c) No Iron, Carbonated, and d) No Iron, Not Carbonated Concrete Monoliths.

Concrete-Sediment Half-Cell Experiments to Determine the Diffusion of Iodine and Rhenium into Fractured Concrete

The previous discussion suggests the potential significance of microcrack formation within the concrete monoliths on the diffusion of I and Re. A comparable set of sediment-concrete half-cell experiments were conducted at 4%, 7%, and 15% sediment moisture content to evaluate the effect of fractures within concrete wasteforms on contaminant diffusion. Two sets of concrete monoliths were prepared with iron and without iron, half the monoliths were carbonated by submersion in a saturated sodium bicarbonate solution for seven days. The resulting compositions consisted of four concrete specimens: 1) no iron, carbonated, 2) no iron, not carbonated, 3) iron, carbonated, and 4) iron, not carbonated. The monoliths were encased in shrink-wrap and struck with a hammer to prevent the formation of rubble. Each fractured core possessed a single fracture extending the length of the core, perpendicular to the concrete-sediment interface. Sediment-concrete half-cells were prepared as previously described with I and Re-bearing sediment.

The concentration profiles for I and Re in the sediment half-cells are presented in Figure 3.5. The concentration profiles display a uniform distribution throughout the length of the half-cells. There is no measurable effect of moisture content on the concentration distribution of I or Re throughout the sediments half-cell; the concentration of I contained in the sediment half-cells was generally less than the concentration of Re (Figure 3.5 and Table 3.3).

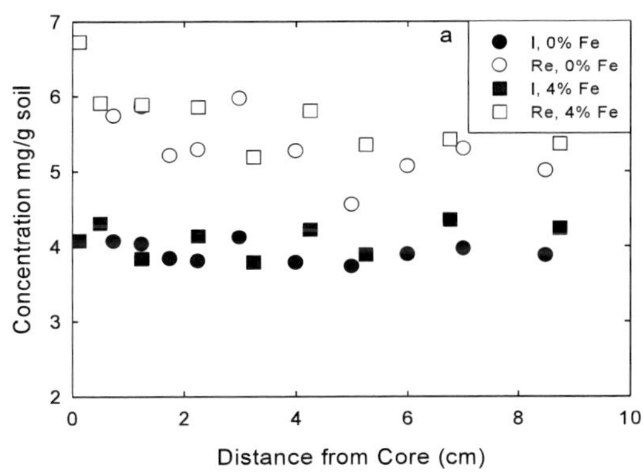

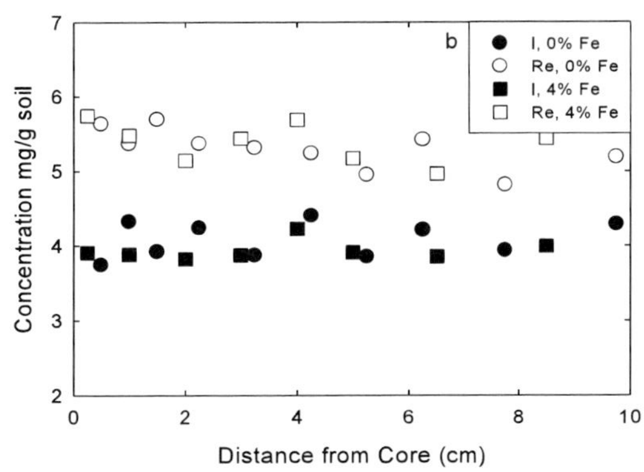

Figure 3.5. (continued).

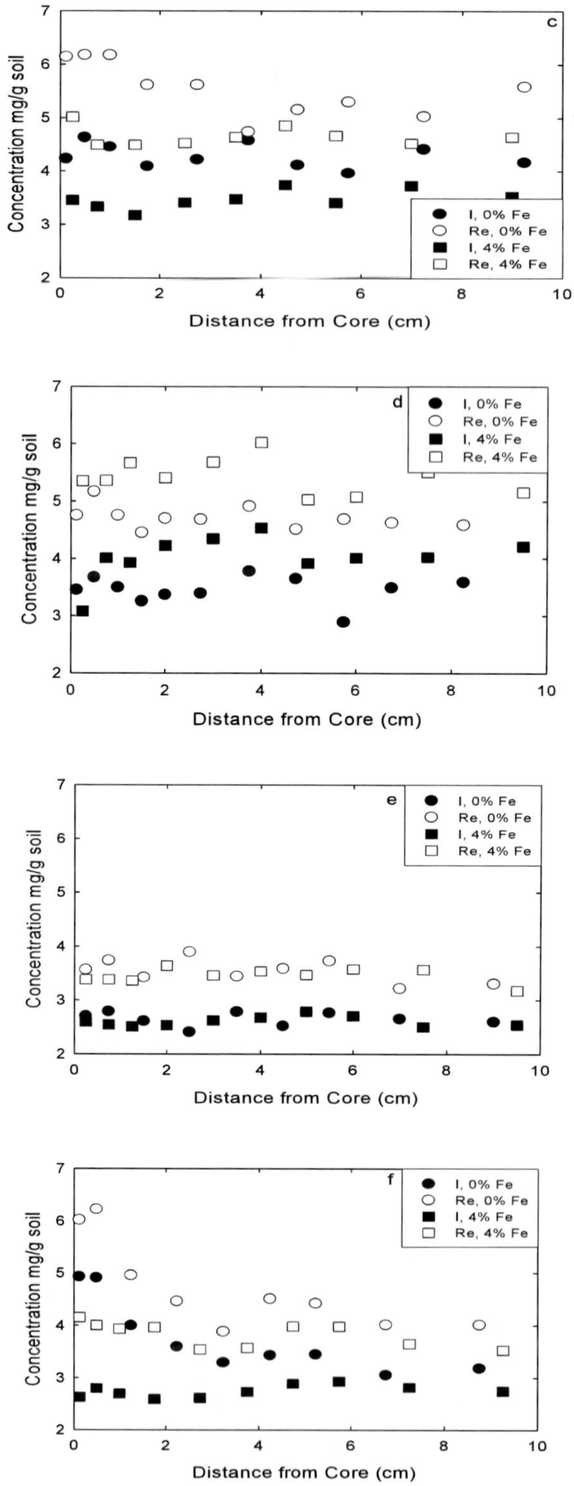

Figure 3.5. Concentration Profiles for Sediment Half-Cell I and Re Profiles. a) 4% moisture content with non-carbonated concrete monoliths, b) 4% moisture content with carbonated concrete monoliths, c) 7% moisture content with non-carbonated concrete monoliths, d) 7% moisture content with carbonated concrete monoliths, e) 15% moisture content with non-carbonated concrete monoliths.

The fracture within the concrete half-cell presents a possible unobstructed path for contaminant migration. It was, therefore, postulated that I and Re would migrate through the fracture and then into the concrete monolith at the fracture interface, in addition to the concrete-sediment interface. As such, one half of the concrete monolith was sectioned parallel to the concrete-sediment interface; the second half was sectioned parallel to the fracture (perpendicular to the concrete-sediment interface).

Figure 3.6 and Table 3.4 present the mass of I and Re per mass of concrete for the concrete monolith half that was sectioned parallel to the concrete-sediment interface. The concentrations of I and Re quantified within the monoliths at 7% and 15% moisture contents, which were not carbonated, were lower at the interface than that measured at 4% moisture content. At 4% moisture content, the concentration of I and Re were greatest within the first 5 mm and displayed an asymptotic decrease in concentration to ~1 mg/g within 17 mm. This behavior is comparable with that observed in solid concrete monoliths (Figure 3.3 and Figure 3.4).

There was no measurable effect of iron on the interfacial concentration or diffusion of I or Re irrespective of sediment moisture content. This is contrary to results observed for solid concrete monoliths which suggested the presence of iron in concrete monoliths may exhibit a slight decrease in the diffusion of Re and may result in the formation of microcracks which could facilitate diffusion. Differences in the diffusion of I and Re between the solid and fractured concrete monoliths suggests that I and Re may be migrating through the fracture and diffusing into the concrete monolith at the fracture interface, in addition to diffusion occurring at the sediment-concrete interface.

Diffusion of I and Re at 7% moisture content is greater in the presence of iron for non-carbonated monoliths. In the absence of iron, the interfacial concentration was ~7 mg/g concrete and decreased linearly over the first 15 mm. In the presence of iron, the interfacial concentration was ~11 mg/g concrete and also displayed a linear decrease over the first 20 mm. The interfacial concentration of Re was also greater in the presence of iron, 16 mg/g concrete, but displayed a rapid asymptotic decrease in concentration over the first 10 mm. A similar asymptotic decrease in concentration was observed for Re in the absence of iron, with an interfacial concentration of ~9 mg/g concrete. At 15% moisture content the initial concentration of I and Re range from 10 to 17 mg/g, which decreases rapidly over the first 5 mm to a relative steady-state concentration of ~8 mg/g that is sustained throughout the extent of the monolith. There was no measurable effect of iron on the interfacial concentration or diffusion of I or Re.

In carbonated monoliths, the interfacial concentration of I and Re at 4%, 7%, and 15% moisture content were comparable, ~ 10 – 12 mg/g concrete. The concentration of I and Re displayed a linear decrease in concentration over the length of the monolith at all moisture contents, irrespective of iron content. At 4% moisture content, there was no measureable effect of iron on the diffusion of I or Re. The concentration decreased linearly over 20 mm of the concrete monolith. At 7%, the diffusion of I and Re displayed a sustained diffusion over the first ~5 mm of the concrete monolith followed by linear diffusion over the subsequent 15 mm of concrete monolith. The diffusion of I and Re were greatest in the absence of iron. This agrees with previous results indicating that the inclusion of iron will reduce I and Re to species which are less soluble and more tightly retained within the concrete monolith (Wellman et al. 2008).

Table 3.3. Concentration of Iodine and Rhenium (mg/g) for Sediment Half-Cell Profiles at 4%, 7%, and 15% Moisture Content

Distance From Core, mm	I, mg/g	Re, mg/g	Distance From Core, mm	I, mg/g	Re, mg/g	Distance From Core, mm	I, mg/g	Re, mg/g	Distance From Core, mm	I, mg/g	Re, mg/g
4% Sediment--NC 0% Iron			4% Sediment--NC 4% Iron			4% Sediment--C 0% Iron			4% Sediment--C-4% Iron		
10.5	4.0	5.1	8.75	4.2	5.4	9.75	4.3	5.2	8.5	4.0	5.4
8.5	3.9	5.0	6.75	4.4	5.4	7.75	3.9	4.8	6.5	3.9	5.0
7	4.0	5.3	5.25	3.9	5.4	6.25	4.2	5.4	5	3.9	5.2
6	3.9	5.1	4.25	4.2	5.8	5.25	3.8	4.9	4	4.2	5.7
5	3.7	4.6	3.25	3.8	5.2	4.25	4.4	5.2	3	3.9	5.4
4	3.8	5.3	2.25	4.1	5.9	3.25	3.9	5.3	2	3.8	5.1
3	4.1	6.0	1.25	3.8	5.9	2.25	4.2	5.4	1	3.9	5.5
2.25	3.8	5.3	0.5	4.3	5.9	1.5	3.9	5.7	0.25	3.9	5.7
1.75	3.8	5.2	0.125	4.1	6.7	1	4.3	5.4			
1.25	4.0	5.9				0.5	3.7	5.6			
0.75	4.1	5.7				0.125	3.7	5.9			
0.25	4.4	5.5									
7% Sediment--NC 0% Iron			7% Sediment--NC 4% Iron			7% Sediment--C 0% Iron			7% Sediment--C-4% Iron		
9.25	4.2	5.6	9	3.5	4.6	10.25	3.5	4.4	9.5	4.2	5.2
7.25	4.4	5.0	7	3.7	4.5	8.25	3.6	4.6	7.5	4.0	5.5
5.75	4.0	5.3	5.5	3.4	4.7	6.75	3.5	4.6	6	4.0	5.1
4.75	4.1	5.2	4.5	3.7	4.9	5.75	2.9	4.7	5	3.9	5.0
3.75	4.6	4.7	3.5	3.5	4.6	4.75	3.6	4.5	4	4.5	6.0
2.75	4.2	5.6	2.5	3.4	4.5	3.75	3.8	4.9	3	4.3	5.7
1.75	4.1	5.6	1.5	3.2	4.5	2.75	3.4	4.7	2	4.2	5.4
1	4.5	6.2	0.75	3.3	4.5	2	3.4	4.7	1.25	3.9	5.7
0.5	4.6	6.2	0.25	3.5	5.0	1.5	3.2	4.4	0.75	4.0	5.4
0.125	4.2	6.1				1	3.5	4.7	0.25	3.1	5.4
						0.5	3.7	5.2			
						0.125	3.4	4.7			
15% Sediment--NC 0% Iron			15% Sediment--NC 4% Iron			15% Sediment--C 0% Iron			15% Sediment--C-4% Iron		
9	2.6	3.3	9.5	2.5	3.2	8.75	3.2	4.0	9.25	2.7	3.5
7	2.6	3.2	7.5	2.5	3.6	6.75	3.1	4.0	7.25	2.8	3.7
5.5	2.8	3.7	6	2.7	3.6	5.25	3.4	4.4	5.75	2.9	4.0
4.5	2.5	3.6	5	2.8	3.5	4.25	3.4	4.5	4.75	2.9	4.0

Distance From Core, mm	I, mg/g	Re, mg/g	Distance From Core, mm	I, mg/g	Re, mg/g	Distance From Core, mm	I, mg/g	Re, mg/g	Distance From Core, mm	I, mg/g	Re, mg/g
3.5	2.8	3.4	4	2.7	3.5	3.25	3.3	3.9	3.75	2.7	3.6
2.5	2.4	3.9	3	2.6	3.5	2.25	3.6	4.5	2.75	2.6	3.5
1.5	2.6	3.4	2	2.5	3.6	1.25	4.0	5.0	1.75	2.6	4.0
0.75	2.8	3.7	1.25	2.5	3.4	0.5	4.9	6.2	1	2.7	3.9
0.25	2.7	3.6	0.75	2.5	3.4	0.125	4.9	6.0	0.5	2.8	4.0
			0.25	2.6	3.4				0.125	2.6	4.2

Table 3.4. Concentration of Iodine and Rhenium (mg/g) for Concrete Half-Cell Profiles at 4%, 7%, and 15% Moisture Content Sectioned Parallel to the Concrete-Sediment Interface

Distance From Interface, mm	I, mg/g	Re, mg/g	Distance From Interface, mm	I, mg/g	Re, mg/g	Distance From Interface, mm	I, mg/g	Re, mg/g	Distance From Interface, mm	I, mg/g	Re, mg/g
4% Sediment--NC 0% Iron			4% Sediment--NC 4% Iron			4% Sediment--C 0% Iron			4% Sediment--C-4% Iron		
0.9	20.11	30.05	1.5	14.76	17.57	1.2	8.55	9.01	0.9	11.81	9.83
2.8	13.17	17.80	4.7	10.26	10.96	3.5	8.74	9.70	3.5	9.10	9.64
5.1	9.95	13.29	7.3	8.58	7.01	6.0	6.88	8.64	5.8	7.93	7.30
7.7	6.90	7.32	10.1	5.69	3.09	8.9	6.32	9.60	8.0	6.19	6.42
10.3	5.35	0.92	13.2	3.70	0.33	11.4	6.03	8.41	10.4	5.32	7.10
12.7	3.74	0.24	16.7	1.57	0.14	14.0	5.10	5.97	13.1	3.90	3.60
15.1	2.12	0.15	22.4	0.12	0.05	16.9	3.94	1.53	15.8	3.19	3.35
17.8	0.80	0.09				19.6	3.03	0.57	18.4	2.28	1.34
7% Sediment--NC 0% Iron			7% Sediment--NC 4% Iron			7% Sediment--C 0% Iron			7% Sediment--C-4% Iron		
1.1	7.04	8.84	1.3	10.66	16.22	1.5	9.08	10.04	1.0	7.15	7.03
3.6	5.92	1.44	4.1	8.77	9.30	4.0	9.65	11.62	3.7	4.64	6.99
6.1	3.54	0.18	6.8	6.24	5.81	6.1	8.73	10.68	6.5	4.03	6.88
8.7	2.90	0.13	9.2	5.20	2.39	8.2	7.45	9.27	9.0	2.91	4.16
11.3	1.30	0.09	11.8	3.24	0.33	10.4	7.28	8.83	11.5	2.24	1.47
13.8	0.50	0.06	14.7	2.21	0.24	13.0	6.06	8.25	13.9	1.64	0.64
16.4	0.07	0.00	17.1	0.66	0.00	15.6	5.08	4.66	16.2	1.04	0.32
19.2	0.00	0.00	20.2	0.28	0.07	18.1	4.54	2.20	18.7	0.62	0.24
15% Sediment--NC 0% Iron			15% Sediment--NC 4% Iron			15% Sediment--C 0% Iron			15% Sediment--C-4% Iron		
1.0	11.65	16.79	1.2	10.12	14.23	1.4	9.42	12.93	1.2	9.26	14.89
3.6	8.09	13.33	3.9	7.45	9.70	4.0	9.71	13.15	3.7	8.09	9.73

Table 3.4. (Continued).

Distance From Interface, mm	I, mg/g	Re, mg/g	Distance From Interface, mm	I, mg/g	Re, mg/g	Distance From Interface, mm	I, mg/g	Re, mg/g	Distance From Interface, mm	I, mg/g	Re, mg/g
6.4	7.69	9.67	6.5	6.96	7.45	6.4	8.47	11.92	6.2	8.26	9.23
9.0	8.55	8.88	9.4	6.75	7.80	8.7	8.34	11.94	8.6	7.49	7.62
11.2	7.62	7.69	12.4	6.75	7.75	11.0	8.10	9.67	11.2	6.63	8.36
13.5	7.05	7.46	15.2	5.78	8.01	13.3	7.29	9.62	13.8	6.01	8.53
16.0	5.62	8.52	18.2	5.50	8.85	15.6	6.84	7.38	16.5	4.98	6.12
18.4	5.62	8.64	20.9	5.18	8.48	18.2	5.85	4.12	19.0	4.74	3.98

The interfacial concentrations of I and Re in the absence of iron were 10 and 9 mg/g concrete, respectively. In the presence of iron the interfacial concentration was 8 mg/g concrete. At 15% moisture content, the interfacial concentrations of Re and I, respectively, were ~14 mg/g concrete and ~9 mg/g concrete, irrespective of iron content. For Re, in the presence of iron concentration rapidly decreased within the first 5 mm and exhibited a very slow decrease in concentration over the length of the monolith to a final concentration of 5 mg Re/g concrete. In the absence of iron, the concentration of Re was sustained at ~12 mg Re/g concrete over the first 10 mm and then exhibited a slow decrease in concentration over the remaining 10 mm of monolith to a final concentration of 5 mg Re/g concrete.

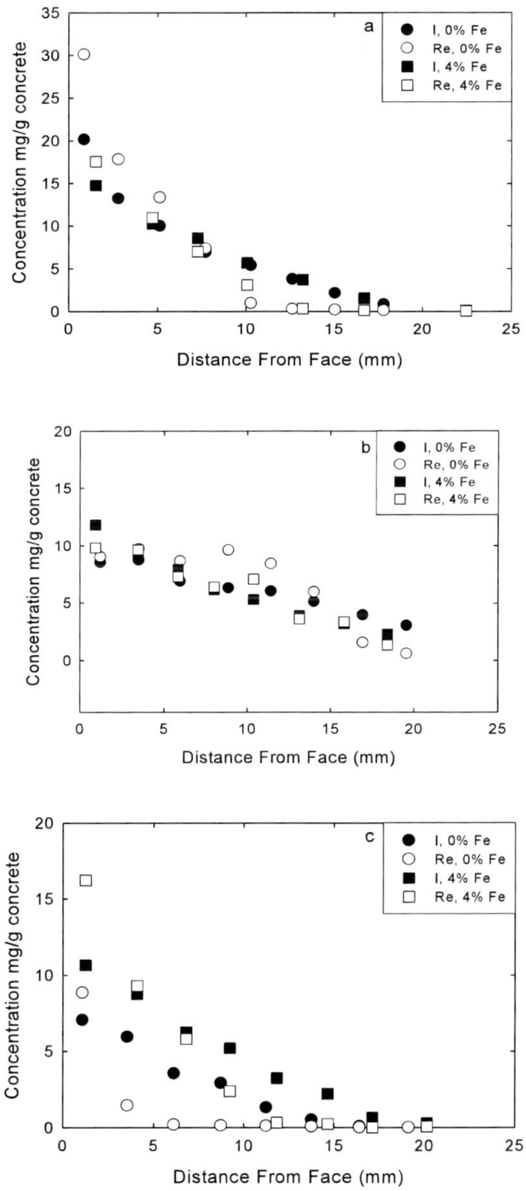

Figure 3.6. (continued).

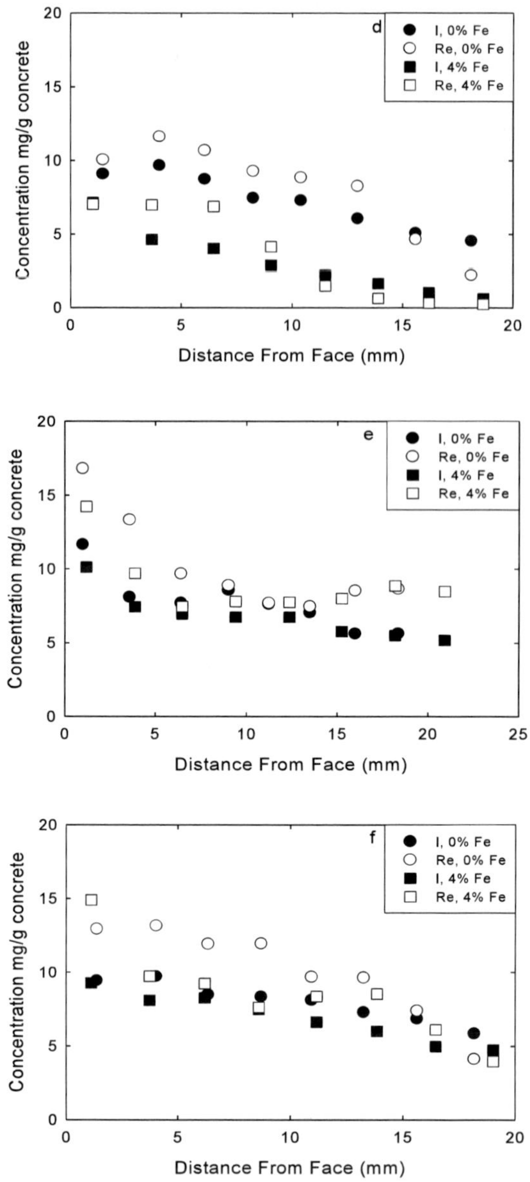

Figure 3.6. I and Re Concentration Profiles Sectioned Parallel to the Concrete-Sediment Interface. a) 4% moisture content, non-carbonated concrete monoliths, b) 4% moisture content, carbonated concrete monoliths, c) 7% moisture content, non-carbonated concrete monoliths, d) 7% moisture content, carbonated concrete monoliths, e) 15% moisture content, non-carbonated concrete monoliths, and f) 15% moisture content, carbonated concrete monoliths.

The inclusion of iron decreased the diffusion of Re within the first 10 mm, but had no effect within the remaining 10 mm of the monolith. There was no measurable effect of iron on the diffusion of iodine. The interfacial concentration of I was 10 mg I/g concrete. The concentration of I displayed a gradual decrease to 4 mg I/g concrete through concrete monoliths with and without iron. The overall diffusion of I and Re is greater in carbonated rather than non-carbonated concrete monoliths. This agrees with previous results (Wellman et al. 2008) which demonstrated carbonation results in the formation of microcracks within concrete monoliths which can facilitate diffusion.

The behavior of I and Re within concrete sections taken perpendicular to the concrete-sediment interface (parallel to the fracture) was markedly different relative to those sectioned parallel to the concrete-sediment interface. Figure 3.7 and Table 3.5 present the mass of I and Re per mass of concrete for the concrete monolith half sectioned perpendicular to the concrete-sediment interface. Diffusion of I and Re in the presence of iron, in non-carbonated cores was comparable at 7% and 15% moisture content, ~6-8 mg/g concrete throughout the monolith profile. At 4% moisture content, however, the diffusion of I and Re initiated at ~8 mg/g and then rapidly decreased to a sustained mass of ~ 2 mg/g throughout the concrete profile. This accords with previous results, suggesting a direct relationship between diffusion and the degree of moisture content (Mattigod et al. 2001; Wellman et al. 2007; Wellman et al. 2006; 2008). The behavior of I and Re in non-carbonated monoliths that do not contain iron is comparable at 4% and 15% moisture content to that observed in the presence of iron. However, at 7% moisture content the concentration of I and Re is ~1 mg/g throughout the monolith profile, substantially less than the sustained concentration of ~5 mg/g measured in the presence of iron. The decreased concentration profile for iodine in the presence of iron agrees with previous findings (Mattigod et al. 2001; Wellman et al. 2007; Wellman et al. 2006; 2008) suggesting that the inclusion of iron results in the formation of microcracks that can facilitate diffusion.

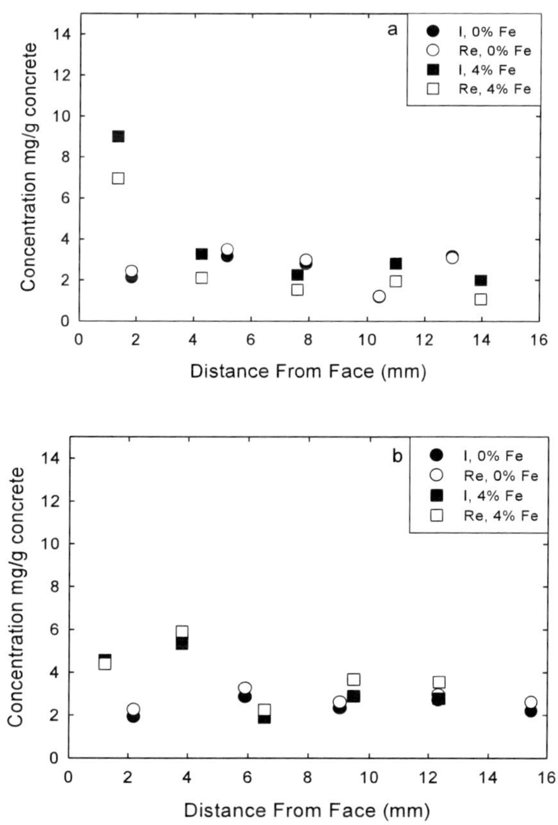

Figure 3.7. (continued).

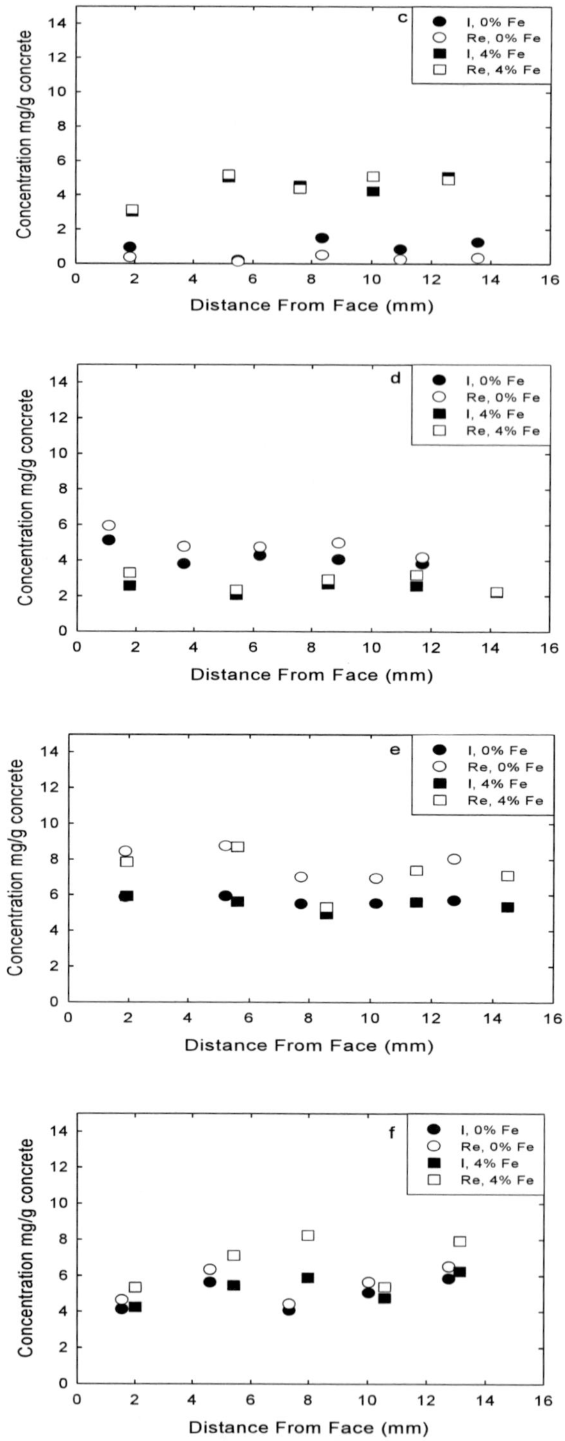

Figure 3.7. I and Re Concentration Profiles Sectioned Perpendicular to the Concrete-Sediment Interface. a) 4% moisture content, non-carbonated concrete monoliths, b) 4% moisture content, carbonated concrete monoliths, c) 7% moisture content, non-carbonated concrete monoliths, d) 7% moisture content, carbonated concrete monoliths, e) 15% moisture content, non-carbonated concrete monoliths, and f) 15% moisture content, carbonated concrete monoliths.

In carbonated concrete monoliths, the behavior of I and Re was comparable to that observed in non-carbonated monoliths at 4% and 15% moisture content. At 7% moisture content, the diffusion of I and Re was greater in the absence of iron than in the presence of iron. Comparable behavior was observed for the other half of this monolith which was sectioned parallel to the concrete-sediment interface. The lack of observable effect at 4% and 15% moisture content suggests that diffusion occurring through the interface parallel to the concrete-sediment interface and the fracture interface result in an increase in the concentration of I and Re within the monoliths, which minimizes the apparent effects of iron and carbonate treatments on diffusion in concrete. Thus, the presence of fractures in concrete waste forms may provide an unobstructed path for contaminant migration. Moreover, the concentration gradient across the fracture interface may serve as an additional plane for contaminant diffusion in concrete wasteforms.

Table 3.5. Concentration of Iodine and Rhenium (mg/g) for Concrete Half-Cell Profiles at 4%, 7%, and 15% Moisture Content Sectioned Perpendicular to the Concrete-Sediment Interface (Parallel to the Concrete Fracture)

Distance From Interface, mm	I, mg/g	Re, mg/g	Distance From Interface, mm	I, mg/g	Re, mg/g	Distance From Interface, mm	I, mg/g	Re, mg/g	Distance From Interface, mm	I, mg/g	Re, mg/g
4% Sediment--NC 0% Iron			4% Sediment--NC 4% Iron			4% Sediment--C 0% Iron			4% Sediment--C-4% Iron		
1.9	2.11	2.41	1.4	9.00	6.96	2.2	1.91	2.24	1.2	4.57	4.40
5.2	3.15	3.48	4.3	3.28	2.11	5.9	2.84	3.24	3.8	5.34	5.90
7.9	2.79	2.97	7.6	2.26	1.53	9.1	2.30	2.59	6.5	1.89	2.25
10.4	1.16	1.19	11.0	2.82	1.95	12.3	2.69	2.95	9.5	2.88	3.68
13.0	3.16	3.08	14.0	2.00	1.07	15.5	2.17	2.59	12.4	2.77	3.55
7% Sediment--NC 0% Iron			7% Sediment--NC 4% Iron			7% Sediment--C 0% Iron			7% Sediment--C-4% Iron		
1.9	0.93	0.35	1.9	3.02	3.13	1.1	5.10	5.92	1.8	2.58	3.31
5.5	0.20	0.10	5.2	5.04	5.20	3.7	3.78	4.75	5.4	2.06	2.34
8.4	1.48	0.48	7.6	4.57	4.37	6.2	4.26	4.72	8.5	2.68	2.93
11.0	0.82	0.23	10.0	4.23	5.11	8.9	4.03	4.96	11.5	2.58	3.20
13.6	1.23	0.33	12.6	5.10	4.90	11.7	3.80	4.15	14.2	2.21	2.25
15% Sediment--NC 0% Iron			15% Sediment--NC 4% Iron			15% Sediment--C 0% Iron			15% Sediment--C-4% Iron		
1.9	5.85	8.42	1.9	5.93	7.85	1.6	4.10	4.63	2.0	4.24	5.35
5.2	5.91	8.73	5.6	5.62	8.70	4.6	5.62	6.32	5.4	5.46	7.12
7.8	5.48	6.99	8.6	4.93	5.31	7.3	4.05	4.40	8.0	5.89	8.24
10.2	5.50	6.92	11.5	5.60	7.39	10.1	5.03	5.63	10.6	4.76	5.38
12.7	5.68	8.03	14.5	5.36	7.10	12.8	5.83	6.50	13.1	6.25	7.93

CONCRETE-SEDIMENT HALF-CELL EXPERIMENTS TO DETERMINE THE EFFECTS OF IRON ON THE DIFFUSION OF TECHNETIUM

Rhenium is typically used as an experimental surrogate for technetium. Re and Tc are manifested as heptavalent cations (+7) of similar size that complex with oxygen to form water soluble anionic species: perrhenate (ReO_4^-) and pertechnetate (TcO_4^-) species. Despite the similarities in cationic size and other chemical properties the standard potential of the Re(VII)/Re(IV) couple is significantly lower than that for Tc(VII)/Tc(IV) (Table 3.6). The oxidation potential of Re is -0.510 V, whereas Tc has a value of -0.738 V, according to (KOTEGOV et al., 1968) and this difference may significantly alter the quantified diffusion in the presence of iron. Therefore, a comparable series of concrete-sediment half-cell tests were prepared with two sets of concrete specimens: 1) no iron, not carbonated and 2) iron, not carbonated.

Table 3.6. Comparison of properties of Tc and Re

M = Tc or Re	Tc	Re
Principal oxidation states:	VII, IV	VII, IV, III
Cation (VII) size (nm)	51[a]	52[a]
Cation (IV) size (nm)	78.5[a]	77[a]
M(VII)—O Å	1.73[f]	1.74[i]
M(IV)—O Å	2.00[b]	1.98[i]
M(IV)—S Å	2.33[e]	2.36[c]
Ehyd (eV)	-2.4[g]	-2.5[g]
Binding energy (eV)	-388.7[g]	-371.1[g]
MO2/MO4- (V)	-0.747[d]	-0.510[h]

Sources: [a]. (Shannon 1976), [b]. (Almahamid et al. 1995), [c]. (Murray et al. 1994), [d]. (Meyer and Arnold 1991), [e]. (Lamfers et al. 1996), [f]. (Faggiani et al. 1980), [g]. (Anderson et al. 2007), [h]. (Kotegov et al. 1968), [i]. (Colton 1965).

The concentration profiles for Tc from concrete-sediment half-cell diffusion experiments are shown in Figure 3.8 and Table 3.7. The Tc concentration profile for sediments at 4% and 7% moisture content exhibited greater concentrations for those in contact with concrete cores without iron (Figure 3.8). In the absence of iron, the concentration of Tc in sediment at 4% moisture content initiated at 7.96 x 10^{-6} mg Tc/g sediment and gradually decreased to 1.96 x 10^{-6} mg Tc/g sediment. In the presence of iron, the concentration of Tc in the sediment at the concrete-sediment interface was 3.29 x 10^{-6} mg Tc/g sediment, 41% less than that quantified in the concrete-sediment half-cell that did not contain iron in the concrete monolith. The concentration decreased over the length of the sediment profile to 5.12 x 10^{-7} mg Tc/g sediment. At 7% moisture content and in the absence of iron, the concentration of Tc in sediment initiated at 1.24 x 10^{-6} mg Tc/g sediment. Contrary to the sediment profile for Tc at 4% moisture content, at 7% the concentration of Tc in the sediment displayed a slight increase as a function of increasing distance from the concrete-sediment interface. The concentration of Tc in the sediment section furthest from the interface, 9.5 cm, was 2.74 x 10^{-7} mg Tc/g sediment. This slight increase through the sediment profile also was observed for the concrete-sediment half-cell at 7% moisture content for concrete containing iron. In the presence of iron, the concentration of Tc initiated at 1.29 x 10^{-6} mg Tc/g sediment and increased over the length of the half-cell to a maximum concentration of 1.89 x 10^{-6} mg Tc/g sediment at 9.5 cm. The concentration of Tc throughout the sediment profile was ~65% lower in the concrete-sediment half-cell that contained iron in the concrete monolith than in the one that did not contain iron at 7% moisture content.

Although the concentration profile for Tc in sediments at 15% moisture content contained slight concentration fluctuations, the concentration profiles were generally invariant throughout the sediment half-cell (Table 3.7). However, the concentration of Tc in the sediment from the concrete-sediment half-cell that contained iron in the concrete monolith was greater than the one that did not contain iron. In the absence of iron, the Tc concentration initiated in the sediment profile at 5.17 x 10^{-6} mg Tc/g sediment, increased to a maximum concentration of 5.45 x 10^{-6} mg Tc/g sediment, and then decreased to 5.19 x 10^{-6} mg Tc/g sediment. In the presence of 4% iron, the Tc concentration initiated in the sediment profile at 6.95 x 10^{-6} mg Tc/g sediment, increased to a maximum concentration of 7.02 x 10^{-6} mg Tc/g sediment, and then decreased to 5.95 x 10^{-6} mg Tc/g sediment.

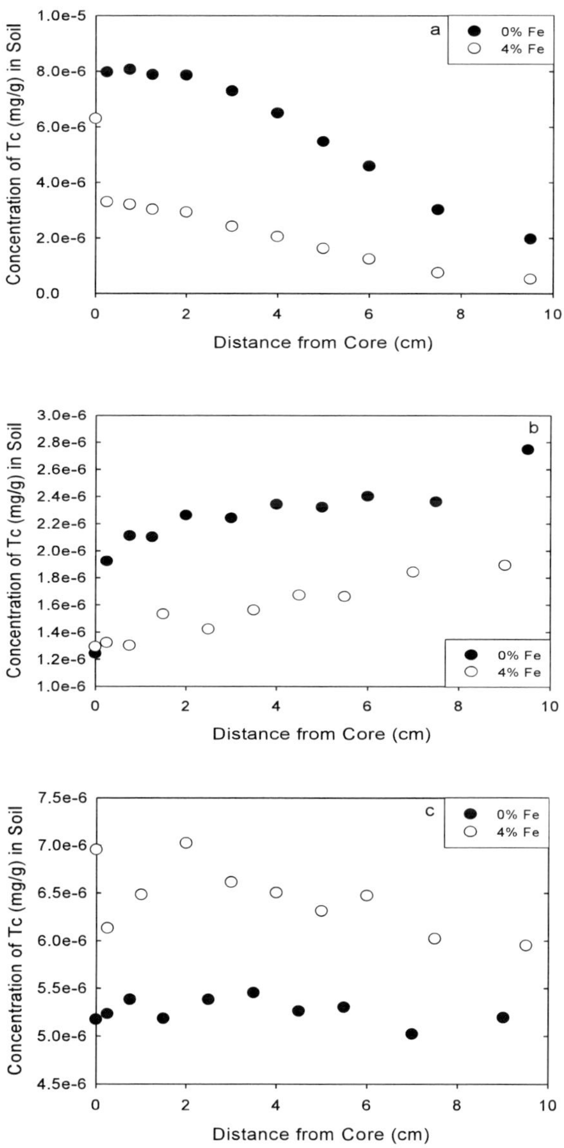

Figure 3.8. Concentration Profiles for Sediment Half-Cell Tc Profiles. a) 4%, b) 7%, and c) 15% moisture content by weight containing non-carbonated concrete monoliths with and without iron.

Table 3.7. Concentration of Technetium (mg/g) for Sediment Half-Cell Profiles at 4%, 7%, and 15% Moisture Content

Distance from Core, cm	4% MC-0% Iron	4% MC-4% Iron	7% MC-0% Iron	7% MC-4% Iron	Distance from Core, cm	15% MC-0% Iron	Distance from Core, cm	15% MC-4% Iron
9.5	1.96E-06	5.12E-07	2.74E-06	1.89E-06	9.0	5.19E-06	9.5	5.95E-06
7.5	3.01E-06	7.46E-07	2.36E-06	1.84E-06	7.0	5.02E-06	7.5	6.02E-06
6	4.57E-06	1.24E-06	2.40E-06	1.66E-06	5.5	5.30E-06	6.0	6.47E-06
5	5.46E-06	1.62E-06	2.32E-06	1.67E-06	4.5	5.26E-06	5.0	6.31E-06
4	6.49E-06	2.04E-06	2.34E-06	1.56E-06	3.5	5.45E-06	4.0	6.50E-06
3	7.28E-06	2.41E-06	2.24E-06	1.42E-06	2.5	5.38E-06	3.0	6.61E-06
2	7.85E-06	2.92E-06	2.26E-06	1.53E-06	1.5	5.18E-06	2.0	7.02E-06
1.25	7.87E-06	3.02E-06	2.10E-06	1.30E-06	0.8	5.38E-06	1.0	6.48E-06
0.75	8.05E-06	3.20E-06	2.11E-06	1.32E-06	0.3	5.23E-06	0.3	6.13E-06
0.25	7.96E-06	3.29E-06	1.92E-06	1.29E-06	0.0	5.17E-06	0.0	6.95E-06
0	N/A	6.29E-06	1.24E-06	N/A				
MC = moisture content								

Table 3.8. Concentration of Technetium (μg/g) for Concrete Half-Cell Profiles at 4%, 7%, and 15% Moisture Content

4% Sediment-0% Iron		4% Sediment-4% Iron		7% Sediment-0% Iron		7% Sediment-4% Iron		15% Sediment-0% Iron		15% Sediment-4% Iron	
Distance from Interface, mm	Tc, μg/g	Distance from Interface, mm	Tc, μg/g	Distance from Interfac e, mm	Tc, μg/g	Distance from Interface, mm	Tc, μg/g	Distance from Interface, mm	Tc, μg/g	Distance from Interface, mm	Tc, μg/g
1.5	0.59	1.2	0.46	1.2	0.25	1.4	0.26	1.4	0.49	1.4	0.3
4.7	0.46	4.3	0.34	3.8	0.3	4.4	0.22	4.2	0.3	4.2	0.18
8.2	0.31	7.4	0.23	6	0.28	7.2	0.19	6.5	0.3	6.9	0.16
11.7	0.24	10.1	0.19	8.4	0.25	10	0.18	8.5	0.28	9.9	0.15
15.2	0.21	12.3	0.19	10.7	0.23	13.1	0.15	10.5	0.26	12.9	0.15
18	0.24	14.5	0.17	12.8	0.21	16.1	0.16	13.1	0.25	15.2	0.19
21.2	0.14	17.2	0.17	15	0.2	18.8	0.18	16.2	0.23	17.6	0.17
		19.7	0.18	17.4	0.19	21.5	0.16	19.3	0.23	20.6	0.17
Total	2.19		1.93		1.91		1.5		2.34		1.47

The total concentration of Tc recovered in the concrete half-cells were comparable for concrete monoliths containing iron and those without iron (Table 3.8). The normalized concentration profiles for concrete half-cells all indicate greater concentrations of Tc for half-cells in which the concrete did not contain iron (Figure 3.9). This agrees with previous findings presented for I and Re suggesting diffusion was limited due to reaction of I and Re with Fe which afforded the reduced, more insoluble species. At 4% moisture content, the concentration profile for Tc in the absence of iron initiates at 0.6 μg Tc/g concrete. The concentration displays a linear decrease over the first 10 mm of the monolith, followed by a more gradual decrease over the subsequent 10 mm of concrete monolith. A similar behavior is observed in the presence of iron, except that the concentration of Tc within the monolith is less than that measured in the presence of iron. The interfacial concentration is 0.46 μg Tc/g concrete and it exhibits a rapid decrease to ~ 0.2 μg Tc/g concrete, which is sustained throughout the remaining 10 mm of monolith.

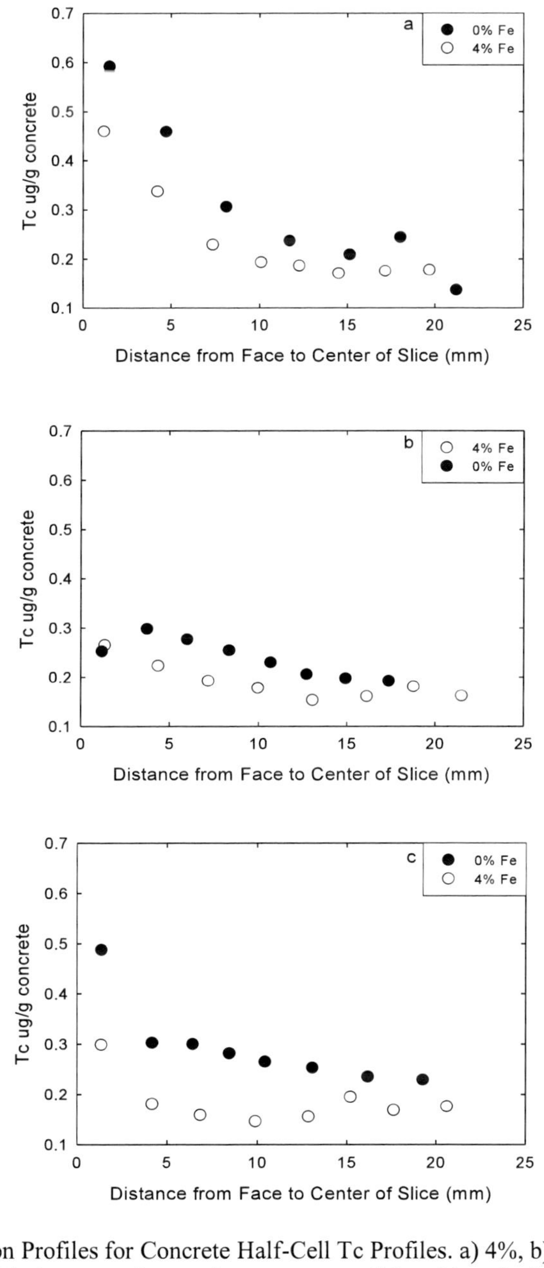

Figure 3.9. Concentration Profiles for Concrete Half-Cell Tc Profiles. a) 4%, b) 7%, and c) 15% moisture content by weight in non-carbonated concrete monoliths with and without iron.

The distribution and behavior of Tc was similar for concrete monoliths at 7% and 15% moisture content. In the absence of iron, the concentration near the interface was ~0.3 μg Tc/g concrete. The concentration of Tc decreased slightly throughout the length of the cores to ~0.2 μg Tc/g concrete. In the presence of iron, the concentration near the interface was ~0.25 - 0.3 μg Tc/g concrete. The concentration again decreased throughout the length of the core, but the decrease was greater than in the absence of iron. In the presence of iron the concentration decreased to a minimum of ~0.15 μg Tc/g concrete, 10 to 15 mm from the concrete-sediment interface. These results contrast with those previous presented by Mattigod et al. (2001), which quantified the diffusion of I and Tc from concrete monoliths containing iron at 4% and 7% moisture contents, sampling at 64 and 169 days. Results by Mattigod et al. (2001) indicated that the inclusion of iron did not significantly influence the diffusion of I or Tc and/or Re. The significantly short test duration for previous half-cell experiments may have limited the observed effect of iron on Tc diffusion.

Carbonation and Micro-Cracking

Four concrete monoliths, from the concrete-sediment half cells, were subjected to scanning electron microscopy studies to evaluate the degree of microcracking induced by carbonation and discern correlations with contaminant diffusion. Scanning electron microscopy studies were performed to identify microcracks with widths < 2–3 μm. Specimen preparation methods, specifically cutting, drying, exposure to solvents, and exposure to vacuum, can cause microcracking. Therefore, stabilization with low-viscosity epoxy resin minimized the extent of damage caused by sample preparation. A small block measuring approximately 20 mm on each side was cut from the central portion of each of the mortar samples. Each block was placed in a plastic cup and impregnated with low-viscosity epoxy resin. After the resin hardened, the embedded mortars were polished using successively finer alumina grits down to a final stage of 0.05 mm. The polished surfaces were carbon coated then examined in backscattered electron mode (BSE) using an ASPEX personal scanning electron microscope (SEM). The relative extent of microcracking can be judged from a comparison of the images. However, it is necessary to note that microcracking can result from any, or a combination, of the preparation steps, which include drying in a low-temperature oven, carbon coating under a vacuum, and SEM studies conducted in a vacuum of about 6 $x10^{-6}$ torr.

Table 3.9. Concrete Monoliths Subject to Carbonation/Micro-Cracking Analysis

NC-4-4-8	Iron, Not Carbonated, 4% Moisture Content
NC-4-0-8	No Iron, Not Carbonated, 4% Moisture Content
C-7-4-8	Iron, Carbonated, 7% Moisture Content
C-7-0-8	No Iron, Carbonated, 7% Moisture Content

Scanning electron microscopy images are presented in Figure 3.10. Results indicate that NC-4-4-8 shows a discontinuous microcracks which are substantially less than 10 μm wide (Figure 3.10a). BSE images of sample NC-4-0-8 indicate the formation of longer microcracks, but which are also substantially less than 10 μm wide (Figure 3.10b). BSE image of sample C-7-4-8 indicates that carbonation of concrete monoliths results in the formation of substantially longer and wider (> 10 μm) microcracks. These microcracks typically pass between aggregate particles (Figure 3.10c). The inclusion of iron in carbonated concrete monoliths results in the formation of short, narrow microcracks branching from the wider, longer microcracks (Figure 3.10d). SEM results clearly display an increase in the length and width of microcracks in concrete monoliths with the inclusion of iron. The formation of branched microcracks in concrete monoliths provides additional paths within the monolith for contaminant diffusion and migration.

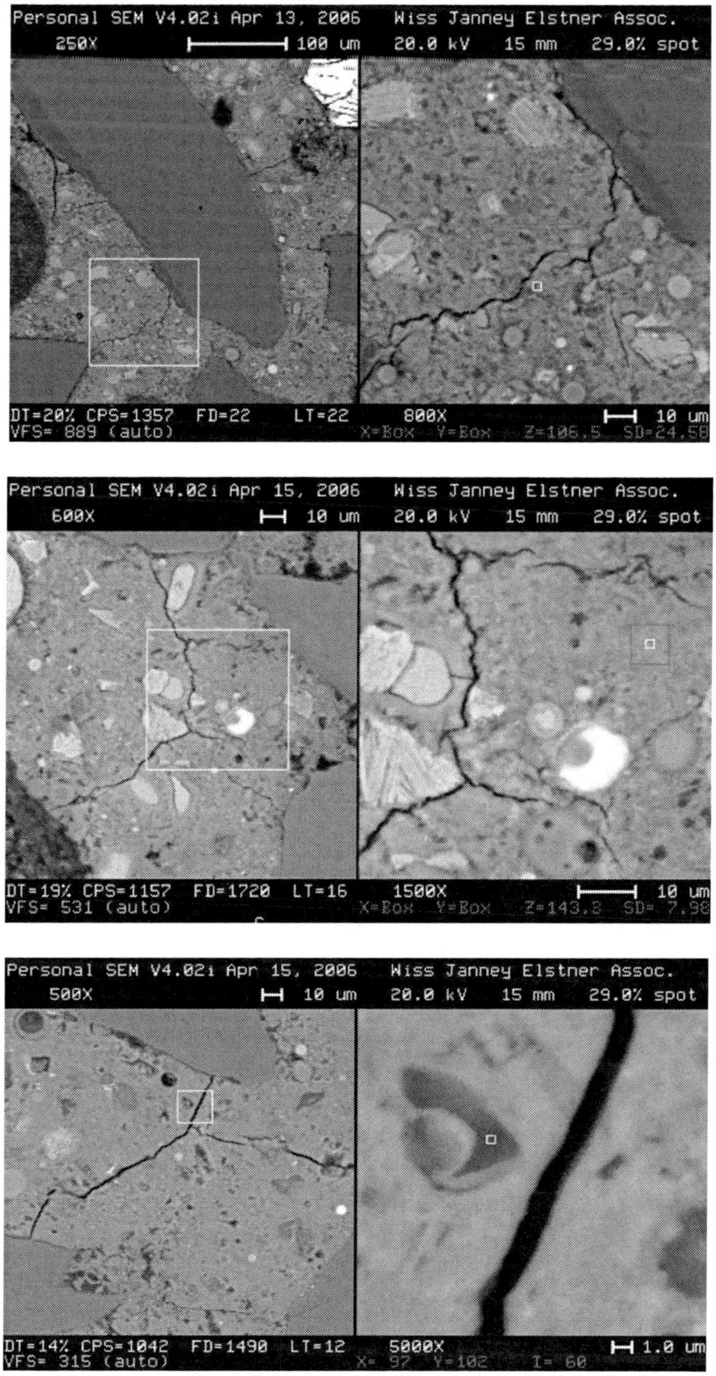

Figure 3.10. (continued).

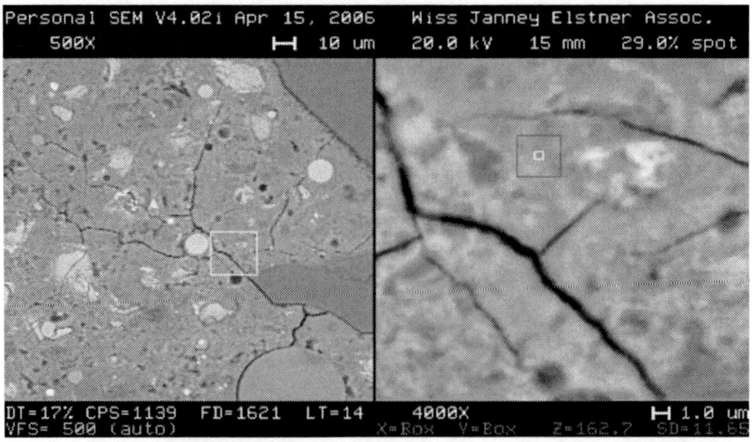

Figure 3.10. a) BSE image of sample NC-4-4-8 shows a discontinuous microcrack; the area enclosed by the white box in the left panel is enlarged on right. The image on the right shows the microcracks are substantially less than 10 μm wide. b) BSE image of sample NC-4-0-8 showing the longer microcracks that are common in this sample. The area enclosed by the white box in the left panel is enlarged on right. The scale bars are above and below the left and right images. c) BSE image of sample C-7-4-8 showing longer and wider microcracks that are common in this specimen. These microcracks typically pass between aggregate particles. d) BSE images of sample C-7-0-8 showing short, narrow microcracks branching from the wider, longer microcracks.

SUMMARY

Concrete-sediment half-cell experiments have been conducted to assess the performance of concrete wasteforms for immobilization of low activity radioactive waste under hydraulically unsaturated conditions. Results presented here, and those previously presented, indicate sediment moisture content, iron content and the degree of carbonation of the concrete monolith provide integrated control on the immobilization of I, Re, and Tc. In general, diffusion increases as a function of water content and the inclusion of iron in non-carbonated concrete monoliths results in the formation of microcracks. In non-carbonated concrete monoliths, the inclusion of iron can reduce I, Re, and Tc migration through reduction of these species to less soluble species, which are better retained within the concrete wasteform. Carbonation of concrete monoliths increases the formation and size of microcracks within the concrete monolith. Iron, carbonated monoliths displayed the formation of large, branched microcracks. The presence of microcracks provides additional paths within the monolith for contaminant diffusion and migration. However, as also illustrated here, changes in one or more of these parameters can result in negating the effects of another. Results presented here underscore the necessity to evaluate wasteform performance under environmentally relevant conditions and to understand the fundamental physical and chemical processes controlling performance. Compilation of data presented here, along with previous results quantifying the diffusion of radionuclides from concrete wasteforms into the surrounding subsurface environment, will allow quantification of the complete diffusion profile and predictive simulation of the long-term performance of concrete wasteforms and the fate contaminants immobilized in cementitious waste forms.

ACKNOWLEDGMENTS

This work was conducted at Pacific Northwest National Laboratory, operated by Battelle Memorial Institute for the U.S. Department of Energy under contract DE-AC05-76RL01830. Funding for this project was provided by Fluor Hanford. The authors wish to thank Eric Clayton for conducting ICP-MS analyses.

REFERENCES

Al-Khayat H, MN Haque, and NI Fattuhi. 2002. "Concrete Carbonation in Arid Climate." *Materials and Structures* 35:421-426.

Almahamid I, JC Bryan, JJ Bucher, AK Burrell, NM Edelstein, EA Hudson, N Kaltsoyannis, WW Lukens, DK Shuh, H Nitsche, and T Reich. 1995. "Electronic and Structural Investigations of Technetium Compounds by X-Ray Absorption Spectroscopy." *Inorganic Chemistry* 34:193-198.

Anderson BE, U Becker, KB Helean, and RC Ewing. 2007. *Perhenate and Pertechnetate Behavior on Iron and Sulfur-Bearing Compounds* Materials Research Society, Warrendale, PA.

ANSI. 1986. *Measurement of the Leachability of Solidified Low-Level Radioactive Wastes Short-Term Test Procedure*, American Nuclear Society, Chicago.

Campbell DH, RD Sturm, and SH Kosmatka. 1991. "Detecting Carbonation." *Concrete Technology Today* 12(1):1-5.

Colton R. 1965. *The Chemistry of Rhenium and Technetium*. Interscience Publishers, London.

Faggiani R, CJL Lock, and J Poce. 1980. "The Structure of Ammonium Pertechnetate at 295, 208 and 141 K." *Acta Crystallographica Section B-Structural Science* 36:231-233.

Garrabrants AC, and DS Kosson. 2003. "Modeling Moisture Transport from a Portland Cement-Based Material During Storage in Reactive and Inert Atmospheres." *Drying Technology* 21(5):775-805.

Garrabrants AC, F Sanchez, C Gervais, P Moszkowicz, and DS Kosson. 2002. "The Effect of Storage in an Inert Atmosphere on the Release of Inorganic Constituents During Intermittent Wetting of a Cement-Based Material." *Journal of Hazardous Materials* B91:159-185.

Garrabrants AC, F Sanchez, and DS Kosson. 2004. "Changes in Constituent Equilibrium Leaching and Pore Water Characteristics of a Portland Cement Mortar as a Result of Carbonation." *Waste Manaagement* 24:19-36.

Gervais C, AC Garrabrants, F Sanchez, R Barna, P Moszkowicz, and DS Kosson. 2004. "The Effects of Carboantion and Drying During Intermittent Leaching on the Release of Inorganic Constituents from a Cement-Based Matrix." *Cemet and Concrete Research* 34:119-131.

Hartmenn T, P Paviet-Hartmann, JB Rubin, MR Fitzsimmons, and KE Sickafus. 1999. "The Effect of Supercritical Carbon Dioxide Treatment on the Leachability and Structure of Cemented Radioactive Waste-Forms." *Waste Management* 19:335-361.

Hills CD, REH Sweeney, and NR Buenfeld. 1999. "Microstrucural Study of Carbonated Cement-Solidified Synthetic Heavy Metal Waste." *Waste Management* 19:325-331.

Kotegov KV, ON Pavlov, and VP Shvedov. 1968. *Technetium*. Academic Press, New York.

Lamfers HJ, A Meetsma, GA Wiegers, and JL DeBoer. 1996. "The Crystal Structure of Some Technetium and Rhenium Dichalcogenides." *Journal of Alloys and Compounds* 241:34-39.

Lange LC, CD Hills, and AB Poole. 1996a. "The Effect of Accelerated Carbonation on the Properties of Cement-Solidified Waste Forms." *Waste Management* 16(8):757-763.

Lange LC, CD Hills, and AB Poole. 1996b. "Preliminary Investigation into the Effects of Carbonation on Cement-Solidifeid Hazardous Wastes." *Environmental Science and Technology* 30(1):25-30.

Mann FM, RJ Puigh II, SH Finfrock, J Freeman, E.J., R Khaleel, DH Bacon, MP Bergeron, PB McGrail, and SK Wurstner. 2001. *Hanford Immobilized Low-Activity Waste Performance Assessment: 2001 Version*, DOE/ORP-2000-24, Rev. B, Pacific Northwest National Laboratory, Richland, WA.

Mattigod SV, GA Whyatt, RJ Serne, PF Martin, KE Schwab, and MI Wood. 2001. *Diffusion and Leaching of Selected Radionuclides (Iodine-129, Technetium-99, and Uranium) through Category 3 Waste Encasement Concrete and Soil Fill Material*, PNNL-13639, PNNL, Richland.

Meyer RE, and WD Arnold. 1991. "The Electrode Potential of the Tc(IV)-Tc(VII) Couple." *Radiochimica Acta* 55:19-22.

Murray HH, SP Kelty, RR Chianelli, and CS Day. 1994. "Structure of Rhenium Disulfide." *Inorganic Chemistry* 34:841-853.

Sanchez F, AC Garrabrants, and DS Kosson. 2003. "Effects of Intermittent Wetting on Concentration Profiles and Release from a Cement-Based Waste Matrix." *Environmental Engineering Science* 20(2):135-153.

Sanchez F, C Gervais, AC Garrabrants, R Barna, and DS Kosson. 2002. "Leaching of Inorganic Contaminants from Cement-Based Waste Materials as a Result of Carbonation During Intermittent Wetting." *Waste Manaagement* 22:249-260.

Serne RJ, JL Conca, VL LeGore, KJ Cantrell, CW Lindenmeier, JA Campbell, JE Amonette, and MI Wood. 1993. *Solid-Waste Leach Characterization and Contaminant-Sediment Interactions*, PNL-8889, Vol. 1, Pacific Northwest Laboratory, Richland, WA.

Serne RJ, RO Lokken, and LJ Criscenti. 1992. "Characterization of Grouted LLW to Support Performance Assessment." *Waste Management* 12:271-287.

Serne RJ, WJ Martin, and VL LeGore. 1995. *Leach Test of Cladding Removal Waste Grout Using Hanford Groundwater*, PNL-10745, Pacific Northwest Laboratory, Richland, Washington.

Serne RJ, WJ Martin, VL LeGore, CW Lindenmeier, SB McLaurine, PFC Martin, and RO Lokken. 1989. *Leach Tests on Grouts Made with Actual and Trace Metal-Spiked Synthetic Phosphate/Sulfate Waste.*, PNL-7121, Pacific Northwest Laboratory, Richland, Washington.

Shannon RD. 1976. "Revised Effective Ionic Radii and Systematic Studies of Interatomic Distances in Halides and Chalcogenides." *Acta Crystallographica, Section A: Foundations of Crystallography* A32:751-767.

Smith RW, and JC Walton. 1991. "The Effects of Calcite Solid Solution Formation on the Transient Release of Radionuclides from Concrete Barriers." *Materials Research Society Symposium Proceedings* 212:403-409.

"Specification for Concrete Encasement for Contact-Handled Category 3 Waste. 1998." In: *Waste Management*, Unpublished Report.

Toxicity Characteristic Leaching Procedure. 1992. Method 1311, Federal Registry.

Venhuis MA, and EJ Reardon. 2001. "Vacuum Method for Carbonation of Cementitious Wasteforms." *Environmental Science and Technology* 35:4120-4125.

Walton JC, S Bin-Shafique, RW Smith, N Gutierrez, and A Tarquin. 1997. "Role of Carbonation in Transient Leaching of Cemtitious Wasteforms." *Environmental Science and Technology* 31(8):2345-2349.

Wellman DM, SV Mattigod, L Powers, KE Parker, LN Clayton, and MI Wood. 2007. *Concrete Property and Radionuclide Migration Tests*, PNNL-17676, Pacific Northwest National Laboratory, Richland, WA.

Wellman DM, SV Mattigod, GA Whyatt, L Powers, KE Parker, LN Clayton, and MI Wood. 2006. *Diffusion of Iodine and Rhenium in Category 3 Waste Encasement Concrete and Soil Fill Material*, PNNL-16268, Pacific Northwest National Laboratory, Richland, WA.

Wellman DM, SV Mattigod, GA Whyatt, L Powers, KE Parker, LN Clayton, and MI Wood. 2008. "Effect of Iron and Carbonation on the Diffusion of Iodine and Rhenium in Waste Encasement Concrete and Soil Fill Material under Hydraulically Unsaturated Conditions." *Applied Geochemistry* 23:2256-2271.

Wood MI, R Khaleel, PD Rittman, AH Lu, S Finfrock, RJ Serne, and KJ Cantrell. 1995. *Performance Assessment for the Disposal of Low-Level Waste in the 218-W-5 Burial Ground*, WHC-EP-0645, Westinghouse Hanford Company, Richland, WA.

In: Radioactive Waste: Sources, Types and Management ISBN: 978-1-62100-188-1
Editors: Satoshi Yuan and Wenxu Hidaka

Chapter 7

HIGHLY RADIOACTIVE WASTE DISPOSAL: A GLOBAL CONCERN

Kenneth A. Rogers
Department of Politics and Geography, Coastal Carolina University,
Conway, SC, US

ABSTRACT

Global inventories of spent nuclear fuel are expected to increase substantially in future years. Because of the widely accepted international principle of whoever generates the waste also is responsible for its safe and secure disposal, the focus for the long-term management of highly radioactive waste has been on national disposal efforts. However, it is unlikely that any national repository for highly radioactive waste will be ready to accept nuclear waste before 2020. As a result, there has been an increased realization that radioactive waste is a global concern that requires more international cooperation.

There are a variety of international actors involved in the management of highly radioactive waste. Clearly, the countries that generate the nuclear waste play a critical role in the management of the waste since they ultimately are responsible for constructing and operating disposal facilities. International conventions help to monitor and regulate national radioactive waste management efforts and establish international norms. International governmental organizations and non-governmental organizations are playing increasingly important roles since they promote research and development, share information, and lobby for various policies affecting nuclear power and highly radioactive waste disposal.

While the international community has made some progress in addressing the issue of safely managing the increasing inventories of highly radioactive waste, much still needs to be done. Ultimately, the safe and effective long-term management of highly radioactive waste must be handled on a global scale in order to reduce safety, environmental and security concerns.

INTRODUCTION

Nuclear generated power is an important global energy source and accounts for approximately 15 percent of the world's total electricity production. As energy demands increase over the next 20 years and concerns over global warming continue, nuclear generated power is projected to increase by as much as 2.6 percent annually to meet the increased electricity needs. (IAEA, 2007: 53) Today, over 10,000 metric tons of highly radioactive waste is produced globally each year as a byproduct of nuclear power generation, spent fuel reprocessing, and nuclear weapons production. (IAEA, 2008: 14) As additional reactors are constructed to meet growing global energy demands, international inventories of spent nuclear fuel are projected to increase almost 40 percent by 2020.[1] Because of the widely accepted principle of whoever generates the radioactive waste also is responsible for its safe and secure disposal, the focus of long-term highly radioactive waste management has been on national disposal efforts. However, it is unlikely that any national repository for highly radioactive waste will be ready to accept the waste before 2020. As a result, there has been an increased realization that radioactive waste is a global concern that requires a global solution. While there has been some international cooperation on highly radioactive waste issues, promoting nuclear waste disposal cooperation at the global level has been slow to develop.

THE ACTORS

A variety of actors are engaged in the long-term management of highly radioactive waste. Clearly, the countries that generate the nuclear waste play a critical role in its management since they ultimately are responsible for its safe and secure disposal. However, it has become increasingly clear that the long-term management of highly radioactive waste cannot be addressed exclusively at the national level. A variety of non-state actors are now being recognized as playing an increasingly important role in nuclear waste management. For example, a number of international conventions and agreements have been concluded that monitor and regulate the management of radioactive waste. In addition, international governmental organizations and non-governmental organizations also are playing an increasingly important role in radioactive waste disposal policy since they monitor national efforts, promote research and development, share information, and lobby for various policies affecting nuclear power and highly radioactive waste disposal.

National Efforts

Today, a number of countries are actively involved in pursuing radioactive waste disposal programs. (See Table 1—Countries with Active Highly Radioactive Waste Management Programs) However, no country has yet completed a geologic disposal facility for civilian nuclear waste. While the United States long has been expected to open the world's first

[1] Estimated totals have been interpolated from IAEA data. See IAEA (2008) Estimation of Global Inventories of Radioactive Waste and Other Radioactive Materials, IAEA-TECDOC-1591, June 2008, pp. 14-15 [http://www-pub.iaea.org/MTCD/publications/PDF/te_1591_web.pdf].

geologic repository for civilian highly radioactive waste at Yucca Mountain, the political realities today are such that the completion of the current project is in doubt. Thus, if Finland and Sweden remain on schedule, they may open the first national repositories for highly radioactive waste. The remaining countries are not scheduled to build repositories until sometime in the distant future. Not all countries believe that disposal is the preferred method of coping with the mounting inventories of spent nuclear fuel; some consider the spent fuel rods as a resource to be reprocessed.[2] While the reprocessing option does reduce the overall amount of radioactive waste that ultimately is generated, it does not eliminate the need for a long-term disposal policy since some highly radioactive waste is a byproduct of the reprocessing cycle.

Table 1. Countries with Active Highly Radioactive Waste Management Programs

Region /Country	Projected Repository Opening
ASIA	
China	earliest 2040
Japan	at earliest 2037
EUROPE	
Belgium	after 2035
Finland	at earliest 2020
France	well after 2020
Germany	no projected date
Russia	to be determined
Spain	to be determined
Sweden	after 2017
Switzerland	after 2050
United Kingdom	to be determined
NORTH AMERICA	
Canada	after 2034
United States	after 2017

Source: Adapted from OCRWM (2001) "Radioactive waste: an international concern" [http://ocrwm.doe.gov/factsheets/doeymp0405.shtml].

There are several features common to those countries that do follow a long-term highly radioactive waste disposal policy. First, countries have been slow to address the problem of the mounting inventories of highly radioactive waste. Second, the "polluter pays" principle is followed (i.e., the nuclear waste generator is responsible for paying for the disposal costs). Third, special disposal funds paid by the nuclear operators have been set up to cover costs for feasibility studies and operating expenses. Fourth, independent agencies have been established to be responsible for the final disposal of nuclear waste. Fifth, the multi-barrier approach is used; using both natural and man-made engineered barriers. Finally, a deep underground geologic repository is the preferred method for disposal.[3]

---

[2] The following countries discussed in this chapter reprocess spent fuel: China, France, Japan, Russia, Switzerland, and the United Kingdom.

[3] For an excellent summary of national efforts, see the IAEA website for national reports submitted for the review meetings of the Joint Convention, the Second Review Meeting of the Contracting Parties to the Joint Convention [http://www-ns.iaea.org/conventions/waste-jointconvention.htm]. Also see Rogers (2009) "Fire in

Asia

China plans to expand its nuclear power electricity generation aggressively over the next several years, and is expected to produce over 6,000 metric tons of highly radioactive waste by 2020. (EIA, 2002: 1) By 2025, the Chinese are projected to be the world's fifth largest annual producer of nuclear waste. Since China has not generated much highly radioactive waste to date, and has committed to reprocessing spent fuel, it is in a good position to plan ahead for the long-term management of nuclear waste. Some preliminary experiments have been conducted on geologic disposal. In all, five potential repository sites, including a proposed underground research laboratory site in the Gobi Desert are being studied. However, China is not expected to have an operational highly radioactive waste disposal facility until 2040 at the earliest. (OCRWM, 2001: 1)

Japan is projected to accumulate about 44,000 metric tons of highly radioactive waste by 2020—the world's fifth largest inventory of spent nuclear fuel. (EIA, 2002: 1) The government announced in 2005 that its official policy is to reprocess spent nuclear fuel. (IAEA, 2006b: 9) Japan's goal is to implement a "closed" nuclear fuel cycle where spent fuel is viewed as a resource to be reprocessed so it can be reused. Because of reprocessing, it is anticipated that a reduced amount of nuclear waste will have to be deposited in the repository. Japan considers deep geologic disposal as the only suitable method for the long-term management of highly radioactive waste. (METI, 2005: B-3) A multi-barrier approach using natural and engineered barriers will be used. The vitrified waste will be placed in canisters, positioned in the repository, and then surrounded by a clay buffer material. (NUMO, 2003b: 2) Four basic principles govern highly radioactive waste disposal in Japan: liability rests with the waste generator, the generation of radioactive waste should be minimized, waste treatment and disposal should be reasonable, and the ultimate disposal should be based on a mutual understanding with the public. (METI, 2008: 10) A disposal facility is not projected to be ready to accept radioactive waste until 2037. (NUMO, 2003a: 1)

Europe

Belgium chose to abandon nuclear generated power in 2003 by prohibiting the construction of any new plants and limiting the operational period of existing plants to 40 years. (FANC, 2006: 7) The 40-year operational limit will not have an impact until 2015. (NEA, 2007a: 15) Because over half of Belgium's electricity is generated by commercial nuclear power, political and economic realities might force the government to reconsider the abandonment of nuclear power. Overall, Belgium is projected to generate about 4,500 metric tons of spent nuclear fuel before the last commercial nuclear power facility is closed. (EIA, 2002: 1; FANC, 2006: 9) Disposal in a deep geologic repository presently is the preferred method for the long-term management of highly radioactive waste. (FANC, 2006: 13-14) A sedimentary clay area in north-east Belgium (Boom Clay) has been identified as a possible repository site. However, no official decision has been made to select the Boom Clay formation for the disposal facility. (FANC, 2006:14) The current focus is to use a multi-barrier design approach using the clay formation as a natural barrier and adding a man-made engineered barrier system to enhance confidence in the feasibility of the repository to contain

the hole: A review of national spent nuclear fuel disposal policy," Progress in Nuclear Energy, Vol. 51, No. 2, March 2009, for a more in-depth review of national nuclear waste disposal programs.

the radioactive material. (FANC, 2006: 15) A national repository for highly radioactive waste is not expected to be ready until sometime after 2035. (OCRWM, 2001: 3)

Finland is projected to produce about 2,800 metric tons of spent nuclear fuel by 2020. (EIA, 2002: 1) In the mid-1990s the government announced that it would stop reprocessing its spent fuel and is now planning to construct and operate a permanent repository for highly radioactive waste. Initial preparations for a permanent disposal option for spent nuclear fuel were begun in the early 1980s. Site characterization studies started in 2004 at the ONKALO underground rock characterization facility at the Olkiluoto nuclear facility to see if the geological formation would be feasible for a repository. (STUK, 2005: 16) The multi-barrier repository design will consist of a network of tunnels constructed at a depth of 400-500 meters. The cast iron and copper disposal canisters will be surrounded by a bentonite backfill to act as a sealant. (STUK, 2005: 17) Finland is not expected to have an operational permanent highly radioactive waste disposal facility until 2020 at the earliest. (OCRWM, 2001: 3)

France relies more on nuclear generated electricity than any other country in the world and is projected to produce about 51,000 metric tons of radioactive waste by 2020—the world's fourth largest accumulation of nuclear waste behind only the United States, Canada and the United Kingdom. (EIA, 2002: 1) Although the French are one of the largest generators of spent nuclear fuel, total waste inventories are lower than would be expected since it reprocesses the spent fuel rods. Since 1994, feasibility studies of a highly radioactive waste repository in the clay Callovo-Oxfordian formation at the Meuse/Haute-Marne underground research laboratory have been conducted. France also has been studying the feasibility of a repository in granite by working with Sweden, Switzerland, Canada, and Finland at their underground research laboratories. (Andra, 2006: 2) Today, France is following a dual strategy in dealing with highly radioactive waste: (1) partitioning and transmutation; and (2) long-term waste management (i.e., disposal). (NEA, 2006c: 17) Partitioning and transmutation would not be radioactive waste disposal options per se, but they would reduce the inventories of spent nuclear fuel. Disposal would take place in a repository based on principle of reversibility—the ability to reverse the process at any time and retrieve the highly radioactive waste. (Andra, 2006: 3) However, France is not expected to have an operational permanent highly radioactive waste disposal facility until sometime well after 2020. (OCRWM, 2001: 3)

Germany officially abandoned nuclear generated power with passage of the 2002 Act to phase out nuclear energy. This was a direct result of growing opposition to nuclear power, which corresponded to the strengthening environmental movement. By 2008, over half of the nuclear power plants in Germany either had been dismantled or were in the process of being dismantled. All six nuclear reactors in the former East Germany have been shut down. (BMU, 2008: 14) As a result, Germany is projected to produce only about 17,500 metric tons of highly radioactive waste by 2020. (EIA, 2002: 1) Feasibility testing of the Gorleben salt dome for a national repository has been put on hold for the next 3-10 years to review design and safety related questions. (BMU, 2008: 16) While there is no set date for completion of an operational permanent highly radioactive waste disposal facility in Germany, the federal government hopes to establish a deep geologic repository by 2030. (BMU, 2005: 14) In Germany, all types of radioactive waste eventually are expected to be stored in a deep underground repository.

The Russian Federation is projected to generate approximately 32,000 metric tons of highly radioactive waste by 2020. (FAEA, 2006: 53; EIA, 2002: 1) Although spent fuel stocks could be higher since Moscow reportedly is planning to double its nuclear energy output by 2020. (WNA, 2009a: 1) Russia does not view spent nuclear fuel as radioactive waste, but instead as a resource to reuse. Spent fuel either is to be reprocessed or stored. In fact, over 12,000 metric tons of spent nuclear fuel has been reprocessed since 1977. (FAEA, 2006: 12-13) The remaining waste either is stored on-site or at a centralized storage facility. According to the 2006 national report on compliance for the Joint Convention on the Safety of Spent Fuel Management and on the Safety of Radioactive Waste Management, "the disposal of spent fuel in the Russian Federation is not envisaged." (FAEA, 2006: 63) However, in spite of the Russian focus on reprocessing, highly radioactive waste as a by-product of the reprocessing cycle still will need to be managed for the long-term. As such, Russia has been investigating the feasibility of salt, granite, clay, and basalt as possible host geologic mediums for a possible future repository; however, there is no projected date for completion of an operational permanent highly radioactive waste disposal facility due to the focus on reprocessing. (OCRWM, 2001: 3)

Spain announced in 2006 that it intended to phase out nuclear generated power and instead rely more on renewable sources such as wind generated power. Further pressure to shut down Spain's nuclear power industry likely will occur as a result of the reported minor radioactive leaks at several nuclear plants and a fire at the Vandellos II plant in 2008. (Reuters, 2008a: 1-2; Reuters 2008b: 1) While Spain is projected to generate about 7,000 metric tons of highly radioactive waste by 2020, the total amount could be lower if nuclear power is phased out over the next several years. (EIA, 2002: 1) Although Spain initially intended to reprocess spent nuclear fuel, it abandoned reprocessing in 2005. The waste producers are responsible for paying for future disposal costs. However, a decision on how to pursue long-term highly radioactive waste management has been postponed until after 2010; instead, Spain is focusing on centralized interim storage. There is no projected date for the completion of an operational permanent highly radioactive waste disposal facility in Spain. (MITYC, 2005: 6; OCRWM, 2001: 3)

Sweden is projected to generate about 7,000 metric tons of spent nuclear fuel by 2020. Agreements for reprocessing with both the UK and France were terminated in the early 1980s after only a small portion of spent fuel was delivered for reprocessing. (MSDS, 2005: 23) After spending a year in a cooling pool, spent nuclear fuel is transported to a centralized interim storage facility where it will be stored for about 30 years before being transported to a permanent repository. Utilities pay a fee into the Nuclear Waste Fund (created to provide sufficient funding for operating the repository). (MSDS, 2005: 20) The repository for highly radioactive waste is envisioned to be a multi-barrier repository design incorporating natural and man-made barriers to ensure that the nuclear waste remains contained. Sweden plans on submitting its license application for a repository for spent nuclear fuel in 2010. (Sweden, 2008: 195) However, an operational disposal facility is not expected to be ready until after 2017 at the earliest. (Thegerström, 2004: 2)

While Switzerland historically has relied on hydroelectric generated power for its electricity, increased energy needs prompted the Swiss government to pursue nuclear power in the 1960s. Switzerland is projected to generate about 3,000 metric tons of spent nuclear fuel by 2020. (EIA, 2002: 1) The Swiss view spent nuclear fuel as a resource that can be reprocessed and no spent fuel has been declared as waste. Initially, Switzerland studied the feasibility of crystalline rock for a potential repository. However, that site was not found suitable; now the Opalinus clay formation in the northern part of the country is being considered. (HSK, 2008: 9) An operational permanent highly radioactive waste disposal facility is not expected to be ready until after 2050. (OCRWM, 2001: 3)

The United Kingdom is projected to produce about 58,000 metric tons of highly radioactive waste by 2020—the world's third largest accumulation of highly radioactive waste. (EIA, 2002: 1) Since the UK reprocesses its own spent nuclear fuel, existing spent fuel inventories have not been designated as radioactive waste for disposal. (Defra, 2006: 65) Nevertheless, in 2006 a special government report called for geologic disposal. (CoRWM, 2006: 111) There is no projected date for completion of an operational permanent highly radioactive waste disposal facility in the United Kingdom. (OCRWM, 2001: 3)

North America

Canada has a strong commitment to nuclear power because of an abundance of uranium. Canada is projected to generate approximately 59,000 metric tons of spent nuclear fuel by 2020—the world's second largest accumulation of spent nuclear fuel. (EIA, 2002: 1) Canada has explored the viability of a deep underground repository (the Underground Research Laboratory) for containing radioactive waste within a granite rock mass. (URL, 2005: 1) A 1998 report concluded that a deep geologic disposal 500 to 1000 meters deep within stable rock would provide a safe long-term storage option for highly radioactive waste. Two main principles govern Canadian radioactive waste disposal: (1) use of multiple engineered barriers to ensure containment and isolation, and (2) procedures to monitor the performance of the engineered barriers. (CNR, 2005: 53) In spite of this progress, however, Canada is not expected to have an operational permanent highly radioactive waste disposal facility until sometime after 2034. (OCRWM, 2001: 3)

The United States is projected to produce almost 83,000 metric tons of highly radioactive waste by 2020—the world's largest accumulation of highly radioactive waste. (EIA, 2002: 1) While the National Academy of Sciences recommended as early as 1957 that highly radioactive waste should be buried deep underground, it was not until 1982 that the U.S. government began the process of identifying a suitable site for the permanent disposal of highly radioactive waste. Five years later, over Nevada's objections, Yucca Mountain was singled out as the only site to be studied for the proposed national repository. In 2008, a license application was filed seeking approval for construction of the repository. The decision on construction approval is expected by 2012. The United States plans to use the multiple barrier principle (i.e., use of natural and man-made engineered barriers) and the concept of defense in depth (i.e., multiple safety functions of isolation, containment, limitation and retardation). While the proposed repository is located in a remote, arid location, there are geologic and environmental concerns that have sparked considerable political controversy. Today, the political realities are such that the viability of the Yucca Mountain Project is in doubt. Energy Secretary Steven Chu announced before a Senate hearing in March 2009 that the Obama Administration no longer was considering Yucca Mountain as an option for disposing of the country's highly radioactive waste. (Murkowski, 2009: 1)

International Conventions Governing Radioactive Waste Disposal

Because nuclear waste disposal is a global concern, there are a number of international conventions and agreements that govern the long-term management of highly radioactive waste (See Table 2 – International Conventions and Agreements Governing Highly Radioactive Waste Disposal).

Table 2. International Conventions and Agreements Relevant for Highly Radioactive Waste Disposal

1960	*Paris Convention on Third Party Liability in the Field of Nuclear Energy* (and subsequent protocols): 14 European countries and Turkey agreed to provide financial protection against damage resulting from a nuclear incident occurring on the territory of a signatory country.
1963	*Vienna Convention on Civil Liability for Nuclear Damage* (and subsequent protocols): established minimum standards to provide financial protection against damage resulting from a nuclear incident.
1971	*Convention Relating to Civil Liability in the Field of Maritime Carriage of Nuclear Materials:* regulates liability for damages suffered as a result of an accident involving nuclear materials and provides a mechanism to resolve conflicts from various conventions involving ship owners' liability.
1972	*Convention on the Prevention of Marine Pollution by Dumping of Wastes and Other Matter* (and 1993 protocol): prohibits the dumping (deliberate disposal) of radioactive waste at sea.
1980	*Convention on the Physical Protection of Nuclear Material* (and 2005 amendment): requires countries to protect the nuclear material within controlled areas that are under constant surveillance with sufficient security to prevent the unauthorized access or removal of the radioactive material.
1986	*Convention on Early Notification of a Nuclear Accident*: requires countries to notify international authorities about any nuclear accident which has the potential for an international transboundary release that could be of radiological safety significance for another country.
1986	*Convention on Assistance in the Case of Nuclear Accident or Radiological Emergency*: sets out a framework for international cooperation to help facilitate timely assistance and support in the event of a nuclear accident or radiological emergency.
1990	*Code of Practice on the International Transboundary Movement of Radioactive Waste*: asks countries to take appropriate steps to ensure that radioactive waste is minimized, managed safely, and disposed of properly. The code also acknowledges that countries have the sovereign right to prohibit the movement of radioactive waste through their territory.
1994	*Convention on Nuclear Safety*: obligates parties to the Convention to take appropriate steps to maintain a high level of nuclear safety.
1997	*Joint Convention on the Safety of Spent Fuel Management and on the Safety of Radioactive Waste Management*: objective is to achieve and maintain a high level of safety worldwide in the management of spent nuclear fuel as well as high-level radioactive waste from military programs that has been transferred to civilian control.
1997	*Convention on Supplementary Compensation for Nuclear Damage*: purpose is to establish a worldwide liability regime by ensuring that liability compensation for a nuclear incident is not less than 300 million Special Drawing Rights (approximately $450 million).
2001	*Code of Conduct on the Safety and Security of Radioactive Sources* (and supplementary guidance): purpose is to encourage the coordination and implementation of national policies to achieve and maintain a high level of safety and security of radioactive sources.

Source: IAEA (2009f) "Treaties, Conventions and Agreements Related to the IAEA's Work" [http://www.iaea.org/Publications/Documents/Treaties/index.html] and IAEA (2009g) "Treaties, Conventions and Agreements under IAEA Auspices" [http://www.iaea.org/Publications/Documents/ Conventions/index.html].

While the earliest conventions entered into force in the 1960s, most agreements have been concluded since the Chernobyl accident in 1986. These treaties and agreements have improved international cooperation in the area of radioactive waste management and have helped to create global norms. For example, there are several generally accepted international principles outlined by international conventions: the generator of the waste ultimately is responsible for its long-term management, a country has a right to refuse the transport of radioactive waste through its territory, and liability for damage caused as a result of an incident involving nuclear materials rests with the operator. However, since nuclear power generating countries are not required to join the conventions, and the agreements generally lack real enforcement mechanisms, the overall impact of these agreements on global radioactive waste management policy has been somewhat limited.

The *Convention on Third Party Liability in the Field of Nuclear Energy* (better known as the Paris Convention) was adopted in 1960 and entered into force in 1968.[4] Fourteen European countries and Turkey have ratified or acceded to the Convention agreeing to provide financial protection against damage resulting from a nuclear incident (including from radioactive waste) occurring on the territory of a signatory country. The Convention has been amended by protocols in 1964, 1982, and 2004 (not yet ratified). (NEA, 2007c: 1) Initially, the Convention set liability at 15 million Special Drawing Rights (SDRs)—approximately $22 million.[5] (NEA, 2008a: Art 7.b) However, upon ratification, the 2004 protocol will set liability at 700 million euro for participating countries. (NEA, 2004: 18) Disputes will be heard by the Steering Committee and if not settled, will be submitted to the European Nuclear Energy Tribunal that was established by the Convention on the Establishment of a Security Control in the Field of Nuclear Energy. (NEA, 2008a: Art 17) That convention initially was concluded to prevent the proliferation of nuclear weapons. However, since that convention has been superseded by other international agreements, the Tribunal now serves to resolve differences concerning the interpretation or application of the Paris Convention and the Brussels Convention on Jurisdiction and the Enforcement of Judgments in Civil and Commercial Matters.[6] Although decisions issued by the Tribunal are binding, no cases have yet been brought before it for adjudication. (NEA, 2007b: 1)

The *Vienna Convention on Civil Liability for Nuclear Damage* was adopted in 1963 and entered into force in 1977. The objective of the Convention is to establish minimum standards to provide financial protection against damage resulting from a nuclear incident. (IAEA, 1996: Preamble) Initially, the Convention set a minimum of five million U.S. dollars and required operators to maintain insurance or other financial security. (IAEA, 1996: Art V.1, Art VII.1) Irrespective of fault, liability of the operator for any nuclear damage is "absolute." (IAEA, 1996: Art IV.1) Three subsequent protocols amending the Convention also have been adopted. The 1992 *Joint Protocol Relating to the Application of the Vienna Convention and the Paris Convention* establishes a link between the Vienna and Paris Conventions and states that either convention will apply to a nuclear incident to the exclusion of the other. (IAEA, 1992, Preamble, Art III.1) The 1997 *Protocol to Amend the 1963 Vienna Convention on Civil Liability for Nuclear Damage* states that a party to the protocol is bound by the Convention and sets liability at not less than 300 million SDRs—approximately $450 million. (IAEA,

---

[4] See NEA (2008a) "Convention on Third Party Liability in the Field of Nuclear Energy of 29th July 1960, as amended by the Additional Protocol of 28th January 1964 and by the Protocol of 16th November 1982," [http://www.nea.fr/html/law]. Three signatory countries (Austria, Luxembourg, and Switzerland) have yet to ratify the Convention.

[5] Special Drawing Rights (SDRs) are defined by the International Monetary Fund in terms of an artificial basket of major currencies used in international trade and finance that include the euro, the pound sterling, the Japanese yen and the U.S. dollar. (IMF, 2009: 1) As of 1 January 2009, one SDR was valued at $1.54.

[6] The Tribunal has seven independent judges appointed for five-year terms.

1997b: Art 7.1.a) A party to the 1997 Protocol but not to the 1963 Treaty is bound by the Vienna Convention as amended. (IAEA, 1997b, Art 19.1) The 1999 *Optional Protocol Concerning the Compulsory Settlement of Disputes to the Vienna Convention on Civil Liability for Nuclear Damage* outlines the adjudication of disputes. Unless parties to the dispute have agreed to another form of settlement, the International Court of Justice generally will have compulsory jurisdiction. (IAEA, 1999: Art I)

The *Convention Relating to Civil Liability in the Field of Maritime Carriage of Nuclear Materials (NUCLEAR)* was adopted in 1971 and entered into force in 1975. The purpose of the Convention is to resolve conflicts with other conventions regulating liability with respect to damages resulting from the maritime carriage of nuclear materials. (IMO, 2002: 1) The Convention aims to ensure "that the operator of a nuclear installation will be exclusively liable for damage caused by a nuclear incident occurring in the course of maritime carriage of nuclear material." (AMLG, 1971: Preamble)

The *Convention on the Prevention of Marine Pollution by Dumping of Wastes and Other Matter* was adopted in 1972 and entered into force in 1975. The Convention prohibits the dumping (deliberate disposal) of highly radioactive waste at sea. (IAEA, 1974: Art IV, Annex 1.6) An amendment to the Convention in 1978 states that disputes shall be submitted by agreement between the parties to the International Court of Justice for adjudication or upon the request of one of the parties to arbitration. (ENTRI, 1978: 2) A subsequent 1993 Protocol prohibits the dumping of even low-level radioactive waste into the ocean. (IMO, 2006: 2)

The *Convention on the Physical Protection of Nuclear Material* was adopted in 1979 and entered into force in 1987. While the amended Convention outlines a number of principles for the physical protection of nuclear materials and facilities, the overlying theme is that the ultimate responsibility for the protection of nuclear materials rests with the state. (IAEA, 2006a: 3-5) Annex I to the Convention establishes levels of physical protection for highly radioactive materials while in use, storage and transport. It requires countries to protect the nuclear materials within controlled areas that are under constant surveillance with sufficient security to prevent the unauthorized access or removal of the radioactive material. The Convention also requires countries to report any unauthorized theft or use of nuclear materials to international authorities. Countries also are required either to prosecute or cooperate in the prosecution of individuals charged with violating the Convention. (IAEA, 1980: 1-13) Initially, signatories to the Convention were asked to resolve disagreements peacefully among themselves, and any party had the right to request arbitration or submit the case to the International Court of Justice for adjudication. However, any party could declare that it did not consider itself bound by either the arbitration or ICJ decision. (IAEA, 1980: Art 17.1-3) Concern over the implications of potential acts of terrorism against nuclear materials and nuclear facilities prompted an amendment in 2005 that made the Convention *legally binding* for parties to protect nuclear materials and facilities against theft and sabotage. (IAEA, 2006a: Art 2A; IAEA, 2005b: 1)

The *Convention on Early Notification of a Nuclear Accident* was adopted in 1986 and entered into force the same year. Prompted by the lack of proper notification by Soviet authorities of the radiation dangers to surrounding countries associated with the Chernobyl nuclear power plant accident in 1986, the Convention requires countries to notify the IAEA "forthwith" about any nuclear accident which has the potential for a transboundary release that could be of radiological significance for another country. (IAEA, 1986a: Art 2 a-b, Art 5.1) All stages of the nuclear cycle (e.g., the reactor, storage, transport of the highly radioactive waste, and disposal) are subject to the provisions of the Convention. (IAEA, 1986a: Art 1.2) Signatories to the Convention are asked to resolve disagreements peacefully among themselves. In the event a dispute cannot be settled within six months, any party has the right to request arbitration or submit the case to the International Court of Justice for adjudication. However, any party can declare that it does not consider itself bound by either the arbitration or the ICJ decision. (IAEA, 1986a: Art 11.1-3)

The *Convention on Assistance in the Case of Nuclear Accident or Radiological Emergency* was adopted in 1986 and entered into force the following year. Adopted after the Chernobyl accident, it sets out a framework for international cooperation to help facilitate timely assistance and support in the event of a nuclear accident or radiological emergency. The Convention outlines provisions for countries to request assistance from other parties to help cope with a radiological emergency, how the assistance will be coordinated, points of contact, reimbursement of costs, privileges and immunities, claims and compensation, and the settlement of disputes. Overall control and direction of third party assistance is vested with the requesting state (IAEA, 1986b: Art 3) Assisting parties are afforded immunity from arrest, detention and legal proceedings. (IAEA, 1986b: Art 8.2, Art 10.2.a) Signatories to the Convention are asked to resolve disagreements peacefully among themselves. In the event a dispute cannot be settled within six months, any party has the right to request arbitration or submit the case to the International Court of Justice for adjudication. However, any party can declare that it does not consider itself bound by either the arbitration or ICJ decision. (IAEA, 1986b: Art 13.2-3)

The *Code of Practice on the International Transboundary Movement of Radioactive Waste* was adopted in 1990 at a general conference of the International Atomic Energy Agency (IAEA) to provide *advisory* guidelines for countries to follow during the transboundary shipment of nuclear waste. The Code acknowledges that countries have the sovereign right to prohibit the transport of radioactive waste through their territory. Moreover, the transboundary movement of radioactive waste should conform to international standards and each country should ensure the transport of the nuclear waste takes place only with the prior notification and consent of all parties affected. Each country also should have sufficient capacity to manage or dispose of the waste safely, and have sufficient liability and compensation protection. (IAEA, 1990: 2-3) The Code asks parties to the agreement to take appropriate steps to ensure that radioactive waste is minimized, managed safely, and disposed of properly. This Code is only advisory in nature and lacks enforcement mechanisms.

The *Convention on Nuclear Safety* was adopted in 1994 and entered into force in 1996. The objectives of the Convention are to achieve a high level of nuclear safety, establish effective protections against radiological hazards, and mitigate consequences from accidents by promoting international cooperation. (IAEA, 1994: Art 1.i-iii) Sometimes referred to as the sister convention to the Joint Convention on the Safety of Spent Fuel Management and on the Safety of Radioactive Waste Management, this convention asks parties to take appropriate steps to maintain a high level of safety for a number of areas such as quality assurance, assessment and verification of safety, radiation protection, and emergency preparedness. There is special attention paid to the siting, design and construction, and operation of nuclear facilities. (IAEA, 1994: Art 17-19) The Convention also states that appropriate steps should be made to minimize the generation of highly radioactive waste. (IAEA, 1994: Art 19.viii) The Convention does not have enforcement mechanisms and there is no compulsory arbitration of disputes. Instead, signatories are asked to resolve disagreements among themselves.

The *Joint Convention on the Safety of Spent Fuel Management and on the Safety of Radioactive Waste Management* (also referred to as the Joint Convention) was adopted in 1997 and entered into force in 2001. The Joint Convention is the most noteworthy international treaty dealing with the disposition of highly radioactive waste because countries are required to report periodically on their progress with radioactive waste management.[7] The required national reports for the periodic review meetings are an important resource for understanding national long-term highly radioactive waste management goals and policies. The objective of the Convention is to achieve a high level of safety worldwide in the management of spent nuclear fuel. (IAEA, 1997a: Art 1.i) High-level radioactive waste from military programs is not covered under this convention unless the nuclear waste has been transferred permanently to civilian control. (IAEA, 1997a: Art 3.iii) The Joint Convention acknowledges that the management of radioactive waste is the responsibility of the generation that produces it, and obligates parties to ensure the safety of any spent fuel management facility. The Convention covers a variety of issues such as the siting, design, construction, and operation of the facility, and the ultimate disposal of the spent fuel. (IAEA, 1997a: Art 4.i-vii) According to the Convention, the transboundary shipment of nuclear waste should occur only with prior notification and consent of the country through which the waste will be shipped. (IAEA, 1997a: Art 27.i) Of note, the Joint Convention prohibits the transboundary movement of radioactive waste south of latitude 60 degrees South (i.e., to Antarctica) for storage or disposal. (IAEA, 1997a: Art 27.2) According to the Convention, disagreements will be arbitrated through normal channels of international law. (IAEA, 1997a: Art 38) A weakness of the Joint Convention is that some countries that produce highly radioactive waste are not party to the Convention.[8]

The *Convention on Supplementary Compensation for Nuclear Damage* was adopted in 1997 and entered into force in 1998. The purpose of this convention is to establish a *global* liability regime by ensuring that liability compensation for a nuclear incident is not less than 300 million SDRs—approximately $450 million. This convention also hopes to promote nuclear safety by encouraging regional and international cooperation. (IAEA, 1998a: Preamble, Art III.1) Only countries that either are party to the Paris or Vienna Conventions on liability, or countries that are party to the Convention on Nuclear Safety and have declared that their national laws comply with provisions of the Annex to this convention will be allowed to accede to the Convention. (IAEA, 1998a: Art XIX.1) In the event a dispute cannot be settled within six months, any party has the right to submit the case for arbitration or refer the case to the International Court of Justice for adjudication. (IAEA, 1998a: Art XVI.2)

The Code of Conduct on the Safety and Security of Radioactive Sources was strengthened after the terrorist events of September 11, 2001. The purpose of the Code is to achieve and maintain a high level of safety and security of radioactive sources by encouraging the development, coordination, and implementation of national policies that will provide an adequate system of regulatory control over radioactive materials. (IAEA, 2005a: Art II.5a.i) The objectives of the Code are to prevent the unauthorized access or damage to, and loss, theft or unauthorized transfer of radioactive sources. (IAEA, 2005a: Art II.5a.ii) Transfers of

---

[7] The Third Review Meeting is scheduled for May 2009.

[8] As of January 2009, 27 countries with nuclear power plants are party to the Convention. (IAEA, 2009d: 1-2).

radioactive sources are supposed to take place only with the prior notification by the exporting state and consent by the importing state. (IAEA 2005a: Art III.23). In 2005, the *Supplementary Guidance on the Import and Export of Radioactive Sources* was added to the Code to provide specific procedures to facilitate the transfer of radioactive materials covered by the Code. In 2006, provisions were added to provide for the periodic exchange of information and lessons learned towards implementing the provisions of the Code. (IAEA, 2009a: 1) The Code is advisory in nature and is not legally binding.

IGOs and NGOs

Global and regional international governmental organizations (IGOs) and non-governmental organizations (NGOs) serve several important functions in the management of highly radioactive waste. First, they both provide a forum for the exchange of information. Second, they both foster research and development on how best to manage radioactive waste. Third, they both collect and publish relevant technical and scientific information to improve cooperation in nuclear waste management. Fourth, IGOs establish international safety standards. Finally, IGOs implement international conventions that monitor and regulate highly radioactive waste.

IGOs

While countries generally are reluctant to surrender national sovereignty over important issues, at times they are willing to allow an IGO some measure of control if membership in the international organization is perceived to be in the country's self interest. Radioactive waste management is one area where international cooperation can provide substantial benefits to participating countries. The International Atomic Energy Agency and the Organization for Economic Cooperation and Development's Nuclear Energy Agency are two principle IGOs that have an important impact on radioactive waste management policy since they provide a forum for communication, disseminate technical reports, monitor highly radioactive waste management practices and policies, and provide some regulation of radioactive waste disposal activities. The United Nations only plays a minor role in nuclear waste management policy.

The International Atomic Energy Agency (IAEA)

The IAEA has taken a leading role in forging an international consensus on the long-term management of highly radioactive waste.[9] The IAEA is an independent international organization with 145 member countries that has a special arrangement with the UN to report annually to the UN General Assembly; and, as needed, to the UN Security Council regarding

[9] See IAEA (2009f) "Treaties, Conventions and Agreements Related to the IAEA's Work" [http://www.iaea.org/Publications/Documents/Treaties/index.html] and IAEA (2009g) "Treaties, Conventions and Agreements Under IAEA Auspices" [http://www.iaea.org/Publications/Documents/Conventions/index.html] and for an excellent resource for international conventions and agreements related to the long-term management of highly radioactive waste.

non-compliance with IAEA regulations. The IAEA's nuclear waste management program concentrates on:

- collecting, reviewing and publishing up-to-date scientific and technical information;
- providing direct services, particularly to developing Member States and the newly independent states of the former Soviet Union, through technical assistance, training, and other programmes to strengthen national infrastructures, including the demonstration of waste management techniques;
- fostering and coordinating research and development through coordinated research programmes;
- establishing safety standards for the management and disposal of wastes by building international consensus;
- implementing international codes of practice and safety conventions;
- providing advisory and peer review services for both developed and developing countries;
- providing a forum for the dissemination and exchange of information at international conferences, symposia and seminars including the promotion of education, public information and training;
- and, developing and implementing special projects, including assessment of the environmental and radiological impacts of waste disposal. (IAEA, 2009e: 4)

The IAEA's Waste Technology Section "covers a broad spectrum of activities from radioactive waste characterization and conditioning to disposal…." and currently is sponsoring the following major ongoing activities: assisting the design and planning of geological repositories, investigating the desirability of retrievability and its technological impact, applying numerical modeling to site characterization and performance assessment, enhancing international cooperation in shared disposal facilities, facilitating training and development of waste disposal technologies, and investigating the disposal of long-lived waste. (IAEA 2009b: 1; IAEA, 2009c: 1-2) The IAEA also issues annual reports that include discussing the safety of nuclear installations, radiation and transport safety, the management of radioactive waste, and nuclear safety. The publications on radioactive waste management are especially useful sources of information. For example, the first important step in exploring the feasibility of a regional or international approach to the disposal of highly radioactive waste was a 1998 IAEA study that assessed the technical, institutional, and economic factors regarding a multinational approach to radioactive waste disposal.[10] The report recognized that the multinational option was a sensitive political issue and that the generation that produces the nuclear waste also should be responsible for its long-term management. However, it did conclude that as long as a consensus was achieved by participating countries, the multinational repository concept does not contradict the principle of who generates the waste is ultimately responsible for its disposal. The report also concluded that the economy of scale

[10] See (IAEA, 1998b) "Technical, institutional, and economic factors important for developing a multinational radioactive waste repository," International Atomic Energy Agency, June 1998, IAEA-TECDOC-1021 [http://www-pub.iaea.org/MTCD/publications/PDF/te_1021_prn.pdf]. The document discusses the legal aspects, safety principles, technical issues, costs and liabilities, institutional aspects and political continuity, ownership of waste materials, ethical aspects, public acceptance, safeguards and RandD requirements associated with the multinational repository option.

that the multinational repository option could offer was a substantial benefit for participating countries that do not generate sufficient amounts of highly radioactive waste to warrant the construction of their own national repository. Finally, the report concluded that the demonstrated safety record of radioactive waste transportation over the years should be sufficient to assuage concerns over impact of nuclear waste transport on public safety and health. (IAEA, 1998b: 19) In 2004, the IAEA again revisited the issue of developing a multinational approach to the disposal of highly radioactive waste.[11] The more substantive 2004 report reiterated the conclusions from the 1998 report as well as discussed a number of additional potential benefits and risks associated with the multinational repository concept. The report concluded that the benefits of a multinational repository can be significant. (IAEA, 2004: 44) While political considerations would make implementation of the multinational option challenging, the potential benefits of a regional or international repository warrant more study of the concept.

The Nuclear Energy Agency (NEA)

The NEA plays an important role in regulating highly radioactive waste since its membership generates most of the spent nuclear fuel produced annually. While the majority of the NEA's 28 member countries are from Europe, other nuclear waste generating countries also are represented (e.g., Australia, Canada, Japan, Korea, Mexico, Turkey, and the United States). Overall, NEA member countries account for about 85% of the world's installed nuclear capacity. (NEA, 2008b: 1) The agency's goal in radioactive waste management is "to assist member countries in the management of radioactive waste and materials…in particular long-lived waste and spent fuel." (NEA, 2006a: 1) The NEA's Radioactive Waste Management Committee promotes international cooperation in radioactive waste management among member countries. The main tasks of the Committee are to:

- provide a forum for the exchange of information related to radioactive waste management;
- develop a common understanding of the basic issues involved and promote common philosophies on possible waste management strategies and alternatives;
- promote state-of-the-art procedures for the safe and secure management of radioactive waste;
- promote the dissemination of information relevant to radioactive waste management;
- and, offer a framework for conducting an international peer review of national activities in radioactive waste management. (NEA, 2006b: 1)

The NEA has four primary objectives for radioactive waste management: foster understanding on the management of radioactive waste; facilitate the elaboration of waste management strategies at both the national and international levels; help to elaborate common

[11] See (IAEA, 2004) "Developing multinational radioactive waste repositories: Infrastructural framework and scenarios of cooperation," International Atomic Energy Agency, October 2004, IAEA-TECDOC-1413 [http://www-pub.iaea.org/MTCD/publications/PDF/te_1413_web.pdf].

regulatory procedures; and promote the application of new science and technology to radioactive waste management. (NEA, 2006a: 1)

The United Nations

The United Nations looks primarily to the IAEA to take the principle role in radioactive waste management. When the UN has addressed the issue of radioactive waste, it has been primarily through the principles of the Rio Declaration on Environment and Development, and Chapter 22 of the Agenda 21 sustainable development program.[12] According to the UN,

> The objective of Chapter 22 is to ensure that radioactive wastes are safely managed, transported, stored and disposed of, with a view to protecting human health and the environment, within the wider framework of an interactive and integrated approach to radioactive waste management and safety. (UN, 2005a: 1)

The UN position on radioactive waste management basically is:

- the countries that produce the nuclear waste are responsible for its safe and secure long-term management;
- radioactive wastes should be disposed of in the territory of the country that generated the waste;
- joint disposal facilities may be appropriate in some circumstances (provided that certain conditions are met);
- all efforts should be made to prohibit the export of radioactive waste to countries that do not have appropriate waste treatment and storage facilities;
- and, the management of radioactive waste should be consistent with international law. (UN, 2005b: 1-2)

While the UN role in the direct supervision over radioactive waste management facilities is limited, the organization does aid international cooperation in nuclear waste disposal efforts by providing a forum for communication and an outlet for the dissemination of technical information and periodic reports on sound radioactive waste management practices.

NGOs

There are a multitude of NGOs active in the highly radioactive waste disposal policy debate. However, their influence on policymaking is somewhat limited since countries are reluctant to surrender sovereignty over the disposition of nuclear waste. Nevertheless, NGOs can, and do, impact the policymaking process by providing information on both sides of the radioactive waste management policy debate. There are two basic categories of NGOs relevant to the management of highly radioactive waste: professional organizations associated with the nuclear industry that exist primarily to promote nuclear energy, and environmental groups who generally are opposed to nuclear power.

---

[12] See (UN, 2004) "Agenda 21: Chapter 22—Safe and Environmentally Sound Management of Radioactive Wastes," United Nations Department of Economic and Social Affairs, Division for Sustainable Development, 15 December 2004 [http://www.un.org/esa/sustdev/documents/agenda21/english/agenda21chapter22.htm].

Professional Organizations

There are a number of professional organizations involved with nuclear waste management.[13] The World Nuclear Association (WNA) is an independent global non-profit private sector organization that promotes nuclear power. The Association's stated objectives are to serve as a global forum and commercial meeting place for the nuclear industry, to provide information on nuclear energy, lobby policy makers, and promote a wider understanding of nuclear energy. (WNA, 2009b: 1) The current WNA membership accounts for 80 percent of the world's population and 85 percent of global nuclear power capacity and radioactive waste generation. (WNA, 2009c: 1) The Association asserts that the sustainability of nuclear energy and the fact that nuclear generated power does not contribute to global warming are major reasons why nuclear power is a sensible option that should continue to be pursued. The WNA provides useful information on member's waste management practices.[14]

The Institute of Nuclear Materials Management (INMM) is another non-governmental organization active in the nuclear waste disposal policy debate. While the membership in the INMM primarily is U.S.-based, Australia, Brazil, Canada and the IAEA are listed as sustaining members. (INMM, 2009d: 1-2) The Institute's mission is to encourage the advancement of nuclear materials management, promote research in the field of nuclear materials management, establish standards consistent with existing professional norms, improve the qualifications of those engaged in nuclear materials management, and increase the dissemination of information through professional contacts and publications. (INMM, 2009a: 1) The INMM promotes "best practices" for a variety of nuclear related areas, including radioactive waste management. (INMM, 2009b: 1) The Institute's journal, *Journal of Nuclear Materials Management* (JNMM), was established to "provide for the exchange of ideas and information related to international safeguard, materials control and accounting, nonproliferation and arms control, physical protection, packaging and transportation, and waste management." (INMM, 2009c: 1)

The World Nuclear Transport Institute (WNTI) will play a role in waste disposal since any repository will require the transport of significant quantities of highly radioactive waste to the facility. The WNTI is a global organization representing the radioactive waste transport industry and is "dedicated to presenting the industry point of view on radioactive material transport matters from an international perspective." (WNTI, 2009b: 1) The Institute was founded to promote international cooperation and understanding of radioactive waste transport issues, and has consultant status with the International Maritime Organization and observer status with the IAEA. (WNTI 2009a: 1) The WNTI provides a forum for discussion and publishes technical information and other research materials on nuclear waste transport issues. (WNTI, 2007: 5)

---

[13] See (Radwaste, 2008) "International organizations," Radwaste.org [http://www.radwaste.org/iorg.htm] for an extensive list of non-governmental organizations involved in the nuclear waste disposal policy debate.

[14] See (WNA, 2006) "International Nuclear Waste Disposal Concepts," World Nuclear Association, September 2006 [http://www.world-nuclear.org/info/inf21.htm].

Environmental Organizations

While environmental groups have been opposed to nuclear power primarily due to concerns about reactor safety, there has been increased attention paid to the long-term environmental impact of spent nuclear fuel storage and disposal. Environmental groups also raise concerns over the safety of transporting significant quantities of highly radioactive waste over long distances to a centralized storage facility or repository; instead advocating leaving the waste on-site at the point of generation. Their opposition to nuclear generated power has had some policy successes. For example, as the environmental movement strengthened in Europe, and opposition to nuclear power intensified, several countries modified their nuclear power policies in response to public pressure. Belgium effectively decided to abandon nuclear power by prohibiting the construction of new reactors and limiting the operational phase of existing reactors to 40 years. Germany announced it was abandoning nuclear power and is now in the process of dismantling its reactors. And, Spain has declared that it was phasing out nuclear power in favor of alternative energy sources. While some environmental groups are genuinely concerned about safety and health concerns related to radioactive waste, other opposition groups hope to discontinue commercial nuclear power by forcing the premature shutdown of reactors due to the limited availability of sufficient on-site storage space.

Most nongovernmental environmental organizations are critical of nuclear power.[15] Today, however, concern over the impact of global warming has prompted some environmentalists to soften their opposition to nuclear power since it does not emit greenhouse gasses. Unless alternative sources of energy become cost-effective and widely available in the near future, the debate over the merits of nuclear power likely will intensify in the environmental community as electricity needs increase and concerns over the impact of global warming intensify.

CONCLUSION

Individual national efforts play the most significant role in ensuring the safe and secure management of highly radioactive waste due to the widely accepted principle of whoever generates the nuclear waste also is responsible for its long-term management. Clearly, countries have a significant self-interest in disposing of the nuclear waste properly since they will have to cope with any adverse effects associated with a radiological accident. However, since no highly radioactive waste disposal facility is expected to be ready for at least another decade, it is apparent that there will be significant quantities of nuclear waste stored above ground on-site near the reactor or in centralized temporary above-ground storage facilities. Even after geologic repositories come on-line in the United States, Finland, and Sweden, it will be difficult to reduce above-ground radioactive waste inventories in those countries quickly since the amount of nuclear waste generated annually will account for much of the waste being transferred to an underground geologic disposal facility. Even more troubling are the mounting inventories of radioactive waste in the developing world. While industrialized countries generate the vast majority of highly radioactive waste today, the projected increase

[15] See (Radwaste, 2007) "Environmental organizations," RadWaste.org [http://www.radwaste.org/ngo.htm] for an extensive list of environmental organizations involved in the radioactive waste disposal policy debate.

in nuclear waste inventories in the developing world is expected to grow at a rate that is twice that of the industrialized world. (See Table 3—World Cumulative Spent Fuel Projections by Region) This trend is expected to continue in future years as the developing world pursues nuclear generated power more aggressively. Sadly, developing countries are the least equipped economically and technologically to cope adequately with mounting inventories of highly radioactive waste.

Table 3. World Cumulative Spent Fuel Projections by Region (Thousand Metric Tons)

	2000	2010	2020
N. America	70.79	106.49	142.83
W. Europe	95.76	127.25	151.99
Asia	27.78	55.90	95.07
Former Soviet Union	26.28	36.15	45.91
E. Europe	4.52	9.20	14.50
Central/South America	2.64	3.85	5.06
Middle East	0.00	0.09	0.41
Africa	0.54	0.90	1.33
World Total	228.31	339.83	457.10

Source: Adapted from EIA (2002) "World Cumulative Spent Fuel Projections by Region and Country" [http://www.eia.doe.gov/cneaf/nuclear/page/forecast/cumfuel.html].

National efforts clearly should remain an important part of any long-term global radioactive waste management policy due to the established international norm that the nuclear waste generator ultimately is responsible for the safe and secure disposal of any highly radioactive waste produced. However, more attention should be paid by the international community to examining the utility of establishing a multinational regional or international repository to cope more effectively with the increasing inventories of highly radioactive waste. While a number of initiatives have been made over the years for examining the multinational repository option, there have been few concrete results. The most promising is the European Support Action Pilot Initiative for European Regional Repositories (SAPIERR), which has brought together representatives from various organizations from 14 European countries with small inventories of highly radioactive waste to study the feasibility of "one or more shared European regional repositories" for radioactive waste. (Ŝtefula, 2006: 1) The multinational option would be especially useful for developing countries since they generally lack the necessary resources to construct and operate a national radioactive waste disposal facility.[16] Moreover, smaller developed countries potentially could benefit from a multinational approach if they do not have a suitable area within their borders for constructing a national repository. In addition, both developing and developed countries with small quantities of nuclear waste could potentially achieve an economy of scale by pursuing a multinational disposal effort. Ultimately, a more coordinated global approach could enhance

[16] If the multinational option is pursued, it is important to ensure that it does not adversely affect the development of existing national disposal programs.

international security by reducing the overall amount of above-ground nuclear waste as well as consolidating it in a more secure underground repository.

International conventions have helped to strengthen radioactive waste management practices by encouraging cooperation in a variety of areas affecting the protection of highly radioactive waste. For example, international agreements have established a number of global norms designed to enhance the safety and security of nuclear materials:

- require the physical protection of nuclear materials (Convention on the Physical Protection of Nuclear Material, and the Code of Conduct on the Safety and Security of Radioactive Sources);
- promote cooperation on the safe management of nuclear waste (Joint Convention on the Safety of Spent Fuel Management and on the Safety of Radioactive Waste Management);
- regulate the transport of nuclear waste (NUCLEAR, Code of Practice on the International Transboundary Movement of Radioactive Waste, and the Code of Conduct on the Safety and Security of Radioactive Sources);
- permit countries to prohibit the transport of nuclear waste across their territory without their permission (Code of Practice on the International Transboundary Movement of Radioactive Waste, the Joint Convention on the Safety of Spent Fuel Management and on the Safety of Radioactive Waste Management, and the Code of Conduct on the Safety and Security of Radioactive Sources);
- regulate where highly radioactive waste can be disposed (Convention on the Prevention of Marine Pollution by Dumping of Wastes and Other Matter, and the Joint Convention on the Safety of Spent Fuel Management and on the Safety of Radioactive Waste Management);
- and, allow for liability in the event of a nuclear waste accident (Paris Convention, Vienna Convention, and the Convention on Supplementary Compensation for Nuclear Damage).

These international norms are a good first step in creating a global regime for the effective long-term management of highly radioactive waste. However, most conventions today ask for voluntary adherence to international norms and lack the necessary enforcement mechanisms to be able to sanction actors who violate the convention's provisions. Another major weakness of some international conventions is that a significant number of countries with nuclear programs have not ratified or acceded to the treaties. This lack of participation does weaken the global impact of the international conventions. International conventions and treaties only will be effective if the individual countries are willing to adhere to the international norms. In order to ensure the long-term safe and secure disposal of highly radioactive waste on a global scale, these international agreements ultimately must evolve into a global regime that has the power to enforce adherence to international norms.

The Paris and Vienna Conventions on third-party liability do have some measure of enforcement since cases can be forwarded to the European Nuclear Energy Tribunal, which has the capability of issuing binding decisions. However, only a handful of countries are party to these conventions, which limits the impact of establishing a truly global regime on liability. The Convention on Supplementary Compensation for Nuclear Damage has made more of a contribution to establishing a global nuclear materials liability regime, but it lacks enforcement mechanisms. The Convention on the Physical Protection of Nuclear Material has a more pronounced impact on radioactive waste management since it is *legally* binding on countries to adequately protect nuclear materials and facilities against theft and sabotage. Given current global security realities, this convention does play an important role in encouraging a more robust standard of protecting nuclear materials. The Joint Convention on the Safety of Spent Fuel Management and on the Safety of Radioactive Waste Management is the most noteworthy international agreement regulating highly radioactive waste since it requires periodic reports on national efforts to manage radioactive waste. These country reports make individual national efforts more transparent and provide an opportunity to review specific national radioactive waste management policies and procedures. Strengthening the enforcement mechanisms of the Joint Convention could begin the process of establishing a truly global regulatory mechanism for the effective and coordinated global long-term management of highly radioactive waste. Ultimately, the Joint Convention could serve as the genesis of a global regime that has the ability not only to monitor, but also to ensure, that signatory countries adhere to established international norms. All countries with nuclear programs would have to be signatories, and non-compliance with the Convention's requirements could be referred to an international organization such as the IAEA for the ultimate disposition of the case.

International governmental organizations and non-governmental organizations can play an important role in monitoring progress of national efforts and fostering international cooperation in radioactive waste management. However, IGOs generally lack the necessary enforcement mechanisms to *require* countries to follow international guidelines. The IAEA takes a leading role in promoting sound radioactive waste management practices and forging an international consensus on how best to approach radioactive waste disposal. The Agency is best positioned to serve as the primary IGO monitoring and regulating highly radioactive waste management activities. The IAEA already has sufficient membership to be effective and currently acts as a clearing house for relevant international conventions. There should be more of a concerted effort by the international community to promote the IAEA as *the* global organization responsible for monitoring and regulating all phases of radioactive waste management efforts with the capability to enact sanctions against parties who are not in compliance with international norms.

The influence of non-governmental organizations on highly radioactive waste policymaking generally is limited. However, professional organizations can play an important role in radioactive waste management since they provide a forum for communication and dissemination of technical information. They can help to provide a wider understanding of nuclear energy. Environmental organizations also can play an important role in the radioactive waste management debate since their opposition can call attention to potential problem areas that need to be addressed.

Promoting cooperation on nuclear waste disposal at the global level has been slow to develop. However, there have been some initiatives in more recent years that appear to signal an increased willingness to promote international cooperation in radioactive waste management efforts.[17] While the international community has made some progress in

[17] For example, The U.S.-sponsored Global Nuclear Energy Partnership (GNEP) and the Russian-sponsored Global Nuclear Infrastructure Initiative (GNII) ultimately could foster international cooperation in radioactive waste

promoting global cooperation in safely managing the increasing inventories of highly radioactive waste, much still needs to be done. In the final analysis, it ultimately will take a global approach to be able to address adequately a truly global problem.

References

AMLG (1971). *Convention relating to Civil Liability in the Field of Maritime Carriage of Nuclear Material*, Admiralty and Maritime Law Guide, International Conventions, 17 December 1971, available at: http://www.admiraltylawguide.com/conven/carriagenuclear 1971.html (accessed 1 February 2009).

Andra (2006). "Dossier 2005: Andra research on the geological disposal of high-level long-lived radioactive wastes," French National Radioactive Waste Management Agency, available at: http://www.andra.fr/publication/produit/D05_264_en.pdf (accessed 5 January 2007).

BMU (2005). *Environmental Policy: Joint Convention on the Safety of Spent Fuel Management and on the Safety of Radioactive Waste Management*, Report Under the Joint Convention by the Government of the Federal Republic of Germany for the Second Review Meeting in May 2006, Federal Ministry for the Environment, Nature Conservation and Nuclear Safety (BMU), September 2005, available at: http://www.bmu.de/files/english/nuclear_safety/application/pdf/2nationaler_bericht_atomenergie_en.pdf (accessed 6 January 2007).

BMU (2008). *Joint Convention on the Safety of Spent Fuel Management and on the Safety of Radioactive Waste Management*, Report of the Federal Republic of Germany for the Third Review Meeting in May 2009, Federal Ministry for the Environment, Nature Conservation and Nuclear Safety (BMU), October 2008, available at: http://www.bmu.de/files/pdfs/allgemein/application/pdf/3nationaler_bericht_atomenergie_en.pdf (accessed 19 February 2009).

CNR (2005). *Canadian National Report for the Joint Convention on the Safety of Spent Fuel Management and on the Safety of Radioactive Waste Management, Second Report*, October 2005, available at: http://www.nuclearsafety.gc.ca/pubs_catalogue/uploads/2005_joint_convention_report_English.pdf (accessed 12 February 2009).

CoRWM (2006). "Final Recommendations," Committee on Radioactive Waste Management, 31 July 2006, available at: http://www.corwm.org.uk/ (accessed 27 December 2006). Also available at: "Third Annual Report 2006," Document 1887, available at: http://www.corwm.org.uk/ (accessed 9 January 2009).

Defra (2006). *The United Kingdom's Second National Report on Compliance with the Obligations of the Joint Convention on the Safety of Spent Fuel Management and on the Safety of Radioactive Waste Management*, Department for Environment, Food and Rural Affairs, February 2006, available at: http://www.defra.gov.uk/environment/radioactivity/government/international/pdf/jointconreport06.pdf (accessed 6 January 2007).

---

management. Most of the major nuclear waste generating countries have signed on to the GNEP, which aims to promote global cooperation in reducing the amount of radioactive waste being generated by recycling spent nuclear fuel in new generation reactors. The GNII proposal intends to construct international nuclear fuel cycle service centers for reprocessing spent fuel from foreign sources.

EIA (2002). *"World Cumulative Spent Fuel Projections by Region and Country,"* Energy Information Administration, 2002, Available from: http://www.eia.doe.gov/cneaf/nuclear/page/forecast/cumfuel.html (accessed 5 January 2009).

ENTRI (1978). *"Amendments to the Convention on the Prevention of Marine Pollution by Dumping of Wastes and Other Matter Concerning Settlement of Disputes,"* Environmental Treaties and Resource Indicators (ENTRI), 12 October 1978, available at: http://sedac.ciesin.columbia.edu/entri/texts/marine.pollution.dumping.wastes.amendments.1978.html (accessed 1 February 2009).

FAEA (2006). *The National Report of the Russian Federation on Compliance with the Obligations of the Joint Convention on the Safety of Spent Fuel Management and the Safety of Radioactive Waste Management*, Prepared for the Second Review Meeting in Frames of the Joint Convention on the Safety of Spent Fuel Management and the Safety of Radioactive Waste Management, Federal Atomic Energy Agency, Moscow 2006, available at: http://www-ns.iaea.org/downloads/rw/conventions/russian-federation-national-report.pdf (accessed 6 January 2007).

FANC (2006). *Second Meeting of the Contracting Parties to the Joint Convention on the Safety of Spent Fuel Management and on the Safety of Radioactive Waste Management*, Federal Agency for Nuclear Control, Kingdom of Belgium, National Report, May 2006, available at: http://www.fanc.fgov.be/download/JointConv_121005noannex.pdf (accessed 11 January 2009).

HSK (2008). *Implementation of the Obligations of the Joint Convention on the Safety of Spent Fuel Management and on the Safety of Radioactive Waste Management*, Second National Report of Switzerland in Accordance with Article 32 of the Convention, Swiss Federal Nuclear Safety Inspectorate (HSK), September 2008, available at: http://www.ensi.ch/fileadmin/deutsch/files/joint-convention_CH_sept-08.pdf (accessed 9 January 2009).

IAEA (1974). *Convention on the Prevention of Marine Pollution by Dumping of Wastes and Other Matter*, International Atomic Energy Agency, Information Circular, INFCIRC/205, 11 June 1974, available at: http://www.iaea.org/Publications/Documents/Infcircs/ Others /inf205.shtml (accessed 10 January 2009).

IAEA (1980). *Convention on the Physical Protection of Nuclear Material*, International Atomic Energy Agency, Information Circular, INFCIRC/274/Rev. 1, May 1980, available at: http://www.iaea.org/Publications/Documents/Infcircs/Others/inf274r1.shtml (accessed 10 January 2009). For an update to the convention see, International Conventions and Agreements, "Convention on the Physical Protection of Nuclear Material," International Atomic Energy Agency, available at: http://www.iaea.org/Publications/Documents/Conventions/cppnm.html (accessed 10 January 2009).

IAEA (1986a). *Convention on Early Notification of a Nuclear Accident*, International Atomic Energy Agency, Information Circular, INFCIRC/335, 18 November 1986, available at: http://www.iaea.org/Publications/Documents/Infcircs/Others/infcirc335.shtml (accessed 10 January 2009).

IAEA (1986b). *Convention on Assistance in the Case of Nuclear Accident or Radiological Emergency*, International Atomic Energy Agency, Information Circular, INFCIRC/336, 18 November 1986, available at: http://www.iaea.org/Publications/Documents/Infcircs/Others/infcirc336.shtml (accessed 10 January 2009).

IAEA (1990). *Code of Practice on the International Transboundary Movement of Radioactive Waste*, International Atomic Energy Agency, Information Circular,

INFCIRC/386, 13 November 1990, available at: http://www.iaea.org/Publications/ Documents/Infcircs/Others/inf386.shtml (accessed 10 January 2009).

IAEA (1992). *Joint Protocol Relating to the Application of the Vienna Convention and the Paris Convention*, International Atomic Energy Agency, Information Circular, INFCIRC/402, May 1992, available at: http://www.iaea.org/Publications/ Documents /Infcircs/Others/inf402.shtml (accessed 10 January 2009).

IAEA (1994). *Convention on Nuclear Safety*, International Atomic Energy Agency, Information Circular, INFCIRC/449, 5 July 1994, available at: http://www.iaea.org/ Publications/Documents/Infcircs/Others/inf449.shtml (accessed 10 January 2009).

IAEA (1996). *Vienna Convention on Civil Liability for Nuclear Damage*, International Atomic Energy Agency, Information Circular, INFCIRC/500, 20 March 1996, available at: http://www.iaea.org/Publications/Documents/Infcircs/1996/inf500.shtml (accessed 9 January 2009).

IAEA (1997a). *Joint Convention on the Safety of Spent Fuel Management and on the Safety of Radioactive Waste Management*, International Atomic Energy Agency, Information Circular, INFCIRC/546, 24 December 1997, available at: http://www.iaea.org/ Publications/Documents/Infcircs/1997/infcirc546.pdf (accessed 10 January 2009).

IAEA (1997b). *Protocol to Amend the 1963 Vienna Convention on Civil Liability for Nuclear Damage* and Annex, International Atomic Energy Agency, available at: http://www.iaea.org/Publications/Documents/Conventions/protamend.html (accessed 10 January 2009).

IAEA (1998a). *Convention on Supplementary Compensation for Nuclear Damage*, International Atomic Energy Agency, available at: http://www.iaea.org/Publications/ Documents/Conventions/supcomp.html, also see IAEA, Information Circular, INFCIRC /567, 22 July 1998, available at: http://www.iaea.org/Publications/Documents/ Infcircs /1998/infcirc567.pdf (accessed 10 January 2009).

IAEA (1998b). "Technical, institutional, and economic factors important for developing a multinational radioactive waste repository, International Atomic Energy Agency, June 1998, IAEA-TECDOC-1021, http://www-pub.iaea.org/MTCD/publications/PDF/te_1021 _prn.pdf (accessed 18 December 2008).

IAEA (1999). "Optional Protocol Concerning the Compulsory Settlement of Disputes to the Vienna Convention On Civil Liability For Nuclear Damage," International Atomic Energy Agency, Information Circular, INFCIRC/%00/Add. 3, 14 October 1999, available at: http://www.iaea.org/Publications/Documents/Infcircs/1999/infcirc500a3. shtml (accessed 10 January 2009).

IAEA (2004). "Developing multinational radioactive waste repositories: Infrastructural framework and scenarios of cooperation, International Atomic Energy Agency, October 2004, IAEA-TECDOC-1413, available at: http://www-pub.iaea.org/MTCD/publications/ PDF/te_1413_web.pdf (accessed 18 December 2008).

IAEA (2005a). *Code of Conduct on the Safety and Security of Radioactive Sources and the Supplementary Guidance on the Import and Export of Radioactive Sources*, International Atomic Energy Agency, INFCIRC/663, 29 December 2005, available at: https://www. iaea.org/Publications/Documents/Infcircs/2005/infcirc663.pdf (accessed 17 March 2009).

IAEA (2005b). *"Convention on the Physical Protection of Nuclear Material,"* International Atomic Energy Agency, International Conventions and Agreements, Background, available at: http://www.iaea.org/Publications/Documents/Conventions/Cppnm.html (accessed 10 January 2009).

IAEA (2006a). *"Amendment to the Convention on the Physical Protection of Nuclear Material,"* IAEA International Law Series No. 2, International Atomic Energy Agency, 2006, available at: http://www-pub.iaea.org/MTCD/publications/PDF/Pub1275_web.pdf (accessed 1 February 2009).

IAEA (2006b). *"Safety of Radioactive Waste Disposal," Proceedings of an International Conference*, Tokyo, 3-7 October 2005, International Atomic Energy Agency, IAEA/NTR/2006, August 2006, available at: http://www-pub.iaea.org/MTCD/publications/PDF/Pub1261_web.pdf (accessed 7 January 2007).

IAEA (2007). *"Energy, Electricity and Nuclear Power: Developments and Projections,"* International Atomic Energy Agency, STI/PUB/1304, December 2007, available at: http://www-pub.iaea.org/MTCD/publications/PDF/Pub1304_web.pdf (accessed 26 January 2009).

IAEA (2008). *"Estimation of Global Inventories of Radioactive Waste and Other Radioactive Materials,"* IAEA-TECDOC-1591, June 2008, pp. 14-15, available at: http://www-pub.iaea.org/MTCD/publications/PDF/te_1591_web.pdf (accessed 1 January 2009).

IAEA (2009a). "Code of Conduct: Code of Conduct on the Safety and Security of Radioactive Sources and Supplementary Guidance on the Import and Export of Radioactive Sources," International Atomic Energy Agency, available at: http://www-ns.iaea.org/tech-areas/radiation-safety/code-of-conduct.htm (accessed 17 March 2009).

IAEA (2009b). "Current and Forthcoming Activities," Waste Technology Section, International Atomic Energy Agency, 2009, available at: www.iaea.org/OurWork/ST/NE/NEFW/wts_home.html (accessed 21 February 2009).

IAEA (2009c). "Geological Waste Disposal," Waste Technology Section, International Atomic Energy Agency, 2009, available at: www.iaea.org/OurWork/ST/NE/NEFW/wts_geologicaldisposal.html (accessed 21 February 2009).

IAEA (2009d). "Joint Convention on the Safety of Spent Fuel Management and on the Safety of Radioactive Waste Management," International Atomic Energy Agency, 26 January 2009, available at: http://www.iaea.org/Publications/Documents/Conventions/jointconv_status.pdf (accessed 7 February 2009).

IAEA (2009e). "Managing Radioactive Waste," International Atomic Energy Agency, Factsheets and FAQs, available at: http://www.iaea.org/Publications/Factsheets/English/manradwa.html (accessed 11 January 2009).

IAEA (2009f). "Treaties, Conventions and Agreements Related to the IAEA's Work," International Atomic Energy Agency, available at: http://www.iaea.org/Publications/Documents/Treaties/index.html (accessed 21 February 2009).

IAEA (2009g). "Treaties, Conventions and Agreements Under IAEA Auspices," International Atomic Energy Agency, available at: http://www.iaea.org/Publications/Documents/Conventions/index.html (accessed 21 February 2009).

IMF (2009). "SDR Valuation," International Monetary Fund, available at: http://www.imf.org/external/np/fin/data/rms_SDRV.aspx (accessed 17 March 2009).

IMO (2002). *Convention Relating to Civil Liability in the Field of Maritime Carriage of Nuclear Material (NUCLEAR), 1971*, International Maritime Organization, 2002, available at: http://www.imo.org/home.asp (accessed 10 January 2009).

IMO (2006). *Convention on the Prevention of Marine Pollution by Dumping of Wastes and Other* Matter, International Maritime Organization, 2006, available at: http://www.imo.org/Conventions/contents.asp?topic_id=258anddoc_id=681#top (accessed 10 January 2009).

INMM (2009a). "*About INMM: INMM Mission Statement*," Institute of Nuclear Materials Management, available at: http://inmm.org/about/mission.cfm (accessed 22 January 2009).

INMM (2009b). "*Global Best Practices for Nuclear Materials Management*,," Institute of Nuclear Materials Management, available at: http://inmm.org/best_practice/index.cfm (accessed 22 January 2009).

INMM (2009c). "Publications: *Journal of Nuclear Materials Management*,," Institute of Nuclear Materials Management, available at: http://inmm.org/publications/journal.cfm (accessed 22 January 2009).

INMM (2009d). "*About INMM: INMM Sustaining Members*," Institute of Nuclear Materials Management, available at: http://inmm.org/about/sustaining_members.cfm (accessed 22 January 2009).

METI (2005). *Joint Convention on the Safety of Spent Fuel Management and on the Safety of Radioactive Waste Management*, National Report of Japan for the Second Review Meeting, October 2005, available at: http://www.meti.go.jp/english/report/index.html (accessed 6 January 2007).

METI (2008). *Joint Convention on the Safety of Spent Fuel Management and on the Safety of Radioactive Waste Management*, National Report of Japan for the Third Review Meeting, October 2008, available at: http://www.meti.go.jp/english/report/data/Japan_mainpages.pdf (accessed 9 January 2009).

MITYC (2005). *Joint Convention on the Safety of Spent Fuel Management and on the Safety of Radioactive Waste Management*, Second Spanish National Report, October 2005, available at: http://www.mityc.es/NR/rdonlyres/BF3E47F5-7861-4A28-8FB3-8F2DBAB183A3/0/ConvencionInforme2_ing.pdf (accessed 6 January 2007).

MSDS (2005). *Sweden's Second National Report Under the Joint Convention on the Safety of Spent Fuel Management and on the Safety of Radioactive Waste Management*, Ministry of Sustainable Development, Ds 2005:44, available at: http://www.sweden.gov.se/content/1/c6/05/40/89/fc570cf2.pdf (accessed 6 January 2007).

Murkowski (2009). Murkowski, Lisa. "Murkowski Urges Administration to Develop Workable Solution to Nuclear Waste Issue," United States Senate Committee on Energy and Natural Resources, Press Releases, 5 March 2009, available at: http://energy.senate.gov/public/index.cfm?FuseAction=PressReleases.DetailandPressRelease_id=0ad22606-a42c-4b5b-9e6c-91be9d0b5a2candMonth=3andYear=2009andParty=1 (accessed 10 March 2009).

NEA (2004). *Protocol to Amend the Convention on Third Party Liability in the Field of Nuclear Energy of 29 July 1960, as Amended by Additional Protocol of 28 January 1964 and by the Protocol of 16 November 1982*, Nuclear Energy Agency, available at: http://www.nea.fr/html/law/paris_convention.pdf (accessed 10 January 2009).

NEA (2006a). "*Radioactive waste management*," Nuclear Energy Agency, available at: http://www.nea.fr/html/rwm/welcome.html (accessed 7 June 2007).

NEA (2006b). "*Radioactive Waste Management Committee*," Nuclear Energy Agency, available at: http://www.nea.fr/html/rwm/rwmc.html (accessed 7 June 2007).

NEA (2006c). "*Safety of Geologic Disposal of High-level and Long-lived Radioactive Waste in France*," Nuclear Energy Agency, NEA No. 6178, 2006, available at: http://www.nea.fr/html/rwm/reports/2006/nea6178-argile.pdf (accessed 9 January 2009).

NEA (2007a). *"Belgium," Nuclear Energy Agency,* available at: http://www.nea.fr/html/general/profiles/belgium.html (accessed 20 January 2009).

NEA (2007b). "European Nuclear Energy Tribunal," Nuclear Energy Agency, 20 November 2007, available at: http://www.nea.fr/html/law/european nuclear-tribunal.html (accessed 3 February 2009).

NEA (2007c). "Paris Convention on Nuclear Third Party Liability: Latest status of ratifications or accessions," Nuclear Energy Agency, available at: http://www.nea.fr/html/law/paris-convention-ratification.html (accessed 10 January 2009).

NEA (2008a). *Convention on Third Party Liability in the Field of Nuclear Energy of 29th July 1960, as amended by the Additional Protocol of 28th January 1964 and by the Protocol of 16th November 1982*, available at: http://www.nea.fr/html/law/nlparis_conv.html (accessed 10 January 2009).

NEA (2008b). "*The Nuclear Energy Agency: The NEA Mission*," Nuclear Energy Agency, available at: http://www.nea.fr/html/nea/flyeren.html (accessed 7 June 2007).

NUMO (2003a). "Outline of Operations: NUMO plants to start operation of the repository (disposal facility). by 2033-2038," Nuclear Waste Management Organization of Japan, available at: http://www.numo.or.jp/en/jigyou/new_eng_tab03.html (accessed 6 January 2007).

NUMO (2003b). "*Outline of Operations: High-level waste arises from the reprocessing of spent fuel used in nuclear power plants*," Nuclear Waste Management Organization of Japan, available at: http://www.numo.or.jp/en/jigyou/new_eng_tab04.html (accessed 6 January 2007).

OCRWM (2001). "*Radioactive waste: an international concern*," Office of Civilian Radioactive Waste Management, U.S. Department of Energy, DOE/YMP-0405, June 2001, available at: http://ocrwm.doe.gov/factsheets/doeymp0405.shtml (accessed 5 January 2009).

Radwaste (2007). "*Environmental organizations*," Radwaste, available at: http://www.radwaste.org/ngo.htm (accessed 21 January 2009).

Radwaste (2008). "International organizations," Radwaste, available at: http://www.radwaste.org/iorg.htm (accessed 21 January 2009).

Reuters (2008a). "Spain *watchdog boosts screening after nuclear leak*," Reuters, 7 May 2008, available at: http://www.reuters.com/article/rbssIndustryMaterialsUtilitiesNews/idUSL0779385920080507 (accessed 3 September 2008).

Reuters (2008b). "UPDATE 1-Fire halts Spain's Vandellos II nuclear plant," Reuters, 24 August 2008, available at: http://www.reuters.com/article/rbssIndustryMaterialsUtilitiesNews/idUSN2447082820080824 (accessed 3 September 2008).

Rogers (2009). Rogers, Kenneth A. "Fire in the hole: A review of national spent nuclear fuel disposal policy," *Progress in Nuclear Energy*, *Vol. 51*, No. 2, March 2009: 281-89.

Ŝtefula (2006). Ŝtefula, Vladan. *Support Action: Pilot Initiative for European Regional Repositories (SAPIERR)*, Final Report, 18 January 2006, available at: http://www.sapierr.net/files1/Documents/SAPIERR%20WP%20D7%20Final%20Report.pdf (accessed 17 February 2009).

STUK (2005). *Joint Convention on the Safety of Spent Fuel Management and on the Safety of Radioactive Waste Management*, 2nd Finnish National Report as referred to in Article 32 of the Convention, STUK-B-YTO 243, October 2005, available at: http://www.stuk.fi /julkaisut/stuk-b/stuk-b-yto243.pdf (accessed 5 January 2007).

Sweden (2008). *Sweden's third national report under the Joint Convention on the safety of spent fuel management and on the safety of radioactive waste management: Swedish implementation of the obligations of the Joint Convention*, Ministry of Environment, Ds 2008:73, available at: http://www.sweden.gov.se/content/1/c6/11/80/03/a5a123ca.pdf (accessed 8 February 2009).

Thegerström (2004). Thegerström, Claes. "Down to Earth and below: Sweden's Plans for Nuclear Waste," *IAEA Bulletin*, 46/1, June 2004, available at: www.iaea.org/ Publications/Magazines/Bulletin/Bull461/article13.pdf (accessed 21 February 2009).

UN (2004). "Safe and Environmentally Sound Management of Radioactive Wastes," available at: http://www.unep.org/Documents.Multilingual/Default.asp?DocumentID=52 andArticleID=70 (accessed 19 January 2009).

UN (2005a). "*Radioactive Waste*," United Nations, 2 August 2005, available at: http://www. un.org/esa/sustdev/sdissues/waste_radioactive/waster.htm (accessed 21 January 2009).

UN (2005b). "*Radioactive Waste: Decisions of the GA and CSD*," United Nations General Assembly, 19th Special Session, New York, 23-27 June 1997, available at: http://www.un.org/esa/sustdev/sdissues/waste_radioactive/waster_decisions.htm (accessed 19 January 2009).

URL (2005). "*Underground Research Laboratory*," Applied Seismology Laboratory, Liverpool University Department of Earth Sciences, available at: http://www.liv.ac.uk/ seismic/research/url/url.html (accessed 5 January 2007).

WNA (2006). "*International Nuclear Waste Disposal Concepts*," World Nuclear Association, September 2006, available at: http://www.world-nuclear.org/info/inf21.htm (accessed 19 January 2009).

WNA (2009a). "*Nuclear Power in Russia*," World Nuclear Association, 25 February 2009, available at: http://www.world-nuclear.org/info/default.aspx?id=366andterms=nuclear+ power+in+russia (accessed 17 March 2009).

WNA (2009b). "Objectives," World Nuclear Association, available at: http://www.world-nuclear.org/about/objectives.html?ekmensel=185bf1b1_12_0_72_1 (accessed 19 January 2009).

WNA (2009c). "*WNA Membership*," World Nuclear Association, available at: http://www. world-nuclear.org/about/membership.html?ekmensel=185bf1b1_12_0_78_4 (accessed 20 January 2009).

WNTI (2007). "*Review 07*" World Nuclear Transport Institute, available at: http://www. wnti.co.uk/UserFiles/File/public/publications/annualreviews/annualreview07.pdf (accessed 20 January 2009).

WNTI (2009a). "*Background information*," World Nuclear Transport Institute, available at: http://www.wnti.co.uk/about-us/background-information (accessed 20 January 2009).

WNTI (2009b). "*How WNTI is organized*," World Nuclear Transport Institute, available at: http://www.wnti.co.uk/about-us/how-wnti-is-organised (accessed 20 January 2009).

INDEX

#

20th century, 84, 127

A

accelerator, 44
access, 26, 46, 60, 66, 192, 194, 196
accessions, 211
accidents, 195
accounting, 201
acid, 56, 58, 87, 88, 89, 92, 93, 94, 98, 102, 117, 130, 131, 133
acidic, 85, 86, 88, 91
actinoid elements (An), viii, 97
activity level, 36, 47
additives, 112, 113
adjudication, 193, 194, 195, 196
advancement, 201
adverse effects, 202
AEA, 209
aerosols, 50
Africa, 203
age, 132, 133, 203, 207
agencies, vii, 187
agent, 148, 149
aggregates, 148
agriculture, 57
aid, 200
air, 148, 149
air quality, 78
alcohol, 156
algorithm, 89, 90
alkaline, 146
alkaline earth metals, ix, 97, 99, 107
alternative, 202
alternative energy, 202
alternatives, 199
amendments, 207
americium, 78
ammonium, 131
amplitude, 18, 22, 28
animations, 28
ANS, 66, 146
APL, 109, 117
aqueous solution, ix, 145, 146
arbitration, 194, 195, 196
Argentina, 50, 51, 59, 73
argon, 50, 83, 110, 117, 118
arid, 146, 147, 191
arms control, 201
arrest, 195
ash, 148, 149
Asia, 188, 203
assessment, 7, 38, 130, 135, 140, 141, 195, 198
asymptotic, 159, 165
atmosphere, 59, 82, 110, 117, 118
atmospheric pressure, 79, 83
atomic nucleus, 121
atoms, 11, 65
Atoms for Peace, 43, 46, 53, 66
attachment, 89
audit, 9
Australia, 199, 201
Austria, 46, 50, 51, 63, 64, 193
authorities, 26, 192, 194
authority, 49
automate, 9, 83
automation, 32, 78, 84
availability, 202
awareness, 7

B

backscattered, 178
bandwidth, 22

barrier, 187, 188, 189, 190, 191
base, 30
basic research, 57
batteries, 8, 9, 10, 38
beams, 48
behavior, x, 145, 147, 155, 158, 159, 165, 171, 173, 176, 177
Beijing, 74
Belarus, 43, 44, 55
Belgium, 43, 46, 57, 187, 188, 202, 207, 211
benefits, 32, 38, 197, 199
beryllium, 43, 44, 65
Best Practice, 210
bias, 18
bicarbonate, 149, 150, 163
binding, 193, 194, 197, 204
binding energy, 124
biopolymers, 79
boric acid, 102
brachytherapy, 58
branching, 178, 180
Brazil, 201
breakdown, 14
breeding, 104
broad spectrum, 198
Brussels, 193
bubbles, 149
buffer, 188
Bulgaria, 43, 55
Bureau of Labor Statistics, 61, 65
burn, 53, 62
by-products, vii, 4

C

cables, viii, 2, 10
CAD, 32
cadmium, viii, 89, 97
calcium, 130, 137
calibration, 29, 37, 88, 90, 91, 94
calixarenes, 88
calorimetry, 39
Canada, 187, 189, 191, 199, 201
candidates, 107
caps, 154
carbon, 178, 181
cartoon, 7
casting, 52, 149, 189
category a, 49
category d, 3
cation, 80, 100, 123
cell, ix, 145, 147, 149, 150, 154, 156, 158, 159, 163, 165, 174, 175, 178, 180
cement, 146, 148, 149
cerium, 85, 89
certification, 143
cesium, 103
channels, 196
chemical, viii, 2, 4, 7, 36, 49, 56, 58, 59, 60, 77, 79, 81, 82, 83, 84, 85, 88, 89, 92, 93, 99, 100, 101, 104, 107, 108, 115, 117, 123, 124, 147, 150, 174, 180
chemical characteristics, 59
chemical properties, 4, 89, 174
chemical reactions, 83
chemicals, 78
chemometric techniques, viii, 77, 85, 94
chemometrics, viii, 77
chernobyl, 192, 194, 195
chernobyl accident, 192, 195
Chicago, 41, 181
children, 4
Chile, 74
China, 44, 55, 78, 187, 188
chloride, 149
chlorine, 109
chromatography, 86
civilian, 186, 192, 196
cladding, 50, 51, 53, 58, 59
classes, 3, 48, 140
classification, 2, 42, 47, 48, 51, 130, 140, 141
clay, 188, 189, 190, 191
cleaning, 48
closure, 54
coal, 4
cobalt, 86, 88
codes, 198
color, 11, 32
commercial, 15, 16, 19, 22, 37, 54, 83, 142, 188, 201, 202
communication, 39, 197, 200, 205
community, x, 185, 202, 203, 205
compensation, 192, 195, 196
complement, 54
compliance, 9, 29, 190, 198, 205
components, 149
composition, 5, 36, 49, 50, 80, 83, 87, 89, 99, 100, 104, 105, 106, 107, 109, 117, 123, 124, 130, 131, 132, 133, 137, 148, 149, 150
compounds, ix, 51, 86, 88, 97, 101, 112, 143
computation, 104
computer, 32, 125
computing, 19
concentrates, 198
concentration, 154, 158, 159, 163, 165, 171, 173, 174, 175, 176, 177

conception, 7
concrete, ix, 145, 146, 147, 148, 149, 150, 154, 156, 158, 159, 163, 164, 165, 170, 171, 172, 173, 174, 175, 176, 177, 178, 180, 203
concrete waste, x, 145, 146, 147, 163, 173, 180
conditioning, 6, 7, 9, 29, 38, 56, 198
conductivity, 56
conference, 195
confidence, 188
configuration, 9, 22, 37
conformity, 86
congress, 66, 67, 75
consensus, 197, 198, 205
consent, 195, 196, 197
conservation, 129
consolidation, 101
constituents, 58, 92, 98, 99, 114
construction, 32, 36, 84, 89, 188, 191, 195, 196, 199, 202
Consumer Price Index (CPI), 61, 65, 74
consumption, 10, 27, 31, 53, 55, 84
containers, 9, 48, 58
contaminant, 86, 147, 149, 163, 165, 173, 178, 179, 180
contaminants, x, 146, 147, 150, 180
contamination, 6, 47, 84, 140
continuity, 198
control, 180, 192, 195, 196, 197, 201
convention, 193, 195, 196, 204, 205, 206, 207, 210, 211
cooling, 43, 59, 104, 154, 190
cooperation, x, 185, 186, 192, 195, 196, 197, 198, 199, 200, 201, 204, 205, 208
coordination, 192, 196
copper, 189
correlation, 9, 19, 94, 138, 157
correlations, 178
corrosion, 36, 47, 50
cost, vii, viii, 1, 2, 3, 8, 10, 12, 25, 27, 32, 40, 43, 44, 61, 62, 63, 64, 65, 66, 78, 83, 84, 146, 202
cost-effective, 202
costs, 187, 190, 195, 198
country of origin, 55
covering, 34
cracking, 159
crust, 130
crystal structure, 110, 111, 119
crystalline, 99, 100, 191
crystallization, 117, 120
current ratio, 80, 82
cycles, 146
cyclohexanone, 87
Czech Republic, 55, 59

D

damages, 192, 194
danger, 8, 141, 142
dangerousness, 2
data analysis, 85
data processing, 90, 92, 93
data structure, 27
database, 9, 27, 29, 38
decay, 2, 3, 4, 5, 14, 48, 59, 60, 63, 138
decisions, 193, 204, 212
decommissioning and storage of radioactive waste, vii, 1
decomposition, 36
decontamination, 49, 60
defects, 9
defence, 115
defense, 191
dehydrate, 100
Denmark, 62, 63
Department of Energy, 54, 68, 69, 70, 181, 211
Department of Labor, 74
deposits, 131, 132, 133, 134, 139, 140, 141, 142, 143
depth, 16, 121, 127, 158, 159, 188, 189, 191
derivatives, 89
destruction, 81
detectable, 14, 102
detection, vii, 1, 7, 8, 9, 12, 13, 15, 20, 24, 33, 51, 78, 79, 83, 134
Detector Mesh for Nuclear Repositories, vii, 1, 9, 39
detention, 195
developing countries, 2, 198, 203
deviation, 28
diamond, 154
diffraction, 20, 110, 119
diffusion, ix, 8, 145, 146, 147, 150, 154, 156, 157, 158, 159, 163, 165, 170, 171, 173, 174, 176, 178, 179, 180
diffusion rates, 159
diffusivity, 147
disaster, 140, 141
discharges, 49, 50, 51
discrimination, 23
dispersion, 52, 110, 130, 135
displacement, 7
disposition, 54, 56, 196, 200, 205
disputes, 194, 195, 196
disseminate, 197
distilled water, 110, 119, 123, 124
distribution, 13, 16, 17, 18, 19, 20, 44, 46, 47, 89, 111, 112, 113, 163, 177
diversity, 81
doping, viii, 41, 42

download, 207
drying, 110, 178
dumping, 192, 194, 207
durability, 117, 123, 124
duration, 154, 178
dykes, 134
dysprosium, 81

E

Earth Science, 212
economics, 32
education, viii, 41, 42, 198
effluents, 49
elaboration, 78, 199
electric charge, 5
electricity, 186, 188, 189, 190, 202
electrochemistry, 98
electrodes, 84, 87, 89
electrolyte, viii, ix, 97, 98, 99, 100, 101, 102, 104, 105, 106, 107, 109, 113, 114, 115, 116, 117, 118, 120, 121, 125
electromagnetic, 5, 33
electron, 5, 18, 110, 147, 178
electron microscopy, 147, 178
electronic structure, 85
electrons, 11, 15, 18
emergency, 140, 192, 195
emergency preparedness, 195
emission, 5, 11, 12, 119
emitters, 4, 48
employment, viii, 77, 85, 86, 92
encapsulation, 57, 60
energy, 4, 5, 8, 12, 13, 14, 23, 32, 33, 34, 64, 65, 71, 78, 174, 186, 189, 190, 200, 201, 202, 205, 210
Energy Information Administration, 207
enforcement, 193, 195, 204, 205
environment, vii, ix, 6, 14, 21, 31, 35, 37, 50, 51, 145, 146, 147, 180, 200, 206
environmental conditions, 147
environmental impact, 51, 202
environmental movement, 189, 202
environmental organizations, 202
environmental protection, 5
environmentalists, 202
epoxy, 178
equilibrium, 101, 108, 109
equipment, 3, 21, 29, 32, 48, 58, 59, 60, 78, 118, 120
ethanol, 83
Europe, 60, 188, 199, 202, 203
European Commission, 73, 74
europium, 81
excitation, 5
exclusion, 138, 193
expected probability, 20
experimental condition, 112
expertise, 60
exports, 55
exposure, 10, 12, 51, 60, 78, 131, 134, 135, 136, 140, 146, 178
extraction, viii, 58, 59, 60, 79, 86, 88, 89, 97, 143, 156
extracts, 154, 156

F

fabrication, 54, 57, 58, 78, 98, 148
failure, ix, 145, 146
false alarms, 8
families, vii, 1, 21
fears, 7, 8
federal government, 189
fee, 190
ferric ion, 121
ferrous ion, 117, 121, 122, 123
fiber, 9, 10, 11, 12, 13, 15, 16, 21, 22, 23, 24, 26, 27, 28, 30, 31, 34, 37, 148
fibers, 9, 10, 11, 12, 13, 14, 23, 25, 27, 28, 30, 34, 35, 36, 37, 39, 148
filters, 48, 59
filtration, ix, 59, 97, 112, 116, 117, 118, 120, 121, 122
finance, 193
financial, 192, 193
Finland, 187, 189, 202
fire, 190
fission, vii, viii, 2, 5, 9, 41, 42, 44, 48, 50, 55, 56, 57, 58, 59, 65, 79, 97, 99, 125
fission products (FP), viii, 97, 99
flexibility, 8, 10
flight, 40
flow, ix, 145, 146, 147
fluctuations, 16, 28, 37, 138, 175
focusing, 190
foils, 58
force, 84, 188, 192, 193, 194, 195, 196
formation, 65, 82, 84, 107, 111, 113, 114, 115, 130, 147, 150, 159, 163, 165, 171, 178, 180, 188, 189, 191
formula, 100, 101, 107, 110, 114, 138
FPGA-based counting system, viii, 2
fracture, 150, 154, 163, 165, 171, 173
fractures, 138, 150, 154, 163, 173
fragility, 12
fragments, 5

France, 43, 44, 46, 51, 55, 56, 57, 78, 187, 189, 190, 211
free volume, 46
freedom, 32, 33, 34
fuel, x, 185, 186, 187, 188, 189, 190, 191, 192, 196, 199, 202, 206, 211, 212
fuel consumption, 55
fuel cycle, 188, 206
fuel management, 196, 212
funds, 60, 187, 190

G

gadolinium, 56, 85, 89, 91
gamma radiation, vii, 1, 12, 24, 130, 134, 135, 136, 138
gamma rays, 5, 9, 12, 14, 23
generation, 186, 188, 195, 196, 198, 201, 202, 206
generators, 189
geochemistry, 146
geology, 81
geometry, 12, 18, 24, 34, 35, 36
Georgia, 43
germanium, 33
Germany, 40, 46, 50, 51, 53, 187, 189, 202, 206
glass transition, 117, 120
glass transition temperature, 117, 120
glasses, 117
Global Nuclear Energy Partnership (GNEP), 205, 206
global scale, x, 185, 204
global security, 205
global warming, 186, 201, 202
government, 188, 189, 190, 191, 206
grain size, 118
graph, 18, 36
graphite, viii, 41, 43, 46, 52, 62, 64, 65, 117
gratings, 20
greed, 192
greenhouse, 202
greenhouse gas, 202
grids, vii, 1, 20
groundwater, ix, 145, 146
groups, 200, 202
growth, viii, 65, 77, 89
guidance, 192
guidelines, 49, 195, 205

H

hafnium, 56
half-life, 4, 5, 48, 58
Halides, 182
hardness, 8, 10, 14, 37
hazards, 195
health, 57, 199, 200, 202
health care, 57
hearing, 191
heat release, 41
heavy metals, 104
height, 8, 32, 62, 63, 120, 121
helium, 56, 65
high-level, 192, 206
high-level radioactive waste (HLW), ix, 97, 99
histogram, 23, 24, 30
history, vii, 1, 7, 9, 29, 38, 60
host, 190
House, 143
human, vii, 9, 31, 200
human health, vii, 200
humidity, 149
Hungary, 55
hydrogen, 36, 44, 62
hydrogen gas, 36
hydrologic, ix, 145, 146
hypothesis, 19, 27

I

icon, 43
ideal, 16
identification, 7, 31, 38
illumination, 34
image, 36, 178, 180
images, 34, 178, 180
immobilization, ix, 49, 98, 100, 101, 102, 104, 115, 180
immunity, 195
implementation, 192, 196, 199, 212
improvements, 32, 37
in situ, 147
inclusion, 147, 150, 159, 165, 171, 178, 180
India, 55, 78
individuals, 194
industrialized countries, 202
industry, 3, 65, 78, 82, 94, 190, 200, 201
inflation, 61, 65, 66
infrastructure, 61
inhibition, 146
inspections, 38
insulation, 19
insurance, 193
integration, 19
integrity, 6, 38
interaction, 146

interface, viii, 2, 9, 10, 28, 29, 77, 83, 150, 154, 159, 163, 165, 171, 173, 174, 177
interference, 88
International Atomic Energy Agency (IAEA), 42, 66, 67, 68, 69, 70, 71, 72, 73, 74, 75, 186, 187, 188, 192, 193, 194, 195, 196, 197, 198, 199, 200, 201, 205, 207, 208, 209, 212
international law, 196, 200
International Monetary Fund (IMF), 193, 209
international standards, 195
international trade, 193
intervention, 44
intrusions, 134
inventories, x, 185, 186, 187, 189, 191, 202, 203, 206
investigations, 181
investment, 25
iodine, 48, 50, 59, 147, 149, 150, 154, 158, 159, 161, 163, 166, 167, 170, 171, 173, 182, 183
ion-exchange, 84, 100, 101
ionization, 5, 48, 79, 82, 83, 84
ionizing radiation, 4, 32
ions, 79, 81, 82, 84, 85, 89, 92, 117, 123
Iraq, 43
iron, ix, 97, 98, 102, 104, 115, 117, 121, 130, 137, 147, 148, 149, 150, 155, 156, 159, 163, 165, 169, 170, 171, 173, 174, 175, 176, 177, 178, 180, 189
irradiation, 14, 37, 48, 50, 58, 60, 65, 83, 84
isolation, 5, 48, 57, 191
isotope, 44, 50, 58, 59, 65
issues, 8, 12, 15, 37, 66, 78, 140, 186, 196, 197, 198, 199, 201

J

Japan, ix, 44, 96, 97, 104, 109, 110, 111, 115, 117, 118, 119, 120, 121, 124, 125, 126, 187, 188, 199, 210, 211
Japanese, 193
judges, 193
jurisdiction, 194

K

Kazakhstan, 55
kinetics, 92
Korea, 103, 199

L

landscapes, 130, 134, 138, 143
lanthanide, 85, 89, 125
lanthanum, 81, 85
Latvia, 43
law, 193, 196, 200, 210, 211
laws, 7, 196
leach, ix, 145, 146, 147, 150, 154
lead, viii, ix, 5, 37, 54, 56, 58, 60, 65, 77, 79, 89, 97, 99, 102, 115
leakage, 6, 7, 36
leaks, vii, 1, 7, 9, 190
legal aspects, 198
lifetime, 53
ligand, 84, 86, 89
light, vii, 1, 9, 10, 11, 12, 14, 16, 17, 20, 22, 23, 24, 40, 63
light transmission, 14
limitation, 146, 191
linear, 165, 176
liquid phase, 131
liquids, 49, 59
lobby, x, 185, 186, 201
location, 191
logging, 4
London, 181
long distance, 202
low-level, 194
low-temperature, 178
LTA, 100, 101
lung disease, 65
Luxembourg, 193
Luxemburg, 73

M

magnesium, 130, 137
magnitude, 83, 84, 89, 138, 140, 147
majority, 59, 138, 199, 202
man, 61, 62, 63, 64, 187, 188, 190, 191
management, vii, x, 2, 7, 39, 48, 51, 54, 58, 78, 185, 186, 188, 189, 190, 192, 196, 197, 198, 199, 200, 201, 202, 203, 204, 205, 206, 210, 212
management practices, 197, 200, 201, 204, 205
manipulation, 8
man-made, 187, 188, 190, 191
manufacturing, 120, 122
mapping, 110, 114
maritime, 194
mass, viii, ix, 5, 10, 44, 52, 54, 77, 79, 80, 81, 82, 83, 95, 120, 129, 131, 142, 145, 146, 147, 154, 156, 157, 165, 171, 191
mass loss, 156
mass spectrometry, 95, 154

materials, viii, ix, 3, 4, 5, 21, 32, 41, 47, 48, 50, 56, 60, 61, 62, 63, 64, 66, 88, 125, 145, 146, 192, 193, 194, 196, 198, 199, 201, 204, 205
matrix, 5, 90, 94
matter, 4, 5, 79, 83
measurement, 7, 14, 20, 21, 83, 120, 122
measurements, 7, 8, 21, 22, 23, 24, 38, 39, 82, 83, 87, 89, 90, 92, 93, 94, 134
mechanical properties, 30
media, 85, 88, 92, 93
medical, 2, 43, 58
medicine, vii, 57, 58
melt, viii, 56, 97, 98, 99, 117, 119, 122
melting temperature, 117, 119, 122
melts, 125
membership, 197, 199, 201, 205, 212
membranes, 84, 86
metal complexes, 79
metal ion, 80, 81, 89, 107
metal salts, 79
metals, ix, 60, 79, 80, 81, 82, 83, 85, 86, 88, 89, 90, 91, 97, 98, 99, 104, 107
meter, 32, 134
methanol, 80, 83
methodology, 91, 92, 94
metric, 186, 188, 189, 190, 191
Mexico, 199
Miami, 124
microcracking, 178
microscope, 110, 178
microscopy, 147, 178
Microsoft, 28
Middle East, 203
migration, 27, 130, 134, 146, 147, 165, 173, 179, 180
military, 192, 196
Ministry of Environment, 212
mission, 201, 210
Missouri, 115
mixing, 49, 59, 100, 104, 109, 117, 150
mixture analysis, 91
modeling, 198
models, 8, 18, 25, 90, 91, 94
moderators, 65
modifications, 12, 78, 148
modules, 8, 27, 31, 38, 56
moisture, ix, 145, 147, 149, 150, 154, 155, 156, 157, 158, 159, 163, 164, 165, 169, 170, 171, 172, 173, 174, 175, 176, 177, 180
moisture content, ix, 145, 147, 150, 154, 155, 156, 157, 158, 159, 163, 164, 165, 169, 170, 171, 172, 173, 174, 175, 176, 177, 180
molds, 52, 149
mole, 83
molecular oxygen, 82
molecular weight, 105, 106, 110, 111
molecules, 50
molybdenum, 59
mosaic, 134
Moscow, 95, 96, 190, 207
Moses, 70
motion control, 30
movement, 192, 195, 196
multidimensional, 85, 89, 92
multiplication, 18, 20, 22
multivariate calibration, 90

N

National Academy of Sciences, 191
National Aeronautics and Space Administration, 61, 74
National Research Council, 71, 73
NATO, 70, 71, 125
natural, 187, 188, 190, 191
neodymium, 81, 85, 88, 89, 91
neptunium, 93, 94
Netherlands, 46, 50, 57, 125
network, 189
neutral, 5, 81, 84, 86
neutrinos, 5
neutron activation analysis, viii, 41, 42, 48
neutron capture therapy, viii, 41, 42
neutrons, 5, 9, 14, 41, 65
Nevada, 191
New York, 182, 212
NGOs, 197, 200
NIR, 94
nitrates, 82
nitrogen, 82
noble gases, 50, 51, 59
nongovernmental, 202
non-profit, 201
nonproliferation, 201
normal, 196
normalization, 149
normalization constant, 18
norms, x, 185, 193, 201, 204, 205
North America, 191
North Korea, 43
Norway, 43, 46, 50, 55, 71, 90
nuclear, x, 185, 186, 187, 188, 189, 190, 191, 192, 193, 194, 195, 196, 197, 198, 199, 200, 201, 202, 203, 204, 205, 206, 207, 211, 212
nuclear energy, 189, 190, 200, 201, 205
Nuclear Energy Agency (NEA), 188, 189, 193, 197, 199, 210, 211

nuclear material, 192, 193, 194, 201, 204, 205
nuclear power, x, 185, 186, 188, 189, 190, 191, 193, 194, 196, 200, 201, 202, 211
nuclear power plant, 189, 194, 196, 211
nuclear program, 204, 205
nuclear reactor, 189
nuclear weapons, 2, 186, 193
nuclei, 4, 5
nucleus, 5, 121
nuclides, ix, 60, 84, 127

O

Obama, 191
Obama Administration, 191
obligations, 212
oil, 4
on-line, 202
operating range, 18
operating system, 28
operations, viii, 2, 3, 6, 7, 9, 10, 28, 31, 38, 56, 61, 62, 63, 64
operator, 193, 194
opposition, 189, 202, 205
optical fiber, 9, 12, 92
ordinary geiger-Muller counters, vii, 1
Organization for Economic Cooperation and Development, 197
ownership, 198
oxalate, 92
oxidation, 59, 92, 174
oxygen, 50, 62, 82, 123, 124, 173

P

Pacific, 125, 145, 181, 182, 183
packaging, 201
parallel, 52, 82, 154, 165, 171, 173
parents, 4
Paris, 192, 193, 196, 204, 208, 211
partial least squares regression, 89
particles, 148, 149, 150, 178, 180
pathways, 47, 51
peak concentration, 159
peer, 198, 199
peer review, 198, 199
perchlorate, 81
periodic, 154, 196, 197, 200, 205
permission, 204
permit, 5, 6, 204
persuasion, 7
PFC, 182
pH values, 146
Philadelphia, 96
phosphate, ix, 97, 98, 102, 104, 107, 108, 109, 110, 112, 115, 116, 117, 125, 182
phosphates, ix, 97, 101, 102, 103, 104, 107, 109, 110, 114, 117, 125
phosphorous, 124
photons, vii, 1, 9, 10, 11, 12, 13, 14, 15, 16, 19, 20, 24
physical properties, 99, 115, 130
physical structure, vii, 1
physics, 44
planning, 189, 190, 198
plants, 27, 30, 41, 188, 189, 190, 196, 211
plasma, 154
plastic, 178
plasticized PVC, 88
plasticizer, 84
platform, 21, 25, 27, 28, 30, 31, 32, 33, 34, 37, 89
platinum, 99, 104, 116
play, x, 185, 186, 201, 202, 205
playing, x, 4, 185, 186
PLS, 90, 92, 93, 94
plutonium, 44, 53, 56, 65, 78, 92, 93, 94, 104
Poland, 46, 55, 59
polarity, 23
policy, 51, 186, 187, 188, 193, 197, 200, 201, 202, 203, 211
policy makers, 201
pollution, 207
poly(vinylchloride), 84
polymer, 84
polymeric membranes, 84
polymers, 79
polyvinyl chloride, 149
population, 3, 4, 7, 201
pore, 146
Portugal, 41, 46, 50, 51
positive correlation, 124
positron, 5
potassium, 130, 138
potential benefits, 199
power, x, 185, 186, 188, 189, 190, 191, 193, 194, 196, 200, 201, 202, 203, 204, 211, 212
power generation, vii, 41, 186
power plants, 2, 50, 51, 54, 56, 66, 189, 196, 211
praseodymium, 85
precipitation, ix, 97, 125
preparation, viii, 77, 78, 83, 86, 88, 89, 91, 110, 117, 149, 150, 178
preparedness, 195
preservation, 38, 63
pressure, 190, 202

principles, 188, 191, 193, 194, 198, 200
private sector, 201
probability, 13, 15, 19, 20, 28
process control, 29, 57
producers, 59, 190
production, 186
profit, 201
program, 198, 200
programming, 25
project, vii, 1, 9, 11, 38, 62, 181, 187
proliferation, 56, 59, 60, 193
prolonged electrolytic operation, viii, 97, 99
propagation, 47
proportionality, 24
protection, 11, 48, 51, 192, 193, 194, 195, 201, 204
protocol, 146, 192, 193
prototype, vii, 1, 9, 12, 14, 15, 16, 19, 22, 25, 27, 29, 30, 32, 34, 36, 44
prototypes, 37, 38, 44
PUB, 209
public, 188, 198, 199, 202, 210, 212
public awareness, 38
public health, 32
public interest, 32
public safety, 199
publishing, 96, 198
purification, 82
purity, 109, 116, 117
PVC, 84, 85, 87, 149, 150, 156

Q

qualifications, 201
quality assurance, 29, 195
quality control, 29
quantification, viii, x, 77, 84, 85, 91, 145, 147, 180
quartz, 130, 137

R

radiation, vii, viii, 1, 2, 5, 6, 7, 8, 9, 10, 12, 14, 23, 31, 32, 33, 37, 39, 43, 51, 57, 60, 65, 78, 88, 92, 121, 130, 134, 138, 140, 141, 142, 143, 194, 195, 198, 209
radiation damage, 39
radiation detectors, viii, 2, 7, 10, 31
radio, 135, 142
radioactive contamination, ix, 127, 130, 134, 142
radioactive isotopes, 60
radioactive waste, vii, viii, ix, x, 1, 2, 3, 4, 6, 7, 9, 38, 39, 40, 41, 42, 47, 49, 50, 51, 57, 58, 60, 62, 63, 64, 66, 94, 96, 97, 99, 142, 180, 185, 186, 187, 188, 189, 190, 191, 192, 193, 194, 195, 196, 197, 198, 199, 200, 201, 202, 203, 204, 205, 206, 208, 212
radiography, viii, 4, 41, 42
radioisotope, viii, 3, 4, 41, 42, 57, 58, 59
radiological, 192, 194, 195, 198, 202
radionuclides, ix, 145, 146, 147, 180
radium, 138
range, 165
rare earth elements, viii, ix, 77, 85, 97, 99, 102, 107
ratification, 193, 211
raw materials, 117, 122
reactant, 111
reaction zone, 109
reactions, 65, 107
reading, 34, 92
reagents, 83, 117
real time, 8
recommendations, 5, 7, 140
recovery, 58, 111, 120
recycling, ix, 56, 65, 97, 99, 206
redundancy, 9, 11
redundant data transmission, viii, 2, 10
reflectivity, 20
refractive index, 11
regenerate, 107, 120
regeneration, 49
regional, 196, 197, 198, 203
registry, 146, 183
regression, viii, 77, 90, 91, 92, 93, 94
regression method, 92
regression model, 90, 91
regulation, 197
regulations, 143, 198
regulatory bodies, 9
rehabilitation, 141, 142, 143
reimbursement, 195
relationship, 159, 171
relevance, 57
reliability, 10, 115
reparation, 178
reprocessing, 4, 48, 54, 55, 56, 57, 78, 85, 89, 92, 93, 96, 98, 103, 115, 125, 142, 186, 187, 188, 189, 190, 206, 211
requirements, 10, 11, 29, 38, 49, 54, 61, 78, 198, 205
research and development, x, 185, 186, 197, 198
residues, 56, 99
resin, 178
resins, 36, 48, 49
resistance, 84
resolution, vii, viii, 2, 8, 10, 21, 33, 82, 83
resources, 61, 203
response, 13, 18, 85, 87, 88, 89, 91, 92, 202

response time, 89, 91
restrictions, 62, 64
retardation, 191
rhenium, 147, 150, 154, 159
Richland, 145, 182, 183
rings, 100
Rio Declaration, 200
risk, 2, 56, 58, 59, 60, 78, 140
risks, 7, 38, 59, 199
rods, 43, 44, 48, 51, 52, 56, 187, 189
Romania, 59
room temperature, 87, 93
root, 35
routes, 59
rules, 7
Russia, ix, 43, 46, 54, 55, 60, 70, 77, 78, 97, 98, 104, 127, 135, 140, 143, 187, 190, 212
Russian, 190, 205, 207

S

sabotage, 194, 205
safeguard, 201
safeguards, 198
safety, x, 2, 11, 32, 38, 44, 51, 78, 138, 141, 143, 185, 189, 191, 192, 195, 196, 197, 198, 200, 202, 204, 206, 212
salt, 189, 190
salt concentration, 83
salts, 109, 114, 115, 125
samarium, 85, 89, 91
sample, 159, 178, 180
sanctions, 205
sand, 150
SAP, 103
scaling, 24
scanning electron, 178
scanning electron microscopy (SEM), 147, 178, 179
scanning process, 137
science, 85, 200
scope, 14, 15, 16
SDRs, 193, 196
security, x, 185, 192, 193, 194, 196, 204, 205
Security Council, 197
sediment, ix, 145, 146, 147, 149, 150, 154, 156, 157, 158, 159, 163, 165, 171, 173, 174, 175, 178, 180
seed, 58
segregation, 2, 62, 64
seismic, 212
selective sensors, 84, 85, 86
selectivity, 79, 85, 88, 89
self-interest, 202
semi-arid, 146, 147
semiconductor, 33
seminars, 198
senate, 191, 210
sensing, 84, 89
sensitivity, viii, 2, 8, 10, 12, 14, 15, 20, 23, 24, 34, 37, 77, 78, 79, 80, 83, 85, 86, 88, 89
sensors, vii, viii, 1, 2, 8, 9, 10, 13, 15, 16, 19, 20, 21, 23, 24, 25, 30, 31, 36, 37, 38, 77, 84, 85, 86, 87, 88, 89, 90, 91, 92
series, ix, 145, 174
services, 198
shape, vii, 1, 8, 10, 12, 16, 17, 23, 46, 63, 81, 82, 92
showing, 15, 64, 158, 180
signals, vii, 1, 9, 15, 16, 17, 27
signs, 6
silica, 130
silicon, viii, 14, 20, 37, 39, 41, 42
simulation, x, 24, 25, 37, 146, 147, 180
simulations, 12, 13, 23, 24
SiPM photosensors, vii, 1, 9, 18
sites, 188
smoothing, 89
sodium, 100, 114, 115, 130, 149, 150, 163
software, 9, 25, 27, 28, 38, 90
sol-gel, 103
solid waste, 48, 58, 59
solution, vii, viii, 1, 3, 7, 21, 26, 32, 33, 34, 37, 54, 59, 77, 78, 80, 81, 82, 83, 84, 85, 87, 89, 110, 111, 119, 131, 149, 150, 163, 186
solvents, 83, 178
sorption, ix, 97, 99, 100, 101, 117, 118, 119, 122, 123
sorption method, 99, 101
South Africa, 55, 57, 73
South America, 203
sovereignty, 197, 200
Soviet Union, 53, 54, 198, 203
Spain, 187, 190, 202, 211
speciation, x, 145
species, ix, 3, 4, 5, 85, 92, 93, 145, 146, 147, 165, 173, 176, 180
specific gravity, 130
specifications, 9, 29, 148, 149, 150
spectroscopy, viii, 77, 86, 92, 94
spectrum, 198
speed, 149, 154
speed of light, 27
spending, 190
spent nuclear fuel, x, 185, 186, 187, 188, 189, 190, 191, 192, 196, 199, 202, 206, 211
St. Petersburg, 77
stability, 30, 37, 38, 83, 88
stabilization, 178

stages, 194
standards, 143, 192, 193, 195, 197, 198, 201
state, 5, 32, 44, 58, 117, 119, 121, 133, 135, 140, 141, 142, 165, 186, 194, 195, 197, 199
states, 92, 174, 193, 194, 195, 198
statistics, 35
steel, 36, 48, 51, 56, 57, 62, 63, 110, 148, 149
storage, vii, viii, 1, 2, 3, 5, 6, 7, 8, 10, 26, 27, 29, 32, 35, 37, 38, 48, 49, 51, 55, 56, 59, 63, 64, 65, 78, 190, 191, 194, 196, 200, 202
strategies, 199
streams, 146
strength, 146
stress, 94
strontium, 103, 112, 123
structure, viii, 21, 22, 27, 28, 30, 31, 32, 41, 42, 60, 61, 84, 111, 112, 115, 116, 120, 123, 134
subtraction, 20
sulfate, 81, 148, 149
supervision, 200
suppliers, 55
surface layer, 130
surveillance, 48, 60, 192, 194
sustainability, 201
sustainable development, 200
Sweden, 187, 189, 190, 202, 210, 212
swelling, 36, 65
Switzerland, 43, 110, 117, 187, 189, 190, 193, 207

T

tanks, 6, 49, 59
target, 57, 58, 59, 85, 108, 117, 121, 122
technetium, 154, 173
technical assistance, 198
techniques, viii, 77, 78, 79, 83, 84, 89, 91, 96, 98, 198
technologies, 56, 78, 94
technology, vii, 22, 32, 38, 46, 64, 66, 84, 85, 86, 94, 143, 200
temperature, 11, 33, 38, 44, 52, 83, 85, 100, 102, 103, 104, 108, 109, 110, 111, 117, 118, 119, 120, 121, 122, 123, 178
terminals, 14
territory, ix, 127, 134, 135, 140, 141, 142, 143, 192, 193, 195, 200, 204
terrorism, 194
terrorist, 196
test data, 27
testing, 38, 82, 146, 150, 189
tetrahydrofuran, 86
textbook, 143
theft, 194, 196, 205
therapy, viii, 41, 42
thermal expansion, 117, 120
thermal properties, 120
thermal stability, 37
thermodynamic equilibrium, 107, 108
thermodynamic properties, 112
third dimension, 92
third party, 195
thorium, viii, 77, 91, 135, 137, 138
threshold level, 18
Tokyo, 209
total costs, 61
tourmaline, 134
trade, 193
training, viii, 41, 42, 198
transfer, 196
transition metal, 84
translation, 24
transmission, viii, 2, 8, 9, 10, 26, 27, 31
transparent, 205
transport, 6, 51, 193, 194, 195, 198, 201, 204
transportation, 10, 56, 78, 199, 201
treaties, 192, 204
treatment, vii, 4, 6, 9, 29, 38, 49, 51, 56, 58, 96, 100, 102, 103, 104, 147, 155, 188, 200
tundra, 128
Turkey, 192, 193, 199

U

Ukraine, 43, 55
UN General Assembly, 197
Underground Research Laboratory, 191, 212
uniform, 30, 80, 154, 163
united, ix, 67, 69, 70, 71, 97, 98, 186, 187, 189, 191, 197, 199, 200, 202, 206, 210, 212
United Kingdom (UK), 43, 44, 55, 57, 65, 75, 125, 187, 189, 190, 191, 206
United Nations (UN), 67, 69, 197, 200, 212
United States, ix, 70, 71, 97, 98, 186, 187, 189, 191, 199, 202, 210
universal gas constant, 85
universities, 104, 115
uranium, viii, ix, 42, 43, 44, 48, 51, 52, 53, 54, 55, 56, 57, 58, 59, 60, 62, 65, 77, 78, 91, 92, 93, 94, 98, 104, 127, 129, 131, 132, 133, 135, 137, 138, 142, 143, 191
USA, 59, 78, 79, 125, 126, 145, 185
USSR, 79, 96
Uzbekistan, 55

V

vacuum, 117, 178
valence, 81, 100, 117, 121
validation, 90, 91
values, 146
vapor, 156, 159
variability, 157
variables, 36, 38, 90
variations, 34, 55, 137, 138
vector, 60, 90
vegetation, ix, 127, 128, 142
velocity, 122
ventilation, 48, 51
vessels, 6
Vietnam, 55
viscosity, 117, 178
volatility, 59

W

walking, 32
war, 7
Washington, 67, 69, 70, 71, 73, 74, 125, 150, 182
waste, vii, viii, ix, x, 1, 2, 3, 5, 6, 7, 9, 10, 12, 14, 21, 27, 29, 30, 31, 35, 36, 37, 38, 44, 47, 48, 49, 50, 51, 54, 56, 57, 58, 59, 60, 61, 62, 63, 64, 78, 89, 98, 99, 101, 102, 103, 104, 115, 125, 145, 146, 147, 173, 180, 185, 186, 187, 188, 189, 190, 191, 192, 193, 194, 195, 196, 197, 198, 199, 200, 201, 202, 203, 204, 205, 208, 210, 211, 212
waste disposal, vii, x, 3, 60, 61, 185, 186, 187, 188, 189, 190, 191, 192, 197, 198, 200, 201, 202, 203, 205
waste disposal sites, 3
waste management, x, 2, 5, 47, 51, 185, 186, 189, 190, 192, 196, 197, 198, 199, 200, 201, 203, 204, 205, 206, 210, 212
waste treatment, 125, 188, 200
wastes, ix, 145, 146, 198, 200
water, ix, 4, 42, 43, 44, 48, 49, 50, 56, 62, 63, 80, 82, 84, 85, 98, 100, 103, 111, 112, 115, 119, 130, 131, 133, 142, 145, 146, 147, 148, 149, 150, 154, 173, 180
watershed, 128
wavelengths, 21
weakness, vii, 1, 24, 196, 204
weapons, 4, 7, 43, 44, 186, 193
web, 7, 26, 27, 186, 199, 208, 209
weight ratio, 84
Western Europe, 66
Westinghouse, 183
wet-dry, 146, 147
wind, 190
workability, 148
workers, 6
worldwide, viii, 3, 7, 41, 57, 58, 192, 196

X

XRD, 110, 111, 112, 114, 115, 119

Y

yield, 11, 18, 24, 47, 59, 82, 93, 134
ytterbium, 85
yttrium, 82
Yucca Mountain, 187, 191

Z

zeolites, 100
zirconium, 51, 52